INTERTIDAL INVERTEBRATES OF THE CENTRAL CALIFORNIA COAST

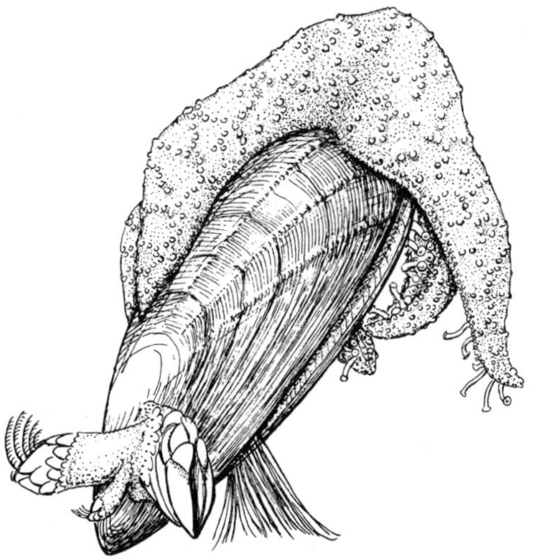

Intertida

UNIVERSITY OF CALIFORNIA PRESS
BERKELEY AND LOS ANGELES 1967

nvertebrates of the Central California Coast

S. F. LIGHT'S "LABORATORY AND FIELD TEXT IN INVERTEBRATE ZOOLOGY," REVISED BY RALPH I. SMITH, FRANK A. PITELKA, DONALD P. ABBOTT, AND FRANCES M. WEESNER, WITH THE ASSISTANCE OF MANY OTHER CONTRIBUTORS

UNIVERSITY OF CALIFORNIA PRESS
BERKELEY AND LOS ANGELES
CALIFORNIA

CAMBRIDGE UNIVERSITY PRESS
LONDON, ENGLAND

FIFTH PRINTING, 1967
PRINTED IN THE UNITED STATES OF AMERICA

S. F. Light

The late Professor S. F. Light (1886-1947) was for twenty-two years a member of the Department of Zoölogy at the University of California, Berkeley. His active interests ranged widely over the field of the invertebrates, and ran the gamut from taxonomy (alcyonarians, scyphozoans, termites, copepods) to the social physiology of termites, their protozoans, symbionts, and caste determination.

Dr. Light gave an extraordinary amount of time and careful thought to his teaching at all levels and exercised a peculiarly pervasive and long-lasting influence on his students' points of view, interests, and habits of thought. His advanced courses were marked by a critical, appreciative, phylogenetic morphology and a critical natural history which insisted on a full realization of the values of sound systematics, keen field observation, and concrete, testable interpretations. The essence of his natural history course was to be found between the lines of its syllabus, which combined a dynamic approach to basic principles (not only of invertebrate zoölogy but of field biology and scientific methodology) with practical aids to the mastery of a specific fauna. It was the result of more than ten years of active contact, virtually the year around, with that fauna and of continual efforts to perfect a teaching approach which aimed at very high and exacting goals. In the present manual, a conscientious updating of the original published syllabus, the revisers have, I believe, been successful in their aim of retaining the values of Light's approach.

Professor Light played his role in biology and academic life in personal contacts rather than in national or university affairs, and profoundly affected the attitudes of many graduate students and associates. Those who knew him will remember the personal characteristics of modesty—extending to a real underestimation of self—, of appreciation of disciplines which lay beyond his own field of study, of exacting criticism in the use of words and ideas— driving him now to caution, now to very forward positions—, of sincere interest in the human relations of his students and assistants, and of a highly developed aesthetic enjoyment of outdoor beauty. Many have felt that although they never really knew the inner man, they sensed vividly the goals and standards for which he lived.

Theodore H. Bullock

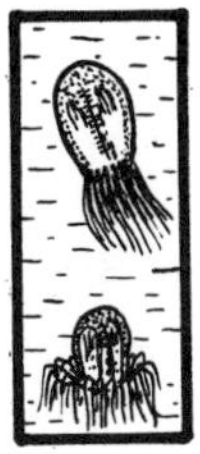

Preface to Second Edition

At the time of Professor Light's death, the first edition of this manual was practically out of print, and only a small amount of the work of revision had been carried out by the author. Since that date the work of revision has proceeded, more slowly than we would have wished, along the lines indicated by present and anticipated needs for a faunal manual of this sort. Recognizing that this manual will rarely, if ever, again be used as the principal text in any course in invertebrate zoölogy, but rather as an adjunct to standard texts or to an instructor's own syllabus, we have generalized much of the introductory material in the various sections and have omitted specific instructions for laboratory procedure in studying particular groups of animals. The section, Field Studies, has been modified to retain the material of general interest, including examples of the problems upon which Dr. Light engaged his classes, but material and schedules pertaining wholly to Dr. Light's conduct of classes at the University of California have been deleted. Some of the keys to fresh-water and terrestrial forms, which are better covered in other works, have also been dropped, and the main emphasis given to the intertidal invertebrate fauna. In most instances, keys and discussions have been completely rewritten rather than simply emended. In an over-all sense, we have edited freely in the effort to achieve a satisfactory and balanced text, reinforced in our efforts by the knowledge of Dr. Light's dissatisfaction with the first edition. Despite these efforts, we may echo Dr. Light's characterization of the work in his preface to the first edition, and state that the second is still "incomplete . . . and continually in process of revision . . ." If we bring this revision to publication without achieving satisfaction in its completeness, we have, at least, adhered to the tradition of the first.

Especially difficult in the revision of keys has been the problem of where to stop. In attempting to achieve a coverage of intertidal animals adequate for the advanced student of general invertebrate zoölogy and marine ecology, we have necessarily stopped short of the treatment required for the specialist. At the same time we have gone beyond the needs of the beginning student. Although we have tried not to sacrifice technical exactness, we have frequently resorted to nontechnical descriptions to make the keys usable by persons new to the field. Since these keys cannot include all the animals which may be encountered in the area covered (the central California coast, roughly from Carmel to Bodega Bay), the student may often be unable to identify his capture by the aid of this book. While making the keys reasonably inclusive of the common forms, we have tried to make them exclusive of rarer forms, and it is our hope that the student who has in hand a species not included in this manual will recognize that it is not treated here, and betake himself to more specialized works, or send the specimen to a competent authority for identification.

Since the publication of the first edition of this manual, certain groups of animals badly in need of revision have received extensive study by persons whose contributions to the second edition are especially noteworthy. Many, although not all, of these contributors are former students of Dr. Light. Cadet Hand has revised the intertidal sea anemones; Donald Abbott has made a noteworthy contribution toward a clear picture of that difficult group, the tunicates; Joel Hedgpeth has contributed a new section on pycnogonids, as well as much general advice; Olga Hartman, a fresh treatment of the polychaetes; Libbie Hyman, an original key to polyclads; Robert Menzies and Milton Miller have furnished a wholly new treatment of isopods; and J. Laurens Barnard has similarly redone the gammarid amphipods. Others who have contributed keys and discussions include Willard Hartman on the sponges; Frank Gwilliam on hydromedusans; Ellsworth C. Dougherty, caprellids; Joan Rattenbury, bryozoans and phoronids; Frances Weesner, asteroids and ophiuroids; Robert L. Usinger, intertidal insects; Irwin M. Newell, marine mites; Joan Steinberg, opisthobranchs and caprellids. Rolf Bolin, of Hopkins Marine Station, has contributed keys to ctenophores and intertidally encountered fishes. Isabella Abbott (Mrs. D. P. Abbott) has provided a key to marine algae. Whenever appropriate, we have given credit in the text where it is due; where authorship of keys and discussions is not indicated, the work is that of the editors, or is carried over from the original manual, or is drawn from so many contributors that specific authorship cannot be decided.

It should be emphasized that we have received a vast amount of unacknowledged help. It is impossible to list these contributors without omitting many, but a few at least must be mentioned. We are grateful to Rudolf Stohler for advice on many points, especially on the molluscs; to M. W. de Laubenfels for advice about sponges; to Wesley R. Coe for advice on nemerteans; to Walter K. Fisher for help with sipunculids; to Clarence R. Shoemaker for checking identifications of amphipods; to A. Myra Keen and Allyn Smith for advice on molluscs; to Raymond C. Osburn for advice on and identification of bryozoans; to Leonie K. Piternick for studies of annelids; to John and Betty Davis for studies of isopods and limpets respectively; to Charles G. Sibley for studies on brachyurans and anomurans; to Harry K. Fritchman for revision of the limpet section in the gastropod key; to Nyven Marchette for work upon the amphipods; to William Newman for studies upon cirripedes and other groups; to Thomas E. Bowman for studies upon which the revision of the hydroid key was based; to Paul L. Illg for advice on copepods; to Frank Filice, upon whose work the revised holothurian key is based; and to James Cannan for contributions to the holothurian key. This list might be greatly extended. Much of the work of revising has been done at Hopkins Marine Station of Stanford University, to whose director, Dr. Blinks, and staff we are grateful for many kindnesses.

For permission to reproduce figures we are indebted to the following: Gilbert M. Smith of Stanford University and the Stanford University Press have kindly given us permission to reproduce figures 129 to 133 of this manual from the original plates of Dr. Smith's *Marine Algae of the Monterey Peninsula.* The University of Chicago Press has permitted us to reproduce our figures 5 and 6 from Ralph Buchsbaum, *Animals without Backbones,* and the McGraw-Hill Publishing Company has permitted us to reproduce our figure 1 from L. H. Hyman, *The Invertebrates: Protozoa through Ctenophora* (1940). Figures previously published in scientific papers by other authors are acknowledged in the text.

The actual work of revision and editing has been made possible only by the efforts of several people with whom it has been a pleasure to work: Frank A. Pitelka bore the main burden of the early stages of revision; Donald P. Abbott has revised the section on Field Studies, and has been instrumental in the over-all organization of the completed revision; Frances M. Weesner has borne the major share of the task of reillustrating and has spent untold effort in checking the various keys. And a special word should be reserved for Theodore H. Bullock, whose faith in the work, advice, and tireless prodding has been invaluable to us all.

Department of Zoölogy,
University of California,
Berkeley

Ralph I. Smith

Excerpts from Preface to First Edition

This volume represents accumulations from fifteen years of teaching the natural history of the invertebrates of the central California coast. . . .

In order to study animals in the field it is necessary to be able to identify them, to recognize them, and to know them by name. A considerable part of the time of the course, therefore, is devoted to the study in the laboratory of animals of the various groups with a view to learning the characteristics important in the identification of the species of these groups. . . . No attempt is made to study taxonomy as such . . . Our end is the prosaic one of learning names for the local assemblage as rapidly and simply as possible.

This end, unpretentious as it is, is by no means easily attained. The invertebrate animals of the Pacific Coast are very imperfectly known. For some there is no monographic account. . . .

Furthermore, such monographs as do exist have a way of getting out of date. Thus, Richardson's monograph on the isopods (1905) is incomplete and badly in need of revision, and even so excellent a monograph as Schmitt's (1921), *The Marine Decapod Crustacea of California,* which we use to very great advantage, contains a number of names that have been changed since its publication.

Finally, modern monographs with keys would not be enough for our purpose. Limited time would still require that these be brought within the range of the study by simplification and limitation of terminology and by limitation of consideration to those species of the various groups significant in our local assemblages. Otherwise it would be impossible for the student to get that familiarity with the fauna as a whole which is one of the greatest values to be obtained in such a study. . . .

The present work, incomplete as it is and continually in process of revision as it is, is the only one known to me which attempts to bring together in more or less completely illustrated keys and lists the information necessary for even a tentative identification of the common invertebrates of this area. The work has been enriched by special studies made by students in the class and by graduate students specializing in the invertebrates. . . .

Under the conditions existing with regard to our knowledge of Pacific Coast invertebrates, changes of name are bound to be the order of the day and a volume such as this is constantly undergoing revision as new works appear or new information is obtained. For its errors, which are numerous, probably beyond even the author's imagination, he accepts full responsibility, consoling himself by the hope that the knowledge of these errors, inevitably forced on the students' attention, may stimulate some of them to undertake corrective investigations such as those mentioned above. . . .

Berkeley, California
March, 1941

S. F. Light

Contents

Introduction

This manual is primarily a guide to the intertidal invertebrate fauna of the central California coast. It may be useful in other West Coast areas, but its effectiveness will lessen with distance. Since it is largely concerned with simple identification and naming of animals and only secondarily with taxonomic principles, it is advisable for the student to understand something of the problems of animal classification.

SCHEMES OF CLASSIFICATION

The purpose of zoölogical classification is to arrange animals into groups on the basis of fundamental similarities and differences which reflect evolutionary relationships. Nearly a million species of animals have been described, and roughly 95 per cent of these are invertebrates. They include a vast, diverse array of types, including all but one of the phyla (Chordata) of the animal kingdom, and even a part of this phylum (the protochordates). The classification of this assemblage is therefore of cardinal importance in the study of invertebrates. It is also a matter of much controversy.

Students and biologists from other fields must be prepared to find that different writers use different systems of classification. These differences are chiefly of two sorts. First, there is the use of different names for the same group, for example, the terms Endoprocta, Kamptozoa, and Calyssozoa refer to the same small group of animals. Second, there is the placing of the same group into different systematic categories, as in the designation of a group as a class by one writer, but as an order by another. Discrepancies of the second sort are so abundant that it is advisable not to have a fixed concept of the taxonomic rank of a particular group, but rather to remember that is it a part of a certain superior group and can be divided into a number of subordinate groups. Thus it is not so important to decide whether the Crustacea represent a subphylum, superclass, or class as it is to know that the group is a major subdivision of the Arthropoda and that it includes the Malacostraca, Copepoda, and so on.

IDENTIFICATION OF INVERTEBRATE ANIMALS

A necessary preliminary to the study of any animal is the determination of its scientific name. A correct determination is especially important if the animal is to be the subject of a scientific investigation.

A scientific name consists of the name of the genus (capitalized), followed by the name of the species (not capitalized), followed by the name of the describer and, if desired, the date of publication of the original description of the species. The name of the describer is properly placed in parentheses if the generic name now in use differs from that used at the time of the original description, thus *Hemigrapsus nudus* (Dana, 1851). For convenience, the name of the describer may often be omitted in our use of scientific names but should always appear on properly labeled specimens. The generic name may be used alone, the specific name never, unless it has already been used with the generic name on the same page. Both generic and specific names should be underlined. This indicates italics to the printer and is of value in picking out the scientific names in manuscript.

Great pains should be taken to spell scientific names correctly. The correlation between careless, unscientific work, and careless use of scientific names is very high.

Common or vernacular names are convenient but have many disadvantages. The rule of priority fixes the correct scientific name, which is universal. Common names are local, since there are no rules to determine which of many such names is the correct name, and the same name may be applied to very different species or types in different regions, or by different persons in the same region. Vernacular names of American birds and certain other groups of vertebrates are relatively uniform owing to the united action of the workers in these fields, but few such authentic common names exist for invertebrates.

It is necessary to make a sharp distinction between these two types of names and to use the vernacular name only after connecting it with the scientific name. Thus, *Hemigrapsus nudus* will be known to, or identifiable by, all zoölogists, but "purple shore crab" would have no meaning, or a different meaning, in areas in which *H. nudus* is not found, or even in certain parts of the range of that particular species.

Few students of zoölogy realize the difficulties involved in identifying with certainty most species of animals, particularly invertebrates. Relatively few species are readily recognized because of distinctive color, pattern, or structure. Such an animal, for example, is the striped shore crab, *Pachygrapsus crassipes*, which is abundant in rocky crevices above low-tide mark along the Pacific Coast. In many other cases, however, identification can only be approximated by the beginner, and in still others even the specialist will find difficulty in making identification. These difficulties are aggravated on the Pacific Coast by the fact that the study of many groups of invertebrates has been greatly neglected. During the period when systematic work was the vogue in zoölogy, there were few zoölogists on this coast. With the change in emphasis in zoölogy, they have largely abandoned this field, and few others have entered it. For some groups such as the amphipods and the littoral copepods, the fauna of the Pacific Coast is largely unknown. In other groups a great amount of work has been accomplished, but much remains to be done.

Some groups, such as the decapod crustaceans, the marine molluscs, and the echinoderms, are fairly well known to systematists. Even here there is need for complete and careful systematic revision and monographing. But when all this has been done, still another step will be necessary if this information is to be available to students of zoölogy or biology who are not specialists in the particular groups, and also to the intelligent laymen. This is the production of manuals of faunas, containing brief diagnostic descriptions and keys to the species, illustrated, if possible, and accompanied by careful definitions of terms. A list of such general publications as may prove useful in conjunction with the present manual is given in the Bibliography.

The keys that form the bulk of this volume represent an attempt to make possible the identification of the common marine intertidal invertebrates of the central California coast. Many keys are incomplete, and no doubt all will need revisions and additions as our knowledge increases. It cannot be too strongly emphasized that keys are shortcuts and often very misleading; that their function is merely to clear the way to an approximation; and that identifications made by them, if to be of scientific value, must be reinforced by reference to the original descriptions or by comparisons with descriptions and illustrations in monographs, if such exist, or by comparisons with authentic named specimens, or by submission to a specialist. Lists of technical publications useful for verification of identifications made with the following keys will be found in the Bibliography.

A NOTE ON THE USE OF THIS MANUAL

Subject matter is grouped according to phyla, classes, orders, or other convenient assemblages. For each group there is usually a general informative introduction, which may be more or less detailed, depending on the needs of the average student and the availability of the information in other general texts. Terms necessary for the use of keys or important outside reference works are explained, and some picture of the group as a whole is attempted.

The keys will be found to vary in their completeness of coverage, geographical range of usefulness, ease of use, and accuracy. Although this is not, perhaps, desirable, it results from differences in numbers of species in various groups, the ease of separating species from each other, the completeness of our knowledge of the group, the professional background of each author, and his success in constructing the key itself. Only use by students up and down the coast will tell us how suitable a given key may be. Editors and contributors alike welcome criticism and information leading to improvement and revision.

Species lists follow most keys. These lists, in general, are of species reliably reported from the central California intertidal. Names of species or genera not included in the keys are marked in the lists by asterisks (*). This has frequently been necessary where certain genera, as among the hydroids, contain numerous and very similar species. In some cases the lists contain only the genera and species identified in the key, omitting a large number of species, as in the Bryozoa. Occasionally we have included species in keys and lists which occur only north or south of the central California region, when such species are distinctive and common. Although this is not consistent, we have felt that such inclusions would be valuable to users of the manual in neighboring areas. Some species lists have been annotated with certain miscellaneous information which may be interesting or helpful. Persons using the book will doubtless add to these notations.

Lists of references are grouped at the end of the manual. These are intended to provide a representative assortment of reading on the biology of each group; they are not complete taxonomic bibliographies, nor do they refer exclusively to forms found in this area.

Phylum Protozoa

Although vastly numerous, most protozoans are so small as to escape notice except in microscopic study of collected material. The familiar *Amoeba, Paramecium,* and *Euglena* of fresh water are so well known that they need not be discussed here. In intertidal collecting, however, we frequently encounter relatively large and conspicuous protozoans which should be mentioned. Foraminifera, shelled relatives of the amoeba, turn up frequently in tide pools. Among sponges, bryozoans, eelgrass roots, and so forth, occur ovoidal brown bodies, 1-3 mm. in diameter, resembling fecal pellets, which are the thin-shelled, single-chambered foraminiferan *Gromia oviformis* (see Arnold, 1951). Among holdfasts of corallines in pools we also encounter spiral chambered foraminiferans of the genus *Discorbis*, whose life history has been described by Myers (1940).

Attached to various objects, the dark flasklike cases of the "bottle-animalcule," *Folliculina*, frequently are puzzling, and are mistaken for egg cases and the like. The animals themselves are large ciliates, often a vivid blue in color, with two or more extensible flattened arms edged with cilia, and are unmistakable once identified. Branching colonies of the stalked ciliate *Zoöthamnium* are likewise rather common and easily recognized. The seashore collector will encounter a great variety of protozoans, very small metazoans, larvae, and other small fry, very engaging to watch, but extraordinarily difficult to identify.

Phylum Porifera

by Willard D. Hartman
Peabody Museum, Yale University

The sponges represent an ancient group of animals which diverged early from the main stem of multicellular forms. Skeletal remains of sponges have been reported from the earliest fossiliferous pre-Cambrian rocks.

The structural plan of sponges is unique among multicellular animals. The body consists of loose aggregations of cells differentiated into ill-defined tissues such as epithelia and mesenchyme, but organs are not formed. A mouth and digestive tract, such as occur in all other free-living Metazoa, are lacking in sponges; instead, the surface of sponges is perforated by numerous openings through which water enters and leaves. The incurrent pores, known as ostia, are small and numerous; the excurrent apertures are larger and relatively few in number. Around the latter are sphincters of myocytes, which function in regulating the size of the openings. These muscle cells appear to respond directly to environmental changes, since nervous elements have not been demonstrated with certainty in sponges.

Water entering a sponge through the ostia passes through a system of canals lined by flagellated collar cells or choanocytes and finally leaves the sponge through the oscula. The canal systems are of varying grades of complexity (fig. 1), ranging from those with a single, large cavity lined by collar cells (e.g., *Leucosolenia eleanor*) to those with numerous subdivisions of the internal channels and with collar cells restricted to spherical or ellipsoidal chambers along the lengths of the channels (e.g., *Leuconia heathi, Aplysilla glacialis*). The latter system is a much more efficient one, causing the water to slow down in its passage through the sponge and thus providing more time for food to be removed. The vast majority of sponges have this complicated plan. The grades of construction of sponges are illustrated in figure 1.

Many sponges have a definite radially symmetrical form. *Leuconia heathi* and *Tethya aurantia* are examples from the California coast.

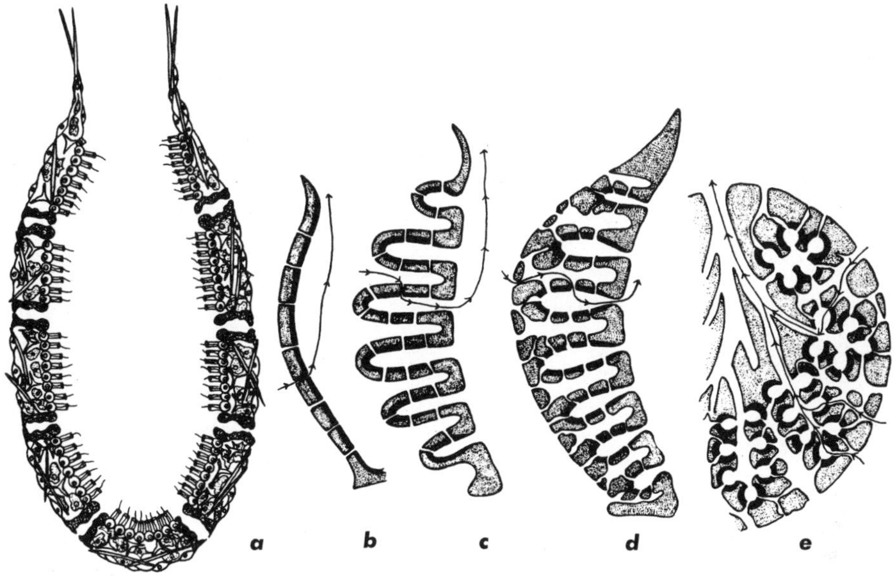

Fig. 1. Diagrams of types of sponge structure. *a*, Diagrammatic vertical section of simplest type of sponge (asconoid type seen in *Leucosolenia*), showing cellular elements; *b-d*, sections of one wall; *b*, asconoid; *c*, primitive syconoid; *d*, developed syconoid such as seen in *Grantia*; *e*, leuconoid, that of all local encrusting sponges. Choanocyte layer shown in heavy black. Arrows indicate course of water. By permission from *The Invertebrates: Protozoa through Ctenophora* by L. H. Hyman, copyright 1940, McGraw-Hill Book Co., Inc.

Others are irregular in shape, encrusting rocks or growing from an encrusting base into irregularly branching colonies. Symmetry of form in sponges is correlated with their vertical distribution in the sea. Deep-sea species tend to have regular and definite shapes; intertidal and shallow water species include a large majority with irregular shapes. It is often difficult to define whether sponges are individuals or colonies. Hyman (1940) points out that physiologically a sponge individual may be considered to be a single osculum with its contributing parts. Thus an encrusting sponge would be termed a colony made up of as many individuals as there are oscula (fig. 2).

The consistency of sponges varies exceedingly, from hard and stony to friable, rubbery, or gelatinous depending upon the nature and arrangement of the skeletal elements. All except a few genera of sponges, such as *Halisarca*, possess some type of skeleton. Indeed, the main subdivisions of the phylum have been based on skeletal characteristics. The skeleton may consist of calcareous or siliceous spicules alone, of spongin fibers alone, or of a combination of siliceous spicules and spongin fibers. One species is reported as possessing both calcareous and siliceous spicules. Elastin fibers are also widely distributed in sponges.

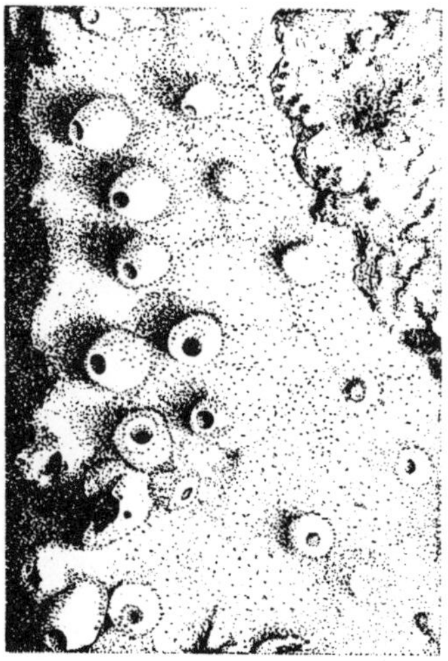

Fig. 2. Surface of part of encrusting sponge showing oscules (after Guberlet).

The feeding mechanism of sponges has been most extensively studied in species of the single family inhabiting fresh waters, the Spongillidae. The food consists of diatoms, small protozoans, bacteria (reported in studies on bath sponges), and small particles of detritus. The smaller particles of food reach the flagellated chambers in the water currents produced by the choanocytes and are ingested by the latter cells. The outer surfaces of the collars of the choanocytes appear to catch the food particles; in the body of the cells the food is transported downward to be picked up by phagocytic amoebocytes in the mesenchyme. The amoebocytes transport the food throughout the colony. In fresh-water sponges it has been demonstrated that the epidermis and the cells lining the incurrent canals also ingest food particles. Larger particles, which cannot find their way through the narrow channels leading to the flagellated chambers, are ingested in this way. Food is predigested in the phagocytes, and the digestive process is completed in the cells (such as scleroblasts, germ cells, or collencytes) which receive food from the wandering cells. The phagocytes also function in excretion, carrying waste material to excurrent canals for release. How the cells differentiate between incurrent and excurrent canals is not known.

Both asexual and sexual modes of reproduction occur in sponges. Fresh-water sponges regularly produce asexual reproductive bodies called gemmules which are aggregations of amoebocytes surrounded by a resistant coat of organic material in which spicules are embedded. The spicules are usually amphidiscs of characteristic shapes or spiny oxeas, and are important in the identification of spongillids. Gemmules carry the sponges through periods of freezing temperatures or of drought and hatch when favorable environmental conditions return. The period of dormancy of fresh-water sponge gemmules can be shortened by artificial refrigeration at temperatures of 50° F for a period of six to eight weeks. Many marine sponges also produce gemmules, and these in some cases develop into flagellated larvae which are indistinguishable from sexually produced larvae.

Sexual reproduction is best known in the calcareous sponges. Eggs develop from archeocytes, which enlarge and take up a position near a layer of choanocytes. Sperm cells develop in groups in the mesenchyme. During fertilization, the sperm, which enter the sponge by way of the water channels, are taken into the cytoplasm of choanocytes or amoebocytes, which transfer the sperm to the ova. In certain Calcarea the flagellated larva, or amphiblastula, arises through a process of inversion of the surfaces of the embryo. Cell divisions up to this point have produced a hollow sphere of cells, with those at the animal pole bearing internally directed flagella. The embryo now turns inside out in a process reminiscent of a similar developmental phenomenon in *Volvox,* and the flagella of the micromeres are brought to the outside. In the Demospongiae the larvae are released as solid stereogastrulae, with an outer flagellated layer and an inner mass of cells already differentiated into choanocytes, scleroblasts, collencytes, and amoebocytes.

It is believed that many deep-sea sponges seldom reproduce sexually, the colonies multiplying chiefly by budding or by the production of gemmules. Local species that reproduce occasionally by budding are *Stelletta clarella* and *Craniella arb.*

Sponges have remarkable powers of regeneration. If a small part of a colony of one of the local species of *Microciona* is pressed through fine (#20) bolting silk into a dish of sea water, a suspension of free cells is obtained. These cells soon form netlike aggregates on the bottom of the dish and eventually, with proper care, small functional sponge colonies appear. Commercial sponge fishermen have taken advantage of the regenerative powers of sponges in the artificial propagation of bath sponges. By cutting a sponge into many pieces and planting these on concrete blocks or attaching rows of them along wires in shallow seas, an increased production of bath sponges has been possible in certain parts of the world.

Some intertidal sponges are quite resistant to desiccation. Colonies of the green-colored *Halichondria panicea* and of the violet-colored *Haliclona permollis* are both regularly found growing in beds of *Mytilus californianus* and *Mitella polymerus* along the California coast. The thin, encrusting, red-colored colonies of *Ophlitaspongia pennata* also occur in mid-tidal areas on the sides and lower surfaces of rocks. Other species, such as *Tethya aurantia* and *Polymastia pachymastia,* are common in offshore waters and can be collected intertidally only at the lowest tides on exposed coasts.

Burton (1949) has recently reported that sponge colonies can move slowly over the substratum. He has observed a tendency for certain littoral encrusting species to settle on rocks as larvae in swarms. As the sponge colonies grow, they slowly move over the surfaces of the rocks, coalescing with neighboring colonies which they happen to meet. The pattern of such a swarm of young colonies is continuously changing as a result of the reorganization of the cells at the periphery of each colony. Similar observations of the rapid reorganization of cells in the periphery of colonies have been made on fresh-water sponges when gemmules were allowed to germinate on a microscope slide and to grow out in a thin layer between the slide and a cover slip.

An interesting family of sponges that has received very little study on the California coast is the Clionidae or the lime-boring sponges. These bore into calcareous material such as mollusc shells, corals, and limestone. One local species is commonly associated with encrusting calcareous algae of the genus *Lithothamnion*, living in a layer under the algal colony and boring up to the surface in places to allow the exit of contractile tubules on which are borne the ostia and oscula. Other common species (to date no adequate taxonomic study of them has been made) bore into abalone or lamellibranch shells, and in these the openings through which the ostial and oscular tubules extend often form regular circular patterns on the surfaces of the shells. If an abalone shell inhabited by a *Cliona* colony is broken, the yellow sponge can be seen filling extensive galleries which it has bored in the interior of the shell. Shells riddled in this way by *Cliona* colonies are greatly weakened, and for this reason the sponge is a nuisance when growing in abundance on oyster beds.

Nudibranchs, limpets, and periwinkles are known to feed on sponges, and it is probable that other gastropods and possibly chitons use them as food, but observations on sponge predators are few.

Many animals find sponge colonies favorable places in which to live. Amphipods, polychaetes, and shrimps commonly live in certain sponge colonies. The masking crabs, *Loxorhynchus crispatus*, often plant parts of sponges on their backs. Other sessile organisms such as hydroids, entoprocts, barnacles, ectoprocts, and tunicates often grow on sponge colonies.

The classification of sponges depends largely upon skeletal characteristics, although these are not always adequate. Embryological studies, life history studies, biochemical characteristics, and cytological details have helped in understanding the relationships of some groups. The phylum Porifera is subdivided into three classes on the basis of the chemical composition and geometrical configuration of the skeletal elements. The classes are characterized as follows:

Class Calcarea (or Calcispongiae): Has spicules (usually triradiate or monaxonid) of calcium carbonate. Spongin absent.

Class Demospongiae: The skeleton consists of spicules of silicon dioxide (laid down in a hydrated form related to opal) or of the iodine- and bromine-containing scleroprotein, spongin, or of both siliceous spicules and spongin. The megascleres are monaxonid or tetractinellid in configuration.

Class Hexactinellida (or Hyalospongiae): The skeleton consists basically of siliceous spicules with three axes (triaxons). This group is not found intertidally, although it is well represented on the continental shelf and slope of California.

There is little agreement among specialists about the subdivision of the classes into orders and families. De Laubenfels (1936) has brought together material for a revision of the entire phylum, and his classification of the class **Demospongiae** is followed here in the main.

DISCUSSION AND GLOSSARY OF SPONGE SPICULES

Some knowledge of the fearsome terminology of sponge spicules is necessary if the taxonomic literature on the group is to be read intelligently. The following discussion of the terms employed is included to aid those who may go further in this field, as well as to make clear certain terms used in the key.

Sponge spicules have an axis of organic material around which calcium carbonate or silicon dioxide is deposited. They have a great variety of shapes and often serve as useful characters in identifying sponges. It is necessary, therefore, to consider spicule structure in greater detail. In general we may differentiate spicules first of all on the basis of size: there are large **megascleres** that form the chief supporting framework of the sponge, and there are smaller **microscleres** scattered throughout the mesenchyme. Spicules are further subdivided on the basis of the number of axes or rays present. Names for spicules are coined by adding the appropriate numerical prefix to the endings **-axon** (when referring to the number of **axes**) or **-actine** (when referring to the number of **rays** or **points**). Thus an important category of spicules consists of **monaxons**, formed by growth along a single axis. If growth occurs in a single direction, the spicule is a **monactinal monaxon** (fig. 3, *a-d*); if in both directions, it is a **diactinal monaxon** (fig. 3, *e-i*). Both **monactines** and **diactines** (the latter also called **rhabds**) are usually megascleres, but in some instances the diactines are small and secondary skeletal elements, which are classed as microscleres. Another important category of spicules includes the **tetraxons** (fig. 3, *j-o*) which have four rays, each pointed in a different direction. The rays may be equal, or one (called the **rhabdome**) may be longer than the others (which are called **clads**), in which case the spicules are referred to as **triaenes** (fig. 3, *k-l*). The **triradiate** spicules common in calcareous sponges are believed to have arisen from tetraxons in which the rhabdome has been lost (fig. 3, *o*). **Triaxons** are formed of three axes crossing at right angles, resulting in six rays or points; hence the synonym, **hexactines**. This category is found only in the hexactinellid sponges.

Microscleres (fig. 3, *p-y*) often have more than four rays; there may be numerous equal rays diverging from a central point. Such types are

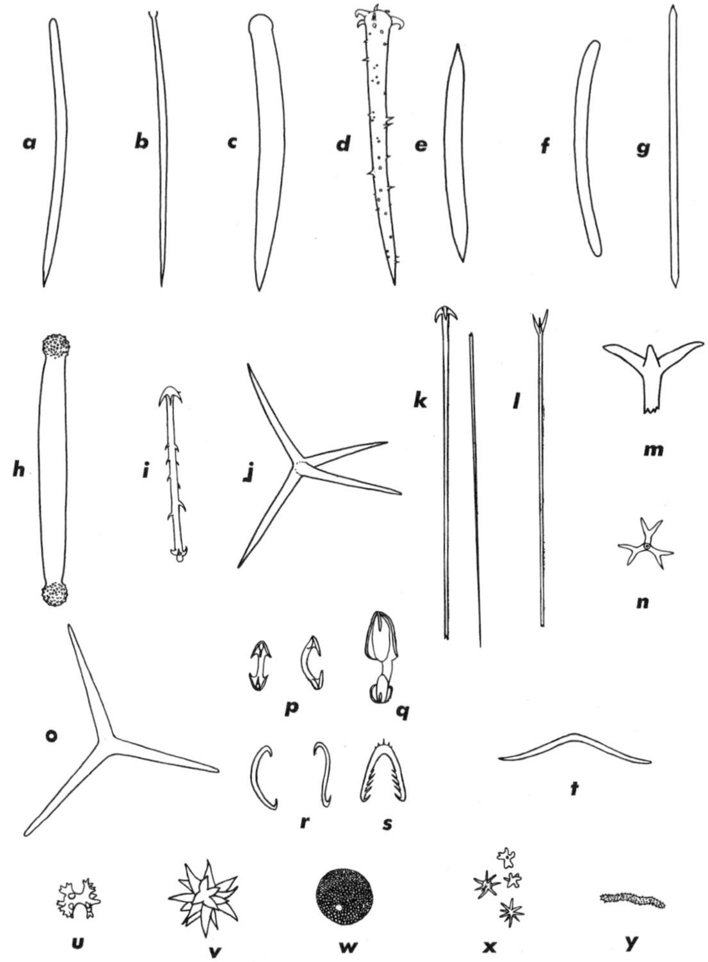

Fig. 3. *a-d*, Monactinal monaxons: *a*, style, *Hymeniacidon sinapium* (x160); *b*, tylostyle, *Cliona celata* (x160); c, subtylostyle, *Ophlitaspongia pennata* (x160); *d*, acanthostyle, *Microciona microjoanna* (x367). *e-i*, diactinal monaxons: e, oxea, *Haliclona permollis* (x300); *f*, strongyle, *Prianos problematicus* (after de Laubenfels); *g*, tornote, *Hymenamphiastra cyanocrypta* (x300); *h*, tylote, *Plocamia karykina* (x160); *i*, cladotylote, *Acarnus erithacus* (x160). *j-o*, Tetraxons: *j*, calthrops, *Poecillastra tenuilaminaris* (x33); *k*, anatriaene, *Craniella arb* (x33); *l*, protriaene, *Craniella arb* (x33); *m*, plagiotriaene, *Stelletta clarella* (x33); *n*, dichotriaene, *Stelletta clarella* (x33); o, triradiate, *Leucosolenia* sp. (x160). *p-y*, Microsceleres: *p*, isochelae (front and side views), *Anaata spongigartina* (x300); *q*, anisochela, *Carmia macginitiei* (x300); *r*, sigmata, *Lissodendoryx noxiosa* (x160); *s*, forceps, *Lissodendoryx noxiosa* (x700); *t*, toxa, *Ophlitaspongia pennata* (x300); *u*, amphiaster, *Hymenamphiastra cyanocrypta* (x367); v, spheraster, *Tethya aurantia californiana* (x160); *w*, sterraster, *Geodia mesotriaena* (x67); *x*, euasters, *Stelletta clarella* (x300); *y*, spiraster, *Cliona lobata* (x367) (Hartman, original).

known as **asters.** Another type of microsclere is formed by concentric growth around a center; such spicules are called **spheres.** Other types of microscleres are C- or S-shaped forms, called **sigmas;** bow-shaped **toxas; chelas,** with recurved hooks, plates, or flukes at each end; and **streptasters,** which are short spiny rods. There follows a more detailed glossary of spicule types of importance in identifying sponges likely to be found intertidally in central California. Examples of species in which each spicule type occurs are given. (Letters before names refer to figure 3.)

Monactinal monaxons

a. **Styles.** Rounded at one end, pointed at the other (e.g., *Hymeniacidon sinapium*).

b. **Tylostyles.** With a distinct knob at one end; pointed at the other end; resembling a marlinspike (e.g., *Cliona celata*).

c. **Subtylostyles.** Tylostyles with indistinct knob at one end: pointed at the other end (e.g., *Ophlitaspongia pennata*).

d. **Acantho-.** A prefix denoting that the spicule is covered with thorny processes (e.g., acanthostyles in *Microciona microjoanna*).

Diactinal monaxons

e. **Oxeas.** Pointed at both ends (e.g., *Haliclona permollis*).

f. **Strongyles.** Rounded at both ends (e.g., *Prianos problematicus*, found in offshore waters of Monterey Bay).

g. **Tornotes.** Lance headed at both ends (e.g., *Hymenamphiastra cyanocrypta*).

h. **Tylotes.** Knobbed at both ends (e.g., *Plocamia karykina*).

i. **Cladotylotes.** Tylotes with more or less recurved clads at each end (e.g., *Acarnus erithacus*).

Tetraxons

j. **Calthrops.** Having the four rays equal or nearly so (e.g., *Poecellastra tenuilaminaris*, an offshore species).

Orthotriaenes. With clads making an angle of about 90° with the axis of the rhabdome (e.g., *Stelletta clarella*).

m. **Plagiotriaenes.** With clads directed forward and making an angle of about 45° with the produced axis of the rhabdome (e.g., *Stelletta clarella*).

l. **Protriaenes.** With clads directed forward as the tines of a fork. The clads make an angle of less than 45° with the produced axis of the rhabdome (e.g., *Craniella arb*).

k. **Anatriaenes.** With the clads directed backwards (e.g., *Craniella arb*).

n. **Dichotriaenes.** With forked clads (e.g., *Stelletta clarella*).

Diaenes. Tetraxons modified through the loss of one clad (e.g., *Craniella arb*).

Monaenes. Modified tetraxons with only one clad (e.g., of rare occurrence in *Craniella arb*).

o. **Triradiates.** Spicules with three rays more or less in the same plane; common in calcareous sponges, such as *Leucosolenia.*

Microscleres

Chelas.

p. **Isochelas.** With equal ends (e.g., *Anaata spongigartina*).

q. **Anisochelas.** With unlike ends (e.g., *Carmia macginitiei*).

r. **Sigmas.** C- or S-shaped (e.g., *Lissodendoryx noxiosa*).

t. **Toxas.** Bow-shaped (e.g., *Ophlitaspongia pennata*).

s. **Forceps.** U-shaped with the arms sometimes crossed; often spined (e.g., *Lissodendoryx noxiosa*).

Streptasters.

y. **Spirasters.** Spirally twisted, spiny rods (e.g., *Cliona lobata*, often found boring in abalone shells).

u. **Amphiasters.** Short rods with spines at each end (e.g., *Hymenamphiastra cyanocrypta*).

x. **Euasters.** With several equal rays radiating from a central point; e.g., *Stelletta clarella*, in which the ends of the rays vary from pointed (**oxeote**) to rounded (**strongylote**).

v. **Spherasters.** With many rays arising from large central spheres e.g., *Tethya aurantia*).

w. **Sterrasters.** Spheres covered with many minute rays (e.g., *Geodia mesotriaena*, an offshore species).

Note on microtechnique.—Detailed studies of sponge anatomy may require the preparation of microscopic sections. For gross preservation 95 per cent alcohol is best; long exposure to formalin leads to maceration in many species of sponges and should be avoided if possible. Fixation in standard histological fixatives is desirable before transferring to alcohol, since often histological work must follow. Thin, free-hand sections are readily made with a razor blade if the specimen has been hardened in alcohol. Sections perpendicular to the surface of the colony as well as tangential sections are useful. Such sections can be directly dehydrated, cleared in xylol, and mounted in piccolyte, or they can be stained with a saturated solution of basic fuchsin or safranin in 95 per cent alcohol before clearing and mounting. In studying sections, the various types of spicules can be observed as well as their arrangement in the skeleton. Other structures to look for are a cortex, often packed with spicules of one category, subdermal spaces, tracts of spicules, and spongin fibers with or without enclosed spicules. In some cases the shapes and dimensions of the flagellated chambers are important.

In sections, certain spicule categories may be overlooked, and it is usually desirable to make slides of spicules which have been freed from the cellular elements of the colony. This can be done by treating a small part of the specimen with sodium hypochlorite (clorox will serve) until the spicules are free. The sample is then washed several times with water, centrifuging between washings to make certain that the minute microscleres have settled. Finally the water can be replaced by 95 per cent alcohol, and the suspension of spicules poured onto a slide and allowed to dry (or the alcohol can be ignited). A drop of xylene is added to the dry preparation, followed by piccolyte and a cover slip. A slide of clean spicules is now available for study and measurement of spicule types.

A KEY TO INTERTIDAL SPONGES, ESPECIALLY OF THE MONTEREY AREA

by Willard D. Hartman and R. I. Smith

The following key may make possible the field identification of a number of sponges having distinctive features of shape, color, or texture; but in many cases verification of the identification must be done in the laboratory. In preparing to identify sponges, notes should first be made of the freshly collected specimen. Shape, size, color, and consistency should be noted; the distribution and sizes of ostia and oscula are also of importance. Complete descriptions of the sponges included in the key as well as others found intertidally in the Monterey area can be found in the monograph of de Laubenfels (1932). Undescribed species are certain to turn up.

1. Spicules calcareous, some always triradiate in structure. If doubt exists about the chemical composition of the spicules, test with 10 per cent acetic acid beneath a cover slip to observe release of bubbles of CO_2; local species are few in number; tubular or cuplike; whitish or pale tan . . . . .Class Calcarea (Calcispongiae) 2
1. Spicules siliceous, seldom triradiate in form. In testing with acid take care that bits of shell or other calcareous foreign matter do not give a spurious positive reaction. Since no Hexactinellida are found intertidally, sponges with siliceous spicules fall in the class Demospongiae, which includes the great majority of local forms . . 5
1. Proper spicules absent; skeleton of branched horny fibers; the surface is almost always raised into little cones; local forms are thin, encrusting, soft, rubbery, or slimy; foreign spicules and sand particles are often embedded in the spongin fibers. . . . . Demospongiae, order Keratosa 4
1. Spicules and spongin fibers absent; in surface view, pores and oscula

are visible; sections show canals and flagellated chambers characteristic of sponges; consistency soft; color pale tan or lavender-gray; might be mistaken for a colonial tunicate or an *Alcyonidium* . . . . *Halisarca sacra*

2. Colonies form dense "spongy" masses of anastomosing white tubes . . . . *Leucosolenia eleanor*
2. Unbranched; cup, vase, or juglike 3

3. Slender, vaselike; white to tan; superficially smooth . . . . *Rhabdodermella nuttingi*
3. Globose or pear-shaped; dirty white; surface spiny; prominent ring of upright spicules about osculum. . . . . . . *Leuconia heathi*

4. Forming thin soft encrustations; surface, showing few pores, is raised into little cones by branched horny fibers; color ranging from yellow to rose . . . . *Aplysilla glacialis*
4. As above, but less common and of a deep purple color which is given off in large amounts when sponge is collected . *Aplysilla polyraphis*

5. More or less globose, or in thick massive encrustations; spicules as seen in a cross section arranged in a markedly radiating fashion . 6
5. Generally encrusting in from thin to quite thick masses; spicules not radially arranged in cross section 10

6. Surface warty, tuberculated, or extended into fingerlike protrusions; external plush of spicules not apparent, at least on the protrusions; external color shows orange or yellow; may be overlaid with film of green algae 7
6. Surface with a plush of projecting spicules; color tan or dirty white 8

7. Globose; surface warty; inner tissue brown, cortical layer orange-yellow . . . . *Tethya aurantia*
7. Hemispherical or encrusting base with fingerlike protrusions; color lemon yellow; interior brown; spicule plush on basal part only . . . . *Polymastia pachymastia*

8. Large biscuitlike masses; color dirty white; heavy plush of spicules (handle with caution!) . . . . *Stelletta clarella*
8. Colonies spherical with a conspicuous osculum; color tan or drab; spicule complement includes microspined sigmas 9

9. Young colonies tan; large colonies drab with a heavy plush of spicules; megascleres include oxeas, anatriaenes, and protriaenes (fig. 3, *k*, *l*) . . . . *Craniella arb*
9. Small, spherical, yellow-tan in color, with ring of upright spicules around osculum; among the megascleres are some shaped like boathooks . . . . *Craniella* sp.

10. Color some shade of red or orange in life 11

10. Color other than reddish or orange in life 16

11. Oscules small (1 mm. or less) or not visible; encrustations generally not more than 1 cm. in thickness; consistency firm or woody . . 12
11. Oscules larger (more than 1 mm.); encrustations relatively thick (as much as several cm.); consistency compressible or rubbery . . 15

12. With subtylostyles (fig. 3, c) as the principal spicules . . . 13
12. Principal spicules styles, tylotes, or acanthostyles (fig. 3, *d*) . 14

13. With subtylostyles and chelas only, the latter often small and easily overlooked; color in life light reddish; stiff, brittle; in small, thin patches on rocks . . . . *Esperiopsis originalis*
13. With subtylostyles and toxas only; color scarlet to orange; pores minute; when out of water, oscules may appear at center of a group of radiating grooves . . . . . . *Ophlitaspongia pennata*

14. Chief spicules styles; subtylostyles and small acanthostyles also present, together with both toxas and chelas; most toxas heavy, deeply arched; small toxas also present; surface finely tuberculate with a low spicule plush; color reddish . *Microciona microjoanna*
14. Chief spicules acanthotylotes and acanthostyles; a thin, brittle, gritty encrustation; surface irregularly pitted or tuberculate; oscules rare; color vermilion to orange . . . . *Isociona lithophoenix*
14. Chief spicules spiny-headed tylotes; consistency firm and woody; oscules small but distinct, regularly scattered over surface; color scarlet; colonies tending to give off mucus when injured. . . . . *Plocamia karykina*

15. Consistency about that of sponge rubber; color salmon pink; oscules irregular or in rows on ridges or on scattered papillae; spicules of dermal layer tylotes, those of interior parts chiefly subtylostyles . . . . *Tedania topsenti*
15. Consistency firm but compressible; color scarlet; chief spicules styles, but most distinctive spicules include tylotes with microspined ends and small palm-tree-shaped spicules with smooth or spiny stems; up to several cm. thick, with large, elevated, craterlike oscules . . . . *Acarnus erithacus*

16. Color cobalt blue; thin; in shaded locations; with amphiasters as microscleres . . . . *Hymenamphiastra cyanocrypta*
16. Color pale lavender or rose-gray 17
16. Color not red, blue, lavender, or rose 18

17. Consistency soft and spongy; forming thin encrustations, with the oscules raised on tubules; oscular openings small (0.5–2 mm. in diameter); spicules oxeas only, with a netlike arrangement . . . . *Haliclona permollis*

17. Of friable consistency; forming encrustations up to 2 cm. thick; oscules raised on low tubules; oscular openings large (2–8 mm. in diameter); spicules are oxeas only, arranged in a regularly reticulate manner; interior of colony orange-tan in color. . *Adocia gellindra*

18. Color lemon yellow; surface irregular or tuberculated; bores in shells of barnacles or molluscs . . . . *Cliona celata*
18. Color nondescript and/or variable, including brown, reddish-brown, yellowish-brown, drab, whitish, or greenish 19

19. Color light brown or tan; oscules with craterlike rims; spicules, oxeas only, arranged in triangular patterns; resembles *Haliclona permollis* (17) but color not lavender . . . . . *Haliclona* sp.
19. Color various; surface more or less broken or tuberculated; oscules without distinct rims, or may not be noticeable 20

20. Color light brown, yellowish, or whitish 21
20. Color otherwise 22

21. Color and thickness variable, but most often light yellowish; consistency spongy, fairly firm but not fibrous; spicules (fig. 4) include slender tylotes, spiny styles, chelas, and sigmas; common; odor strong, distinctive to some persons . . . *Lissodendoryx noxiosa*
21. Rather thick; internal structure distinctly fibrous; chief spicules subtylostyles; distinctive spicules, giant sigmas with serrations (teeth) on outer sides of ends; also present, small sigmas and chelas with unequal ends . . . . *Paresperella psila*
21. Color burnt gold to rich yellow; forms thin encrustations on rocks; spicules are acanthostyles, tylostyles, and minute chelas . . . . *Eurypon asodes*

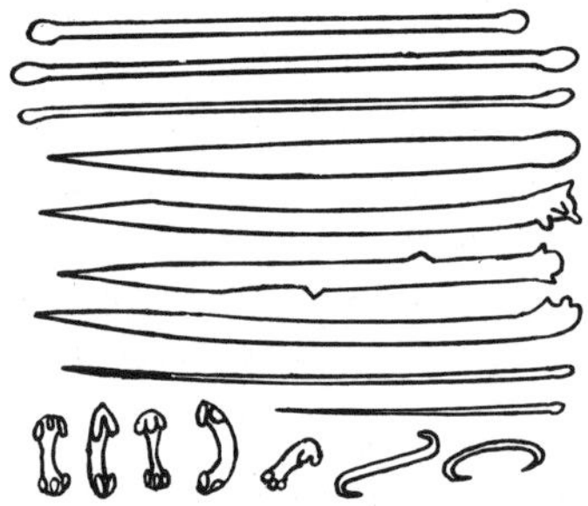

Fig. 4. The spicules of *Lissodendoryx noxiosa*. From above downward; tylotes, styles of three types (smooth, sparsely spiny, and very slender), arcuate chelas (left), and contort sigmas (after de Laubenfels).

22. Greenish, sometimes varying to orange; oscules raised, small, and irregular; spicules oxeas only, but (unlike *Haliclona*) strewn in irregular fashion; structure resembling bread when broken; found in well-lighted situations such as crevices in mussel beds. The "crumb-of-bread" sponge . . . . *Halichondria panicea*
22. Color a deep, rich brown approaching reddish; consistency firm; oscules usually without craterlike rims; spicules include slender subtylostyles, spiny or very spiny styles, and large chelas . . . *Anaata spongigartina*

LIST OF SPECIES OF SPONGES OCCURRING INTERTIDALLY ON THE CENTRAL CALIFORNIA COAST

Class Calcarea (Calcispongiae)
- *Leuconia heathi* (Urban, 1905)
- *Leucosolenia eleanor* Urban, 1905
- **Leucosolenia nautilia* de Laub., 1930
- *Rhabdodermella nuttingi* Urban, 1902

Class Demospongiae
- Subclass Tetraxonida
  - Order Choristida
    - *Craniella* sp. (Undetermined species found at Mission Point, Carmel and Pescadero Point, Monterey Peninsula)
    - *Craniella arb* (de Laub., 1930)
    - **Papyrula saccharis* de Laub., 1930
    - **Penares cortius* de Laub., 1930
    - *Stelletta clarella* de Laub., 1930
  - Order Epipolasida
    - *Tethya aurantia* (Pallas) *californiana* de Laub., 1932
  - Order Hadromerina
    - *Cliona celata* Grant *californiana* de Laub., 1932
    - **Cliona lobata* Hancock, 1849
    - *Polymastia pachymastia* de Laub., 1932
    - **Spheciospongia confoederata* de Laub., 1930
- Subclass Cornacuspongida
  - Order Halichondrina
    - *Halichondria panicea* (Pallas, 1766)
    - **Hymeniacidon ungodon* de Laub., 1932
  - Order Poecilosclerina (Includes all local red sponges except *Aplysilla*)
    - *Acarnus erithacus* de Laub., 1927
    - **Anaata brepha* (de Laub., 1930)
    - *Anaata spongigartina* (de Laub., 1930)
    - **Astylinifer arndti* de Laub., 1930
    - **Carmia macginitiei* (de Laub., 1930) (Found on the breakwater at Elkhorn Slough)
    - **Clathriopsamma pseudonapya* de Laub., 1930
    - **Ectyodoryx parasitica* Lambe, 1893
    - *Esperiopsis originalis* de Laub., 1930
    - *Eurypon asodes* de Laub., 1930
    - *Hymenamphiastra cyanocrypta* de Laub., 1930 (Orange-colored variants are sometimes found)

Isociona lithophoenix (de Laub., 1927)
Lissodendoryx noxiosa de Laub., 1930
Microciona microjoanna de Laub., 1930
Ophlitaspongia pennata (Lambe) *californiana* de Laub., 1932
Paresperella psila de Laub., 1930
Plocamia karykina de Laub., 1927
**Plocamissa igzo* (de Laub., 1932)
Tedania topsenti de Laub., 1930
**Tedania toxicalis* de Laub., 1930
**Tedanione obscurata* de Laub., 1930
**Zygherpe hyaloderma* de Laub., 1932

Order Haplosclerina

Adocia gellindra (de Laub., 1932)
Haliclona permollis (Bowerbank, 1866) [= *H. cinerea* (Grant, 1827)]
**Haliclona lunisimilis* de Laub., 1930
**Sigmadocia edaphus* (de Laub., 1930)
**Xestospongia vanilla* (de Laub., 1930)

Order Keratosa

Aplysilla glacialis (Dybowski, 1880)
Aplysilla polyraphis de Laub., 1930
Halisarca sacra de Laub., 1930
**Leiosella idia* (de Laub., 1932)

Phylum Coelenterata

The coelenterates may be compared to the sponges in that they are an ancient group, relatively simple in structural organization, wholly aquatic in habit, and with their greatest development in the oceans, where they are to be found from shore line to abyssal depths. At the seashore they are confined, with few exceptions, to low-tide level or below, as are the sponges, bryozoans, and tunicates, because like these groups they are not adapted to withstand intertidal exposure. In contrast to these latter groups, which are all filter feeders, all coelenterates are predators. Their continued success seems to be explained by two devices for food getting and defense, tentacles and nematocysts, combined with effective means of distribution by ciliated planula larvae and, in addition, in many groups (Scyphozoa, some Hydrozoa) a free-swimming, sexually mature medusa.

Class HYDROZOA

The class Hydrozoa is abundantly represented in the intertidal by the attached polyp stages, hydroids (fig. 5), of the order Hydroida. These range from tiny individuals to large and showy colonies, varying tremendously in form. The life cycles of hydrozoans characteristically include a sexually mature medusoid stage, which may exist free in the plankton but is more commonly retained upon the polypoid generation as an attached medusoid or as a reduced "sporosac." The presence of free-living medusa stages has led to a curious double taxonomy in the Hydrozoa, since in many cases polyp and medusa of one species have been described under separate generic and specific names, some of which have persisted even after the two forms have been recognized as stages in the life cycle of one species. Thus the tiny tentacleless polyp described as *Microhydra ryderi* is commonly called by this name even though it is now known to give rise to the sexually mature fresh-water medusa *Craspedacusta sowerbyi*, whose name should apply to the entire life cycle. In many cases, however, polyps and medusae exist for which the life cycle has not been worked out, and the separate names have to be used for the time being. In the common but rather atypical marine hydromedusae of the genus *Polyorchis*, the polyp stage has not yet been recognized.

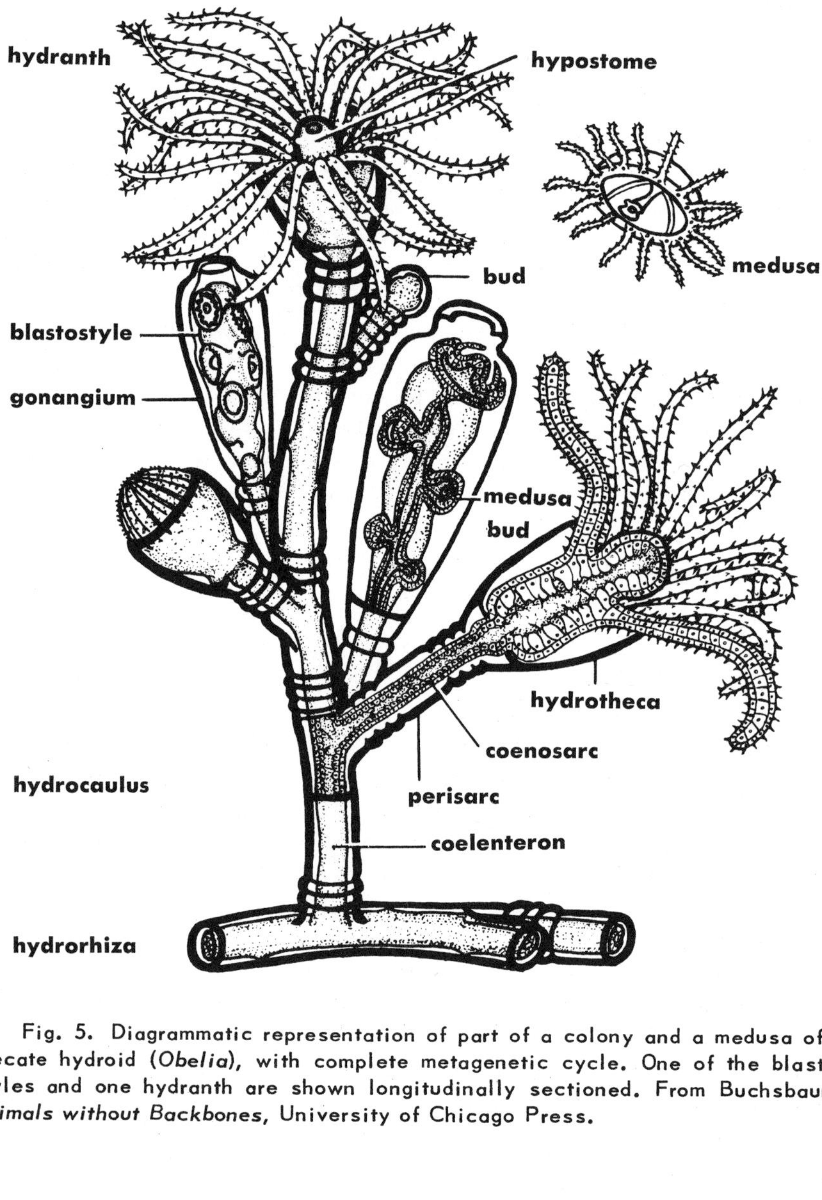

Fig. 5. Diagrammatic representation of part of a colony and a medusa of a thecate hydroid (*Obelia*), with complete metagenetic cycle. One of the blastostyles and one hydranth are shown longitudinally sectioned. From Buchsbaum, *Animals without Backbones*, University of Chicago Press.

The order Hydrocorallina (often divided into two separate orders, Milleporina and Stylasterina) is represented in the intertidal by the lavender encrusting *Allopora porphyra.* Two other orders, Trachylina and Siphonophora, do not live intertidally, but may be frequently cast ashore or collected in harbors. The trachylines are typical hydromedusae with the polyp stage inconspicuous or lacking in the life cycle. The siphonophores are complex swimming colonies, combining both polypoid and medusoid individuals into composite organisms of great beauty, such as *Velella lata* (the "purple sailor") with a flat oval float and sail-like fin (fig. 6), which is often washed ashore in vast numbers.

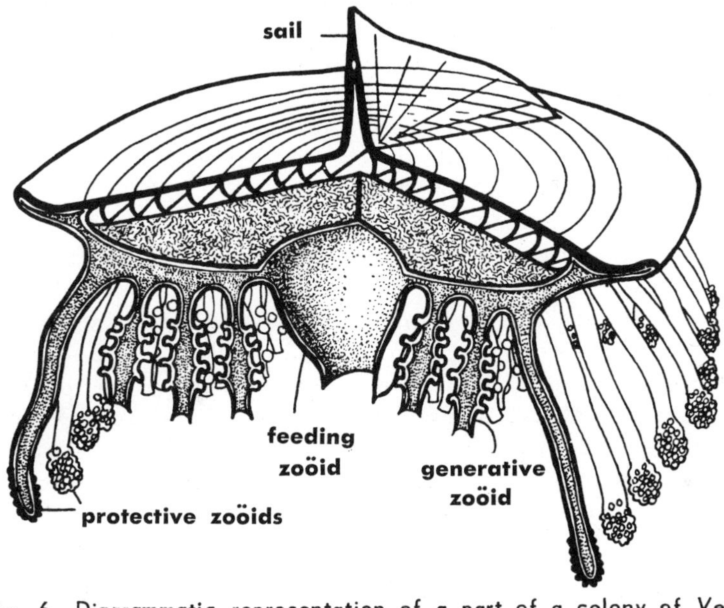

Fig. 6. Diagrammatic representation of a part of a colony of *Vellela*, a purple siphonophore. From Buchsbaum, *Animals without Backbones*, University of Chicago Press.

The order Hydroida includes an array of local intertidal forms of such varied form and structure that a detailed account is desirable. Certain terms are widely used for the parts of the hydroid colony. Unfortunately, however, they by no means always have the same meaning nor are they always consistently applied. We shall use these terms as they are used by C. McLean Fraser (1947), although somewhat different terms, as noted below, may be employed by other authors. Figure 5 illustrates the basic terms as they are used in the key.

The hydroid colony comprises essentially a continuous cellular tubular part, the coenosarc. The coenosarc consists of a layer of ectoderm or epidermis, separated by a thin layer of noncellular mesoglea from an inner layer of endoderm (gastrodermis) surrounding the continuous central cavity, the coelenteron or gastrovascular cavity. Surrounding the coenosarc more or less completely is a thin, chitinous, noncellular layer, the perisarc or periderm, secreted by the epidermis.

The colony arises by budding of new individuals from an original zoöid, which developed from a zygote. It consists, therefore, of a number of individuals known in general as zoöids. Actually no definite limits can be set to the individuals, and much of the colony is made up of connecting and rootlike branches.

There is always a certain amount of polymorphism, and the zoöids may be of several different types, named on the basis of the function for which they are specialized: nutritive (gastrozoöids), generative (gonozoöids), or defensive (dactylozoöids). The term hydranth is used to designate the terminal part of a nutritive (vegetative) zoöid but does not include any associated perisarcal structures such as the hydrotheca. The hydranth is therefore entirely coenosarcal, consisting of the body, hypostome, and tentacles. The term trophosome is used to include all the nongenerative zoöids—typically nutritive zoöids—of the colony together with any perisarcal structures associated with them.

The term gonosome is used to include all the specialized generative zoöids of the colony, with the perisarcal structures associated with them, and is thus in contrast to the term trophosome which refers to the rest of the colony. In thecate hydroids (fig. 5) the gonosome would include those asexual generative zoöids, the blastostyles, which are specialized to produce the sexual zoöids by budding, together with the coverings of the blastostyles known as gonangia (Fraser) or gonothecae, and the sexual zoöids to which the blastostyles give rise by budding. Fraser uses the term gonosome in this sense, but Pratt restricts it to the medusoid stage (sexual generative zoöids). The sexual zoöids are termed gonophores by some or are referred to in general as the medusoid stage or generation, in contrast to the hydroid stage or generation. However, Fraser uses the term gonophore synonymously with blastostyle, including often the budding sexual zoöids and protective coverings. Others would call this a gonagium, a term that Fraser uses as synonymous with gonotheca. Such an element of a colony (blastostyle with buds and protective covering, if any) is often spoken of as a "fruiting body." The term gonophore is used, therefore, with the most diverse meanings, and we must be certain to know with which of these meanings it is used in each case. This necessity of using terms whose meanings differ with the author, while annoying for the moment, affords very excellent intellectual experience.

The sexual zoöids produce gametes from which, by fertilization, arise the zygotes that develop into the initial zoöids from which new colonies arise by budding. All the zoöids of a given colony are derived from a single zygote; hence the sexual zoöids of a colony are commonly all of the same sex, and we speak of the colony as being male or female.

Typical or highly developed sexual zoöids are free-swimming medusae. All intermediate conditions are found between medusae with a long free life on the one hand, and sessile, completely reduced, sexual zoöids which have lost all remnants of medusa structure, on the other. The latter have the appearance of being gonads of the asexual generative zoöids (blastostyles) on which they arise. These extremely reduced sexual zoöids are spoken of as sporosacs, or, as mentioned above, as gonophores. Such may be seen in *Clava* among athecate hydroids (fig. 7, *c*), and in thecate hydroids in *Campanularia*. Intermediate types, sessile but still showing vestiges of medusa structure, are termed medusoids, such as are seen in *Gonothyraea* (fig. 8, *b*) in contrast to the fully developed

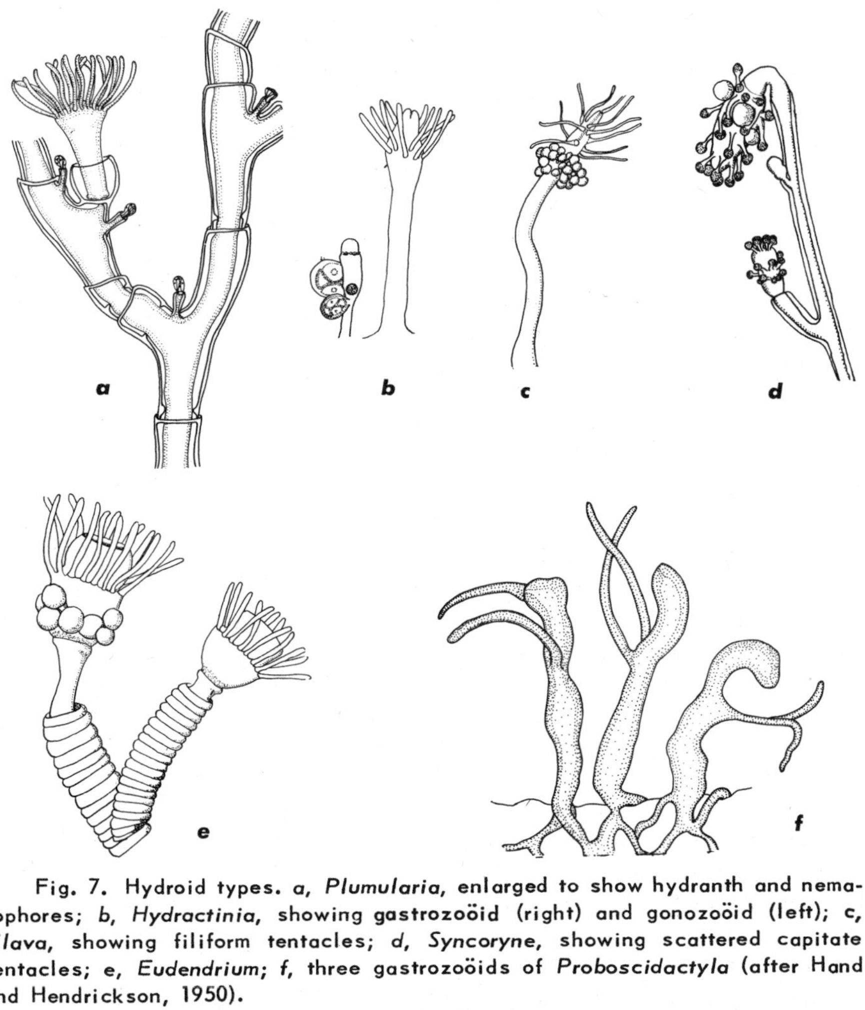

Fig. 7. Hydroid types. *a, Plumularia,* enlarged to show hydranth and nematophores; *b, Hydractinia,* showing gastrozoöid (right) and gonozoöid (left); *c, Clava,* showing filiform tentacles; *d, Syncoryne,* showing scattered capitate tentacles; *e, Eudendrium; f,* three gastrozoöids of *Proboscidactyla* (after Hand and Hendrickson, 1950).

medusae of *Obelia* or the sporosacs of *Campanularia* or *Clava.* In *Gonothyraea* the ripe medusoids are extruded from the gonothecal aperture by continued growth of the blastostyle.

In certain genera of thecate hydroids the fruiting body is made even more complex by complicated protective structures derived from other elements of the colony. The corbula in *Aglaophenia* (fig. 8, *d*) is an example.

The tentacles of the hydranth may form a single whorl as in *Obelia* (fig. 5), or two whorls as in *Tubularia,* or may be scattered over the hypostome, the elevation that bears the mouth, as in *Syncoryne* (fig. 7, *d*). They may be filiform (slender) as in *Clava* and *Obelia* or capitate (knobbed) as in *Syncoryne.* In one large group of thecate hydroids, the plumularians (family Plumulariidae), there are present, in addition to the typical hydranths, numerous small individuals that function entirely for defence and are known as nematophores (fig. 7, *a*). The small branches that bear the hydranths and nematophores are known as hydrocladia.

The hydrotheca may be stalked (*Obelia,* fig. 5) or sessile (*Sertularia,* fig. 8, *f*; *Abietinaria,* fig. 8, *a*; *Plumularia,* fig. 7, *a*; *Aglaophenia,* fig. 8, *e*). They may be uniserial (Plumulariidae, fig. 7, *a*) or biserial (Sertulariidae, fig. 8, *f*). When biserial, they may be opposite (*Sertularia,* fig. 8, *f*) or alternate (*Abietinaria,* fig. 8, *a*: *Sertularia*).

For convenience in the use of the following key and other works of reference, a glossary of terms commonly applied to hydroids follows, but it should be borne in mind that some authors apply different shades of meaning.

Glossary of Hydrozoa

Adnate. One side of structure growing attached, such as the mesial nematophores of *Aglaophenia.*

Annulated. Possessing a ringed appearance, such as the pedicel of *Eudendrium.*

Athecate. The hydranth is not protected by a chitinous periderm (hydrotheca); characteristic of gymnoblastic hydroids.

Blastostyle. Modified zoöid (gonozoöid) from which the medusoids are budded; may be protected by a surrounding theca (gonotheca or gonangium).

Calyptoblastea. The order of thecate hydroids, having each hydranth protected by a cuplike hydrotheca.

Campanulate. Bell-shaped.

Capitate tentacle. Short tentacle with a terminal knob, which is usually studded with nematocysts, as in *Syncoryne.*

Castellated. Margin of hydrothecae possessing squared teeth, giving the appearance of the battlements of a castle, as in certain campanularids.

Fig. 8. Hydroid types. *a, Abietinaria,* greatly enlarged; *b, Gonothyraea clarkii,* two gonophores: left, mature with protruding medusoids; right, immature; c-e, *Aglaophenia struthionides:* c, colony with six scattered **corbulae**: *d,* a corbula much enlarged; e, part of a hydrocladium much enlarged; *f, Sertularia furcata* (*b-f* after Fraser, 1937).

Clavate. Club-shaped, gradually thickened near one end.

Colony. The hydroid colony is a polymorphic system, consisting of at least the gastrozoöids (hydranths) or food-securing individuals, and the gonozoöids (gonangia) or asexual reproductive individuals which by budding produce free or attached medusoids. The various zoöids form a unit arising from a single planula larva and connected by means of stems and branches; rootlike structures attach the colony to substrate.

Corbula. A protective covering of the gonangia; in *Aglaophenia* the corbula is composed of several pairs of leaves which are outgrowths of the highly modified hydrocladium.

Dactylozoöid. A mouthless polyp armed with nematocysts, and serving for protection or to aid in food catching.

Distal tentacles. The circlet of tentacles farthest from the stalk of gymnoblasts that possess two circles of tentacles.

Fascicled. Hydroids in which several stems of the colony are closely bound together into a bundle or fascicle.

Filiform tentacle. Long threadlike tentacle, over which the nematocysts are usually evenly distributed.

Flexuous. Stem not straight but having a slight zigzag or sinuous appearance.

Gastrozoöid. A nutritive polyp; a hydranth.

Gonangium. A variously used term: (1) the entire asexual reproducing individual of the hydroid colony; in calyptoblasteans composed of the protecting gonotheca, the blastostyle, and the medusoid forms which bud from the blastostyle and are termed gonophores; (2) the theca surrounding the blastostyle and medusa buds; this use of the terms equals "gonotheca."

Gonophores. Sexual individuals, which in calyptoblasteans arise asexually by budding from the blastostyle, being either free medusae or reduced to sporosacs; in gymnoblasteans they may arise from the stolons, hydranths, pedicels, and hydrocauli. However, Fraser calls the blastostyle, which bears the sexual zoöids, a gonophore.

Gonotheca. The peridermal protective structure surrounding the blastostyle and gonophores in calyptoblasteans.

Gonozoöid. Another term applied to any individual of the colony which buds off sexual individuals such as free medusae or the various degrees of reduced, attached sporosacs.

Gymnoblastea. The order of athecate hydroids, lacking protective thecae about the hydranths.

Hydranth. The hydranth (= gastrozoöid) is the feeding member of the colony, having a terminal mouth surrounded by tentacles; may be either sessile or stalked.

Hydrocaulus. The main stem of the colony.

Hydrocladium (-ia). Lateral branch(es) growing from the hydrocaulus.

Hydrorhiza. The rootlike structure, which may be more or less simple or a mass of tangled tubes attaching the colony to a substrate.

Hydrotheca. The peridermal, cuplike structure that surrounds the hydranth, and into which the hydranth may be drawn when disturbed.

Internode. Part of the stem or branch between the nodes.

Medusoid form. Inclusive term applied to the various types of sexual zoöids (gonophores); it includes: (1) **free medusae** with velum, tentacles, radial canals, and manubrium (*Obelia*); (2) **gonophores (eumedusoid)** that resemble developing medusae but have no tentacles or other marginal structures, and that in some forms break away but soon die, since they lack a mouth; however, in most cases they remain attached (*Tubularia*); (3) **gonophores (cryptomedusoid)** that are simple gastodermal layers in the wall (*Clava*); and (4) the most reduced type in which the sex cells ripen directly on the sides of the blastostyle (*Sertularella*). Types 2, 3, and 4 are usually called **sporosacs**. The condition in *Hydra* is not unlike that in *Sertularella*, but there is no differentiated blastostyle.

Nematophore. The small, highly modified zoöid (dactylozoöid) of protective function which is characteristic of the family Plumulariidae.

Nodes. Joints of the stems and branches.

Operculum. A lid closing the hydrotheca or gonotheca when the zoöid is drawn in (Calyptoblastea).

Pedicel. The stalk that bears a hydrotheca.

Pinnate. Stems and lateral branches arranged in a featherlike pattern (*Aglaophenia*).

Proximal tentacles. The circlet of tentacles nearer the stalk in gymnoblasts that have two whorls of tentacles.

Retrorse. Bent backward or downward.

Septal ridges. Regular ridges of diagnostic value that occur in the internodes.

Sessile. Hydrotheca or other structure attached directly to stem or branch of colony; lacking a pedicel.

Sporosacs. Reduced gonophores that do not develop into free medusae but remain attached and produce the gametes.

Stolon. Tubular or rootlike processes that extend over the substrate; the stolons are a part of the hydrorhizal system.

Thecate. Hydroids that possess a protective, cuplike hydrotheca which surrounds the hydranth as in the Calyptoblastea.

Truncate. Having the tip squared off, as in the teeth of the hydrothecal margin.

Zoöid. A general term applied to any of the several types of individuals or "persons" of the hydroid colony.

Key to Hydroid Stages of Hydrozoa

1. Hydranths athecate. . . . . Suborder Gymnoblastea 2
1. Hydranths thecate . . . . Suborder Calyptoblastea 12
1. Hydranths concealed in a stony, vivid purple, encrusting growth on rocks (order Hydrocorallina) . . . . *Allopora porphyra*

2. Hydranth with single basal whorl of filiform tentacles . . . 4
2. Hydranth with tentacles scattered 9
2. Hydranth with only 2 tentacles; on sabellid tubes (fig. 7, *f*) . . . . *Proboscidactyla* (= *Lar*) sp.
2. Hydranth with 2 whorls of filiform tentacles; hydranths the largest of any of our hydroids; medusoids borne on long blastostyles just above basal whorl of tentacles . . . . *Tubularia* 3

3. Colony unbranched, about 1 in. in length; distal tentacles less numerous than proximal; in rocky intertidal . . *Tubularia marina*
3. Colony branched, growing in immense bushy clusters, up to 6 in. in length; proximal and distal tentacles nearly equal in number, 20 to 24; on wharf pilings in extreme low-tide zone . *Tubularia crocea*

4. Stalks of hydranths without perisarc, unbranched, rising from a spiny calcareous crust (fig. 7, *b*) . . . . . . *Hydractinia* spp.
4. Stalks of hydranths with a perisarc (except *Clava*), usually branched, not rising from a calcareous crust 5

5. Hydranth with trumpet-shaped hypostome; perisarc stiff, brown, closely and sharply annulated (fig. 7, *e*) . . . . *Eudendrium* spp.
5. Hydranth with a conical hypostome; perisarc usually not sharply annulated, or if so only on the smaller branches 6

6. Colony very small, stems not at all or only very slightly branched . . . . *Perigonimus repens*
6. Colony larger (up to several cm. in length); stems branched . . 7

7. Hydranths relatively large, orange-colored, conspicuous . . . . *Garveia annulata*
7. Hydranths smaller, not orange-colored 8

8. Hydranths very small; colony finely branched; annulations of perisarc almost entirely lacking . . . . *Bougainvillia mertensi*
8. Hydranths larger, colony with few branches; annulations distinct at base of hydranths . . . . *Bimeria* sp.

9. Tentacles capitate (enlarged at tip); colony small, very slightly branched; medusae arising among tentacles (fig. 7, *d*) . . . . *Syncoryne* 10
9. Tentacles filiform (tapering at tip); sexual zoöids borne below tentacles 11

10. Colony much branched; perisarc irregularly annulated . . . . *Syncoryne eximia*
10. Colony unbranched or slightly branched; perisarc not annulated . . . . *Syncoryne mirabilis*

11. Hydranths small, colony very slightly branched; perisarc present; free medusae . . . . *Turritopsis nutricula*
11. Colony of fair size, irregularly branched; perisarc well developed; sexual zoöids are sporosacs; in fresh or brackish water . . . . *Cordylophora lacustris*
11. Hydranths large, pink, branched, often in masses, arising from filiform hydrorhiza; perisarc absent; sexual zoöids are sporosacs, pink in male, purple in female (fig. 7, *c*) . . . . *Clava leptostyla*

12. Hydrothecae stalked 13
12. Hydrothecae sessile 17

13. Hydrothecae tubular, with operculum of converging segments; colony very small . . . . *Calycella syringa*
13. Hydrothecae reduced to saucer-shaped hydrophores, too small to contain retracted hydranths . . . . *Halecium* spp.
13. Hydrothecae bell or wineglass-shaped; colony larger 14

14. Hydrothecae with very thick walls; gonangia contain large medusoids without mouth or digestive system . . . . *Eucopella* spp.
14. Hydrothecae without excessively thickened walls 15

15. Sexual zoöids sporosacs; blastostyle not projecting from gonotheca . . . . *Campanularia* spp.
15. Sexual zoöids medusiform but sessile; blastostyle and mature medusoids project from gonotheca (fig. 8, *b*) . . *Gonothyraea clarki*
15. Sexual zoöids free medusae (during attached period do not project from gonotheca) 16

16. Medusae globular, with 4 tentacles at time of liberation; colony not much branched. . . . . *Clytia* spp.
16. Medusae flat, with 16 or more tentacles (fig. 9); colony usually more branched . . . . *Obelia* spp.

17. Hydrothecae on 1 side of branches; nematophores present . . 18
17. Hydrothecae in 2 rows on opposite sides of branches; no nematophores 19

18. Colony slender and flexible, colorless; hydrothecae at considerable intervals; margins of hydrothecae not toothed or sculptured (fig. 7, *a*) . . . . *Plumularia* spp.
18. Colony large, stiff, brown or greenish-brown; branches of colony resemble feathers (fig. 8, *c*); hydrothecae close-set with toothed and sculptured margins (fig. 8, *e*); numerous species . . . . *Aglaophenia* spp.

19. Hydrothecae exactly opposite each other; operculum of 2 pieces; colony small, delicate, without branches (fig. 8, *f*) . *Sertularia* spp.
19. Hydrothecae alternate or subalternate 20
20. Hydrotheca with smooth margin; operculum of one piece; colony larger, plumelike, with numerous side branches in one plane (fig. 8, *a*); numerous species . . . . *Abietinaria* spp.
20. Hydrotheca with toothed margin; operculum of 3 or 4 pieces; colony small, without side branches . . . . *Sertularella* spp.

List of Species of Hydroid Stages of Hydrozoa

Order Hydroida

Suborder Athecata (Gymnoblastea)

**Bimeria franciscana* Torrey, 1902
Bougainvillia mertensi Agassiz, 1862
**Candelabrum* sp. (From low intertidal at Pigeon Point)
Clava leptostyla Agassiz, 1862 (Abundant on Fruitvale and Bay Farm Island bridges, Oakland, in spring)
Cordylophora lacustris Allman, 1844 (From Antioch, on San Joaquin River)
Eudendrium sp.
**Eudendrium californicum* Torrey, 1902
Garveia annulata Nutting, 1901
Hydractinia spp.
**Hydractinia laevispina* Fraser, 1922 (On kelp off Coast Guard breakwater in Monterey Harbor)
**Hydractinia milleri* Torrey, 1902
Perigonimus repens (Wright, 1858).
**Proboscidactyla flavicirrata* Brandt, 1834 (Common at Friday Harbor on tubes of *Schizobranchia, Pseudopotamilla*)
**Proboscidactyla occidentalis* Fewkes, 1889 (Found in southern California on *Pseudopotamilla*)
Proboscidactyla sp. (Found in Monterey area on tubes of *Pseudopotamilla ocellata* and *P. intermedia*)
Syncoryne eximia (Allman, 1859)
Syncoryne mirabilis (Agassiz, 1862)
Tubularia crocea (Agassiz, 1862) (Abundant on Fruitvale bridge, Oakland, in summer)
Tubularia marina Torrey, 1902
**Tubularia harrimani* Nutting, 1901 (Sometimes found at Pescadero Point, on Monterey Peninsula)
**Turris neglecta* Lesson, 1843
Turritopsis nutricula McCrady, 1857

Suborder Thecata (Thecophora or Calyptoblastea)

Abietinaria spp.
**Abietinaria abietina* (Linnaeus, 1758)
**Abietinaria amphora* Nutting, 1904
**Abietinaria anguina* (Trask, 1857)
**Abietinaria filicula* (Ellis and Solander, 1786)
**Abietinaria greenei* (Murray, 1860)
**Abietinaria pacifica* Stechow, 1923
Aglaophenia spp.
**Aglaophenia diegensis* Torrey, 1904
**Aglaophenia latirostris* Nutting, 1900 (If ripe, will produce planulae readily when left in sun for approximately an hour.)
**Aglaophenia lophocarpa* Allman, 1877
**Aglaophenia struthionides* (Murray, 1860)
Calycella syringa (Linnaeus, 1767)
Campanularia spp.
**Campanularia denticulata* Clark, 1876
**Campanularia fusiformis* Clark, 1876
**Campanularia gelatinosa* (Pallas, 1766)
**Campanularia urceolata* Clark, 1876
**Campanularia volubilis* (Linnaeus, 1767)
**Campanulina forskalea* (Peron et Lesueur, 1809)
Clytia spp.
**Clytia bakeri* Torrey, 1904
**Clytia edwardsi* (Nutting, 1901)
**Clytia hendersoni* Torrey, 1904
Eucopella spp.
**Eucopella caliculata* (Hincks, 1853)
**Eucopella compressa* (Clark, 1876)
**Eucopella everta* (Clark, 1876) (Abundant on kelp)
Gonothyraea clarki (Marktanner-Turneretscher, 1895)
Halecium spp.
**Halecium annulatum* Torrey, 1902
**Halecium corrugatum* Nutting, 1899
**Halecium tenellum* Hincks, 1861
**Hydrallmania distans* Nutting, 1899
Obelia spp.
**Obelia bicuspidata* Clarke, 1876
**Obelia commissuralis* McCrady, 1858
**Obelia dichotoma* (Linnaeus, 1758)
**Obelia dubia* Nutting, 1901
**Obelia geniculata* (Linnaeus, 1767)
**Obelia longissima* (Pallas, 1766)

**Obelia plicata* Hincks, 1868
Plumularia spp.
**Plumularia alicia* Torrey, 1902
**Plumularia goodei* Nutting, 1900
**Plumularia lagenifera* Allman, 1885
**Plumularia plumularoides* (Clark, 1876)
**Plumularia setacea* (Ellis, 1755)
Sertularella spp.
**Sertularella pinnata* Clark, 1876
**Sertularella rugosa* (Linnaeus, 1758)
**Sertularella turgida* (Trask, 1857)
Sertularia spp.
**Sertularia cornicina* McCrady, 1858
**Sertularia desmoides* Torrey, 1902
**Sertularia furcata* Trask, 1857

Order Hydrocorallina
Allopora porphyra (Fisher, 1931)

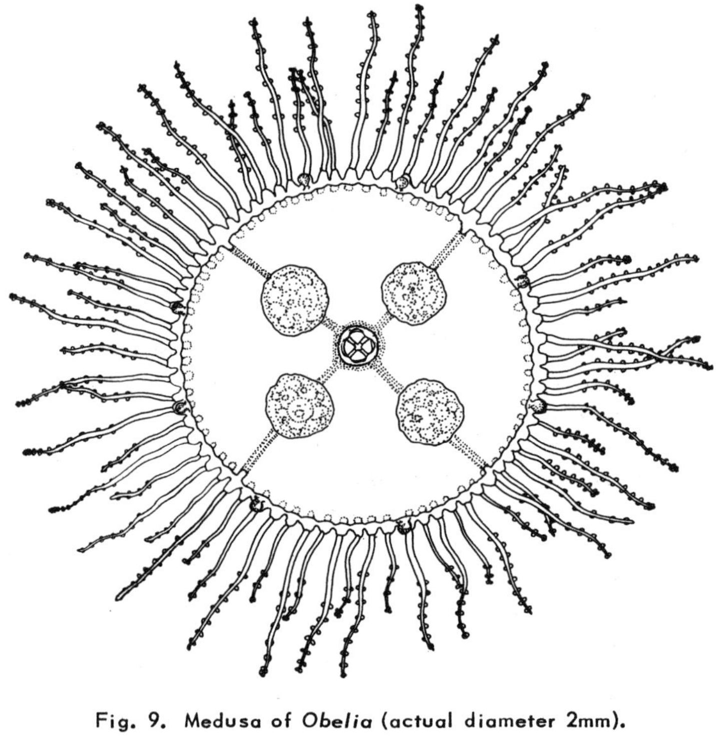

Fig. 9. Medusa of *Obelia* (actual diameter 2mm).

THE HYDROMEDUSAE

by G. F. Gwilliam
University of California, Berkeley

Hydromedusae are the free-swimming sexual stages of the class Hydrozoa as distinct from the Scyphomedusae (class Scyphozoa). The order Siphonophora is not usually included in this grouping because its members are not strictly medusiform, but are complex, highly polymorphic, floating colonies of specialized individuals, both polypoid and medusoid (fig. 6). Hydromedusae may be divided into the following groups on the basis of differences to be noted later:

Order Hydroida
- Suborder Anthomedusae (Gymnoblastea, Athecata)
- Suborder Leptomedusae (Calyptoblastea, Thecophora, Thecata)

Order Trachylina
- Suborder Trachymedusae
- Suborder Narcomedusae

The medusae belonging to the order Hydroida are the most commonly encountered forms in inshore waters and will be considered below in some detail.

The Trachylina are distinguished by the absence or slight development of a polypoid stage and by the possession of endodermal concretions in the lithocysts (these are ectodermal in the Hydroida). The Trachymedusae are difficult to recognize as such at first sight, being very similar to the medusae of the order Hydroida. Often, however, at least some of the tentacles are inserted above the bell margin on the exumbrella, and from the point of insertion an endodermal core courses through the jelly to join the ring canal. In many species there are also adhesive discs on some or all of the tentacles. These characters, however, will not always hold true for the Trachymedusae alone, because older individuals of *Polyorchis* (*Scrippsia*) *pacifica* will possess supramarginally inserted tentacles, and *Cladonema* possesses suckerlike discs on the tentacles, but both of these can be recognized by other characters. The Narcomedusae are readily distinguished by the scalloped bell margin.

The life cycle of the trachylines may, in some cases, be "direct" (i.e., planula-actinula-medusa), or it may include a polyp stage (e.g., *Gonionemus*, *Craspedacusta*). In the majority of the Narcomedusae, however, the life cycle is complicated by the parasitism of the actinula on the parent or on a different genus of narcomedusan (Hyman, 1940). The only medusae occurring in fresh water are the trachylines *Craspedacusta* and *Limnocnida*, and of these only *Craspedacusta* has been reported from the United States.

The Hydroid medusae, as mentioned above, are divided into two suborders, **Anthomedusae**, arising from gymnoblastic hydroids, and **Leptomedusae**, arising from calyptoblastic hydroids. The Anthomedusae are generally deeply bell-shaped and bear the gonads on the manubrium, whereas the Leptomedusae are usually more shallow, tending to be saucer-shaped, and bear the gonads on the radial canals.

The generalized hydromedusan (fig. 10) is usually a free-swimming animal consisting of a gelatinous bell which can be either bell- or saucer-shaped with all gradations between. The outer surface is known as the exumbrella, the inner as the subumbrella. From the center of the subumbrella surface hangs the manubrium, which can be of various lengths, and in some examples is mounted upon a gelatinous peduncle. The oral opening is terminal on the manubrium and is frequently provided with lobes (often spoken of as "lips"), frills, or tentacles which are liberally provided with nematocysts. Where the manubrium joins the bell or peduncle there is usually a gastric cavity. From the gastric cavity arise the radial canals, which course through the bell to the margin where they join the ring canal. The radial canals are usually four in number and occupy the perradii, but other numbers occur (six, eight, numerous). In a few forms the radial canals are branched.

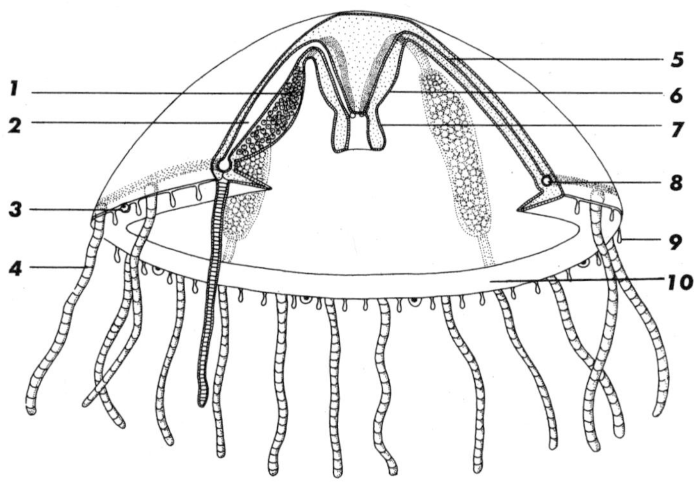

Fig. 10. A generalized hydromedusan with a part of the bell removed. 1, gonad; 2, radial canal; 3, statocyst; 4, tentacle; 5, gastrodermal lamella; 6, peduncle; 7, manubrium; 8, ring canal; 9, cordylus (sense club); 10, velum.

The Hydromedusae are typically craspedote, that is, they possess a velum or membrane that partly closes off the subumbrellar space at the level of the bell margin. This is occasionally rudimentary, as in *Obelia* (fig. 9).

The bell margin (except in Narcomedusae) is usually simple and unscalloped. Tentacles arise from the bell margin and may be simple, few or many in number, occurring singly or in groups, or they may be branched (e. g., *Cladonema*) or rudimentary. The margin may also be provided with specialized sense organs. Chief among these are the ocelli and statocysts (= lithocysts or otocysts). The ocelli occur as dark pigmented spots, usually one on each tentacle bulb. The statocysts are vesicles or open pits containing one or more concretions known as statoliths. Other marginal sensory structures are the sensory clubs (cordyli), cirri, and tactile combs (Hyman, 1940; Mayer, 1910). The gonads are epidermal folds occurring on the radial canals or the manubrium. Medusae are almost always of separate sexes.

The following key is not complete, but includes only the better-known forms encountered in Monterey Bay and a few from other West Coast localities. If a hydromedusan is found that is not included in the key, it usually can be identified by referring to Mayer (1910) or Bigelow (1909).

Medusae are best examined alive, but it is frequently necessary to anesthetize them. For this, a solution of magnesium chloride (73.2 gm. of $MgCl_2 \cdot 6H_2O$ per liter of fresh water) is recommended in the proportion of 30 or 40 per cent added to the water the animals are in. Preservation is best accomplished in 10 per cent formalin. This tends to dissolve the statoliths but leaves the structure of the statocyst intact. For histological work, Bouin's fixative is recommended.

Key to Hydromedusae

1. Gonads borne on the radial canals 2
1. Gonads borne on the manubrium 11

2. At least some tentacles apparently arising above the bell margin, gonads not elongate or fingerlike, nor pendant in the subumbrellar space . . . . Trachymedusae 3
2. Tentacles arising in a more or less single whorl at bell margin . 4

3. Marine; 4 of tentacles *terminating* in suckerlike adhesive disc; other tentacles with adhesive discs located other than terminally . . . . *Vallentinia adherens*
3. Marine; adhesive discs located on aboral surface of all tentacles, but none terminal . . . . *Gonionemus*
3. In fresh water; tentacles without adhesive discs . . . . *Craspedacusta sowerbyi*

4. Gonads fingerlike or filamentous, pendant in subumbrellar space; gonads may arise on peduncle, but always from radial canals . . . . *Polyorchis* 5

4. Gonads not fingerlike and not pendant in subumbrellar space . . 6
5. Peduncle well developed, broadly conical with pendant gonads arising from radial canals on peduncle; radial canals unbranched or only slightly so; large, about 85 mm. high and 60 mm. in diameter; tentacles numerous . . . . *Polyorchis (Scrippsia) pacifica*
5. Smaller, peduncle not as well developed as in *P. pacifica*; well-developed manubrium that hangs almost to velum; radial canals pinnately branched . . . . . . *Polyorchis penicillatus*
6. Radial canals numerous (more than 24), simple; mouth large, surrounded by numerous frilled "lips" . . . . *Aequorea*
6. Radial canals fewer than 24 7
7. Radial canals 4; very small medusae with numerous tentacles, a shallow bell (fig. 9); frequently seen swimming "inside out"; statocysts 8 (2 per quadrant), borne on basal bulbs of tentacles . . . . *Obelia* spp.
7. Radial canals 4; bell deeper than *Obelia*; seldom if ever seen swimming inside out; statocysts otherwise or wanting 8
8. Well-developed peduncle; numerous small tentacles; with 8 adradial lithocysts (2 per quadrant) borne directly on bell margin; common in Monterey Bay . . . . *Eutonina*
8. Peduncle poorly developed or absent 9
9. Lithocysts 1-3 between each 2 tentacles . . . . . . *Phialidium*
9. Lithocysts otherwise, or lacking 10
10. Lithocysts lacking; with cirri and sensory clubs (cordyli) between tentacles . . . . *Laodicea*
10. Tentacles 16, alternating with enclosed lithocysts; at time of release from hydroid only 4 tentacles and 8 lithocysts present . . *Clytia*
11. Tentacles arise in 4 clusters at distal ends of the 4 radial canals . . . . *Bougainvillia*
11. Tentacles arise singly; may be rudimentary or branched . . . 12
12. Tentacles branched 13
12. Tentacles not branched 14
13. Radial canals and tentacles usually 9, with a single well-developed suctorial appendage given off orally on each tentacle; other part (aboral) bifurcated in adult . . . . . . *Cladonema californica*
13. Tentacular appendages 2 (young) or 3 (adult); radial canals and tentacles 7 (rarely 5 or 6); tentacles in adult with numerous branches . . . . *Cladonema myersi*
14. Radial canals branch dichotomously in older specimens; interradial cnidothylacies (nematocyst sacs) on exumbrella . . *Proboscidactyla*
14. Radial canals simple; cnidothylacies lacking. 15

15. Tubular manubrium, surrounded by ringlike gonad; 4 tentacles equally developed . . . . *Sarsia*
15. Gonads in stomach wall, 4 in number, complexly folded . . . 16
16. Two diametrically opposed well-developed tentacles, others rudimentary; often with apical bell projection. . . . . *Stomotoca*
16. Tentacles 4 or more; otherwise resembles *Stomotoca* . . *Turris*

List of Hydromedusae

Anthomedusae

Bougainvillia Lesson, 1843
Cladonema californica Hyman, 1947 (Monterey Bay and Tomales Bay, California)
Cladonema myersi Rees, 1949 (Reported from southern California only)
Polyorchis penicillatus (Eschscholtz, 1829) (From San Francisco Bay)
Polyorchis (Scrippsia) pacifica (Torrey, 1909) (From San Francisco Bay and Monterey Bay)
Proboscidactyla Brandt, 1838
Sarsia Lesson, 1843
Stomotoca L. Agassiz, 1862
Turris Lesson, 1843 (*sens. amend.*)

Leptomedusae

Aequorea Peron and Lesueur, 1809 (Common in Friday Harbor area; luminescent)
Clytia Lamouroux, 1812 (in part, *sens.* Hincks, 1868)
Eutonina Hartlaub, 1897 (Very common at times; has generally been identified as *Eutimium*)
**Halistaura cellularia* (A. Agassiz, 1865) (California to southern Alaska)
Laodicea Lesson, 1843
Obelia Peron and Lesueur, 1809
Phialidium Leuckart, 1856 (Rare in Monterey Bay; common in Friday Harbor region)

Trachymedusae

Craspedacusta sowerbyi Lankester, 1880 (Fresh-water; recently reported from California)
Gonionemus A. Agassiz, 1862 (Abundant in Puget Sound among *Zostera*)
Vallentinia adherens Hyman, 1947 (Common on *Macrocystis* off Hopkins Marine Station, Monterey Bay)

CLASS SCYPHOZOA

The class Scyphozoa comprises chiefly the "true" jellyfish, which generally may be distinguished from the Hydromedusae by their larger size, fringed mouth lobes, scalloped margins, absence of a velum, and by the complex pattern of radial canals. Hydromedusans usually are small and glassy clear, possess a velum, and have four simple radial canals. More fundamental morphological differences exist, for an account of which the student is referred to Hyman, *The Invertebrates: Protozoa through Ctenophora.*

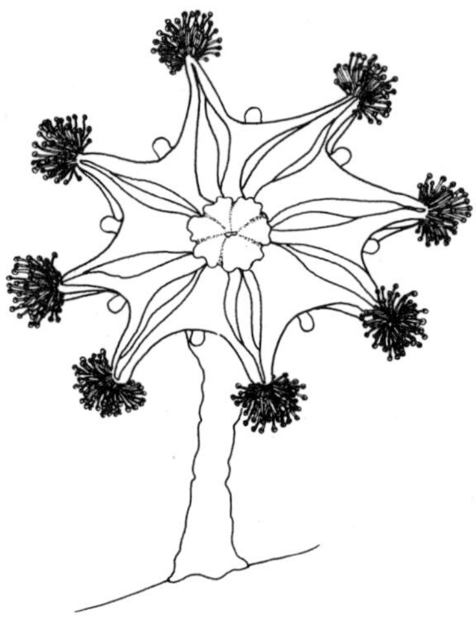

Fig. 11. *Haliclystus* sp., from *Zostera* in False Bay near Friday Harbor, Wash.

Scyphomedusae, being pelagic, are not encountered in the intertidal as adults except when cast ashore, but may be often seen about harbors. Exceptions are the curious stauromedusans (fig. 11), which possess the basic structure of scyphomedusans, but are attached by an aboral stalk to eelgrass, rocks, or algae. In addition, the attached "scyphistoma" larval stages of certain medusae may be encountered occasionally.

Key to the More Common Scyphomedusae

1. Free-swimming medusae, often large 3
1. Small and inconspicuous attached forms with 8 marginal clusters of knobs or short, knobbed tentacles (order Stauromedusae) . . 2

2. Eight lobes equally spaced about margin, each bearing short capitate tentacles . . . . . . *Thaumatoscyphus* and *Haliclystus* spp.
2. Eight lobes (each bearing small knobs) in 4 distinct pairs, separated by 4 deep notches . . . Unident. lucernarid on kelp, Monterey Bay

3. Bell dish-shaped; marginal tentacles very small and numerous; 4 horseshoe-shaped gonads; 4 moderately fringed mouth arms . . . . *Aurellia* 4
3. Bell flat or domed; marginal tentacles large and well developed; mouth arms extremely fringed and prominent 5

4. Bell margin scalloped into 8 lobes . . . . . . *Aurellia aurita*
4. The 8 lobes of bell margin secondarily notched so as to appear as 16 . . . . *Aurellia labiata*
5. Marginal tentacles arranged singly; bell domed; mouth lobes prominent and long 6
5. Marginal tentacles arranged in clusters; bell flatter; mouth lobes extensively fringed 7
6. Eight long marginal tentacles; bell radially patterned in deep purplish-brown . . . . *Pelagia panopyra*
6. Twenty-four marginal tentacles; bell with radial yellow-brown lines . . . . *Chrysaora melanaster*
7. Marginal tentacles in 16 linear groups; 16 marginal sense organs . . . . *Phacellophora camtschatica*
7. Marginal tentacles in 8 crescentic groups; 8 marginal sense organs . . . . *Cyanea capillata*

List of Scyphomedusae

Aurellia aurita (Linnaeus, 1746) (= *Aurelia*)
Aurellia labiata Chamisso and Eysenhardt, 1820
Chrysaora melanaster Brandt, 1838
Cyanea capillata (Linnaeus, 1746)
Pelagia panopyra (Peron and Lesueur, 1807)
Phacellophora camtschatica Brandt, 1838

Stauromedusae. This group is very poorly known on the West Coast. Two species of *Haliclystus* are found in the Friday Harbor region, and we have collected a species of *Thaumatoscyphus* on the Monterey Peninsula and at Pigeon Point. An unidentified lucernarid has been collected on *Macrocystis* outside Monterey Harbor. It is questionable if *Haliclystus stejnegeri* occurs in this area.

Class ANTHOZOA

The Anthozoa are the largest class of coelenterates, numbering more than 6,000 species, many of which are extremely massive. The group reaches its greatest and most diversified development in the coral-reef areas of tropical and subtropical seas, where several orders not found in this region occur. Locally, the class is chiefly represented by a fair number of sea anemones (order Actiniaria), by a few octocorals (sea pens and their kin), and by a single species of solitary stony coral (order Madreporaria).

Octocorals are distinguished by having eight pinnately branched tentacles, and by being colonial. Sea anemones are not colonial, although they may occur in dense aggregations. Stony corals are typically colonial, although our sole intertidal representative is a solitary type. The distinctive features of the class Anthozoa may be illustrated by a sea anemone (fig. 12). The oral end is broadened into a disc, in the center of which the mouth opens into a deep gullet (actinopharynx). The digestive cavity is divided by numerous radially placed vertical partitions known as mesenteries (sometimes called septa, although this term may also be used for the calcareous sclerosepta of the corals, which are not the same thing). The edges of the mesenteries are thickened into glandular and digestive regions known as mesenterial filaments, which may at their lower ends be continued as threadlike acontia (of diagnostic importance and present only in some sea anemones). The oral disc is surrounded by a circlet of hollow tentacles used in food getting.

No trace of the medusoid stage remains in the life cycle of any anthozoan. Gametes are produced by "endodermal" gonads on the mesenteries,

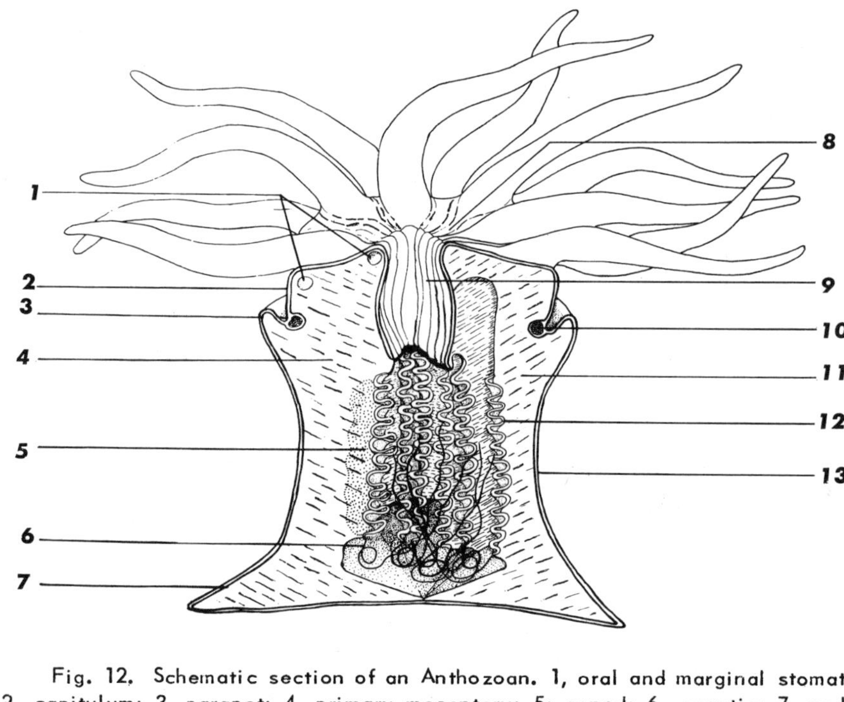

Fig. 12. Schematic section of an Anthozoan. 1, oral and marginal stomata; 2, capitulum; 3, parapet; 4, primary mesentery; 5; gonad; 6, acontia; 7, pedal disc; 8, oral disc; 9, gullet; 10, sphincter; 11, secondary mesentery; 12, mesenteric filament; 13, column.

a planula larva is developed, and after some wandering this settles down to grow into a new polyp. In *Epiactis* the larvae are protected in pockets about the outside of the column of the parent, where they appear as if in the process of being budded off asexually. Although this is not true in *Epiactis*, a number of anemones do reproduce asexually. *Metridium senile*, as it moves about, often leaves behind small mounds of tissue from the edges of the pedal disc, a process called "pedal laceration." From these bits of tissue, small anemones develop. The dense aggregations of *Anthopleura elegantissima* also arise asexually, but in this case by a more or less equal longitudinal fission of the entire animal.

Key to the Anthozoa

by Cadet Hand
University of California, Berkeley

1. Tentacles simple, neither branched nor pinnate, usually numerous . . . . Subclass Zoantharia (Hexacorallia) 2
1. Tentacles pinnately branched and 8 in number . . . . Subclass Alcyonaria (Octocorallia) 20

2. Possessing septate calcareous skeleton. (Only local representative is *Balanophyllia elegans*. The solitary polyps are a bright transparent orange in life) . . . . Order Madreporaria
2. Without skeleton . . . . Order Actiniaria 3

3. Base unattached; elongate and slender (up to 2.5 in. long by about ¼ in. in diameter); 10 tentacles . . . . *Halcampa* sp.
3. Base attached to substrate 4

4. With 24 tentacles; column with cuticular sheath; tentacles orange, sometimes with brown markings; on open coast, frequently in kelp holdfasts . . . . *Cactosoma arenaria*
4. With more than 24 tentacles 5

5. Without acontia 11
5. With acontia 6

6. Margin of tentacle-bearing disc usually deeply frilled or lobed when extended; tentacles short and very numerous; very little tentacle-free area around the mouth; in harbors and bays, commonly on pilings . . . . *Metridium senile*
6. Margin of disc circular and not frilled or lobed when extended; rarely more than 100 tentacles present; a fairly large area of disc tentacle-free 7

7. The tentacles (directives) closest to each end of mouth slit with yellow bases; color of the column variable (cream, gray, light green);

column transparent when extended and frequently with vertical white stripes . . . . *Diadumene* sp.

7. Directive tentacles not marked differently than others . . . 8

8. Column transparent olive green, brownish-green, or other shade of green when extended; sometimes orange gonads are visible in the lower parts of the column; vertical stripes of orange, yellow, or white frequently present on column; along bay shores, under rocks or on pilings . . . . *Diadumene luciae*

8. Column white, pink, or orange; not as above 9

9. Column and tentacles flesh to pale salmon-colored; upper parts of column may show tints of green; long and slender when extended; on floats and pilings, stones, or oyster shells; in bays . . . . *Diadumene leucolena*

9. Column orange, yellow, or reddish 10

10. In extension column about as tall as wide or slightly taller than wide; up to 1 cm. in diameter; orange or yellow, sometimes reddish; under rocks or ledges on open coasts . . . . . . *Metridium* sp.

10. In extension column at least twice as tall as wide; column never more than 5 mm. in diameter; color a transparent light orange or yellow; along edges of tidal channels where sand has been deposited among algal holdfasts; . . . . *Diadumene* sp.

11. Tentacles capitate; color variable (white, pink, red, brown, lavender) . . . . *Corynactis californica*

11. Tentacles not capitate 12

12. Column with tubercles capable of holding bits of sand, shells, and so on 13

12. Column smooth, without tubercles 18

13. With white to yellow marginal spherules containing very numerous atrichous nematocysts (spherules at top of column just outside of tentacles); column green, yellow, or white 14

13. No marginal spherules; column a shade of red, sometimes red interrupted by green or brownish-green patches 16

14. Tubercles on upper 2/3 of column only; column white or pink on lower 2/3, usually black or gray on upper 1/3; lives buried in sand or gravel and attached below the surface to rocks; tentacles frequently brightly colored; body capable of great elongation . . . . *Anthopleura artemisia*

14. Whole column covered with tubercles; column usually green . . 15

15. Tubercles irregular, compound, not conspicuously in longitudinal rows; tentacles uniform in color, not pink-tipped; large, solitary anemones . . . . *Anthopleura xanthogrammica*

15. Tubercles round, arranged in longitudinal rows; tentacles frequently

tipped with pink or purple; column green to yellow or white; small to medium-sized anemones; usually in aggregated masses, often attached to rocks in sand . . . . . *Anthopleura elegantissima*

16. Column densely covered with tubercles which are strongly adherent; usually buried in gravel or shell debris; column covered with adhering gravel, and so on . . . . *Tealia coriacea*
16. Column not densely covered with tubercles and seldom with adhering foreign material 17

17. Column bright scarlet; tubercles white, few, in regular longitudinal rows . . . . *Tealia lofotensis*
17. Column red, frequently marked with irregular patches of green or brownish-green; tubercles absent or very few, weak, scattered . . . .*Tealia crassicornis*

18. Young anemones usually attached on the column of the adult; diameter of column greater than height; edge of base marked with white lines . . . . *Epiactis prolifera*
18. Edge of base not marked with white lines 19

19. Column several times longer than broad (up to 7 in. long by 1/2 in. in diameter); in mud or sand, attached to stones, worm tubes, or shells; estuarine, Elkhorn Slough . . . . Undescribed isanthid
19. Column as tall as broad; color red or red varied with green or greenish-brown patches; large (up to about 9 cm. diameter) . . . . *Tealia crassicornis*

20. Low, encrusting, ramifying growths on rocks, shells, and so on; individual polyps usually contracted to small pinkish mounds, tentacles rarely visible in the field; sometimes mistaken for small simple ascidians . . . . Order Stolonifera
(There are two local intertidal forms, of which the more common, at least, appears to be a species of *Clavularia*)

20. Erect, featherlike; rooted in soft substrates by a fleshy base; usually subtidal but occasionally washed up or seen in protected shallow water areas; the sea pens . . . . . Order Pennatulacea 21

21. Distinctly thick and fleshy . . . .*Leioptilus* spp.
21. Elongate and slender 22

22. Rough to the touch; axis strong and brittle . . *Stylatula elongata*
22. Smooth to the touch; axis thin and flexible . *Acanthoptilum gracile*

List of Anthozoa

Subclass Zoantharia (= Hexacorallia)
  Order Madreporaria
    Balanophyllia elegans Verrill, 1864
  Order Actiniaria
    Anthopleura artemisia (Pickering in Dana, 1848) (Formerly *Evactis*)
    Anthopleura elegantissima (Brandt, 1835) (Formerly *Bunodactis* or *Cribrina*)
    Anthopleura xanthogrammica (Brandt, 1835)
    Cactosoma arenaria Carlgren, 1931
    Corynactis californica Carlgren, 1936 (Possesses very large nematocysts, excellent for class study)
    Diadumene leucolena (Verrill, 1866) (= *Sagartia leucolena*)
    Diadumene luciae (Verrill, 1898) (= *Sagartia luciae*; characteristically among barnacles in bays and estuaries)
    Diadumene sp. (7) Hand (in MS)
    Diadumene sp. (10) Hand (in MS)
    Epiactis prolifera Verrill, 1869
    Halcampa sp. Hand (in MS)
    Metridium senile (Linnaeus, 1767) [= *M. fimbriatum* McMurrich and *M. dianthus* (Ellis)]
    Metridium sp. Hand (in MS)
    Tealia coriacea (Cuvier, 1798)
    Tealia crassicornis (Müller, 1776)
    Tealia lofotensis (Danielssen, 1890)
    (Undescribed isanthid) Hand (in MS) (A species from Elkhorn Slough, referred to as *Harenactis attenuata* Torrey by MacGinitie, 1935; not found in past several years)
Subclass Alcyonaria (= Octocorallia)
  Order Stolonifera
    Clavularia sp.
  Order Pennatulacea
    Acanthoptilum gracile (Gabb, 1863)
    Leioptilus sp. [Specimens washed up at Monterey appear to be either *L. quadrangularis* (Moroff) or *L. guerneyi* (Gray)]
    Stylatula elongata (Gabb, 1863)

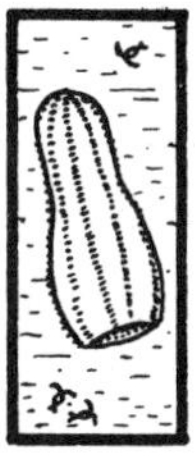

Phylum Ctenophora

The ctenophores or comb jellies (fig. 13) are transparent jellylike animals belonging to a small and entirely marine phylum which resembles the coelenterates in appearance, but which is set apart by a number of important differences. These include the absence of nematocysts and of tentacles about the mouth, the possession of a unique aboral sense organ and of a marked biradial symmetry, and locomotion by means of eight meridional rows of ciliary platelets (the comb rows) which give the phylum its name. All are micropredators, often capturing their prey by means of the colloblasts (glue cells or lasso cells) on long, paired tentacles. Some members of this group glow brilliantly when disturbed in darkness. The commonest local species is *Pleurobrachia bachei*, often cast up on beaches, and variously referred to as sea gooseberries, sea walnuts, or cats' eyes.

Fig. 13. A ctenophore.

Key to the Ctenophora of California

by Rolf Bolin
Hopkins Marine Station of Stanford University, Pacific Grove

1. Body resembling empty sack; neither tentacles nor oral lobes present . . . . Class Nuda, Order Beroida 2
1. Body not sacklike; tentacles (in some species minute and almost impossible to detect) and sometimes oral lobes present . . . . Class Tentaculata 3
2. Side branches of meridional canals forming anastomosing network;

body basically triangular in shape, widest at mouth . *Beroë forskali*
2. Side branches of meridional canals ending in blind twigs; body basically ovoid in shape, mouth somewhat constricted *Beroë cucumis*

3. Oral lobes (liplike flaps on either side of mouth) well developed and conspicuous . . . . Order Lobata 4
3. Oral lobes not developed . . . . Order Cydippida 5

4. General body surface smooth and colorless . *Bolinopsis microptera*
4. Surface of body bearing large conspicuous papillae tipped with orange or brown . . . . *Leucothea* sp.

5. Body much flattened laterally . . . . *Mertensia ovum*
5. Body subcircular in cross section 6

6. Body almost spherical . . . . *Pleurobrachia bachei*
6. Body oval, markedly longer than wide 7

7. Ciliary plates ending about midway between oral pole and equator; tentacular sheaths in contact with stomach throughout most of their length . . . . *Hormiphora* sp.
7. Ciliary plates ending much nearer oral pole than equator; tentacular sheaths not in contact with stomach . . *Euplokamis californiensis*

List of Species

Beroë cucumis Fabricius, 1780
Beroë forskali Milne-Edwards, 1841 (Luminesces beautifully after a period in darkness)
Bolinopsis microptera (A. Agassiz, 1865)
Euplokamis californiensis Torrey, 1904
Hormiphora sp.
Leucothea sp.
Mertensia ovum (Fabricius, 1780)
Pleurobrachia bachei A. Agassiz, 1860 (By far the commonest species)

Phylum Platyhelminthes

The phylum Platyhelminthes is the lowest of the truly bilateral phyla or "Bilateria." Two of the three classes, the **Trematoda** or flukes and the **Cestoda** or tapeworms, are parasitic and will not be discussed in this manual. The third class, the **Turbellaria**, is composed of predaceous free-living forms, with a few parasitic members.

The class **Turbellaria** contains five groups or orders: (1) **Acoela**, small marine forms without an intestine; one species common in the Monterey region. (2) **Rhabdocoela**, small, typically narrow forms in which the intestine is a single straight tube; common in fresh water, but represented in the local intertidal only by two parasitic species. (3) **Alloeocoela**, small, mostly marine forms, sometimes lumped with the **Rhabdocoela**, from which they differ in having a more complex intestine and other features; these are not likely to be encountered locally, or have been overlooked. (4) **Tricladida**, in which the intestine has three main branches, one forward from the pharynx and two backward; common locally in fresh water and soil, with marine representatives scarce. (5) **Polycladida**, a large marine group abundant locally, having thin, leaflike bodies of good size, intestine with many branches; frequently active and colorful.

The only conspicuous local acoel is the orange *Polychoerus carmelensis*, of the *Ulva*-filled high-tide pools of the Monterey region.

Rhabdocoels are commonly encountered in fresh-water collecting, but the only readily obtained intertidal representative of the order is *Syndesmis*, parasitic in the intestines of sea urchins.

We have not encountered triclads in marine collecting on the central California coast, but members of the order are common in fresh water and in certain situations on land. Students interested in the group should consult the papers of Hyman listed in the Bibliography. The chief local forms are included in the list that follows the key to polyclads.

The polyclads of the Pacific Coast are an abundant but difficult group. Fortunately we have been able, at the last moment, to receive a key to the group from Dr. Hyman, which provides for the first time a reliable synopsis. For descriptive details and for the less common species, the student should consult Hyman's valuable monograph (1953).

Key to the More Common Polyclads from Monterey Bay to Vancouver Island

by Libbie H. Hyman
American Museum of Natural History

Those seriously desiring to identify polyclads must realize that exact identifications are impossible in many cases except by study of serial sagittal sections of the copulatory apparatuses. Eye arrangement cannot be determined satisfactorily except on fixed, dehydrated, and cleared specimens. Such features as the penis stylet and Lang's vesicle may be seen by pressing live specimens between slides. Color is variable and often depends on ingested food.

1. Ventral surface without a sucking disc; tentacles when present of nuchal type (over brain region, well back from anterior margin) . 2
1. With sucker on ventral surface; tentacles when present situated at anterior margin 21

2. Completely devoid of eyes; small, brown, elliptical . *Plehnia caeca*
2. With eyes 3

3. Band of eyes present along all, or anterior part, of body margin; eyes also present elsewhere 4
3. Without marginal eyes; eyes limited to clusters over brain region 9

4. Marginal band of eyes limited to anterior 1/3–1/2 of body margin; with nuchal tentacles 5
4. Marginal band of eyes completely encircling margin; with or without nuchal tentacles 6

5. Small, anterior end narrowed; cerebral eyes few . . . . *Stylochus franciscanus*
5. Large, oval; cerebral eyes numerous . . . *Stylochus tripartitus*

6. With prominent nuchal tentacles 7
6. Nuchal tentacles wanting in adult, present in young; adult very large . . . . *Stylochus atentaculatus*

7. Very large, oval, no posterior notch 8
7. Small, with notch for gonopores in center of posterior margin . . . . *Stylochus exiguus*

8. With vaginal duct . . . . *Kaburakia excelsa*

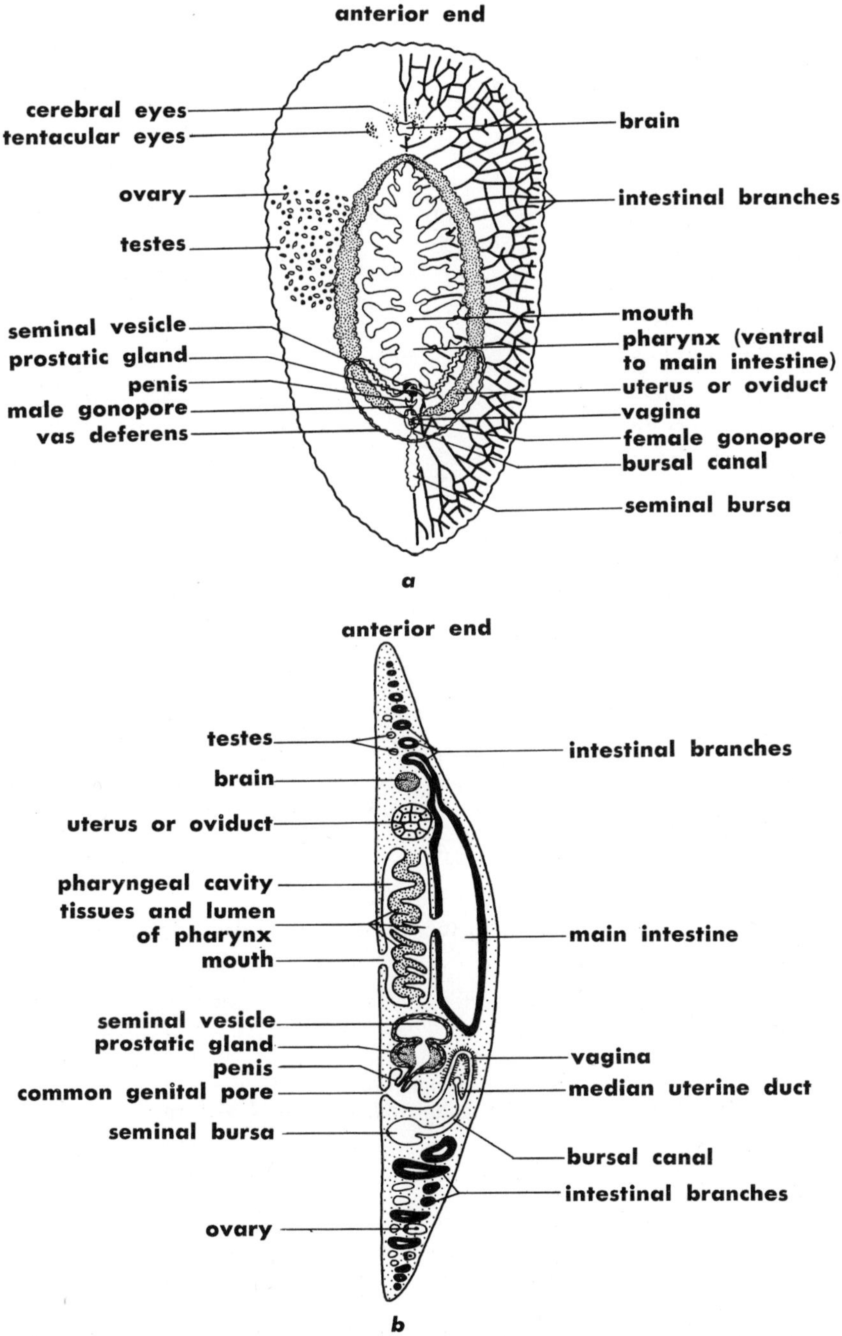

Fig. 14. *a*, diagrammatic ventral view of *Leptoplana timida* Heath and McGregor (redrawn from Bresslau after Heath and McGregor); *b*, diagrammatic median longitudinal section through *Stylochoplana agilis* Lang (redrawn from Bresslau after Lang).

8. Without vaginal duct . . . . *Stylochus californicus*

9. With nuchal tentacles 10
9. Without nuchal tentacles 12

10. Small, of cuneate form. . . . . *Stylochoplana gracilis*
10. Larger, oval 11

11. With digestive branches conspicuously outlined in chocolate brown on lighter ground of greenish tint . . . . *Alloioplana californica*
11. Digestive branches not so outlined . . *Pseudostylochus burchami*

12. With penis stylet 13
12. Without penis stylet 14

13. Penis stylet straight, relatively short; common gonopore; Lang's vesicle very long . . . . *Notoplana inquieta*
13. Penis stylet very long, coiled; gonopores widely separated; Lang's vesicle reduced . . . . *Notoplana longastyletta*

14. With strongly developed spermiducal bulbs 15
14. Without spermiducal bulbs. 16

15. With common large gonopore and large conical penis papilla, often seen protruding from gonopore in preserved specimens . . . . *Freemania litoricola*
15. With separate inconspicuous gonopores and small hidden penis papilla . . . . *Leptoplana chloranota*

16. With red mark on dorsal surface; Puget Sound . *Notoplana sanguinea*
16. Without red mark, grayish or brownish, plain or mottled . . . 17

17. Lang's vesicle very long . . . . *Notoplana rupicola*
17. Lang's vesicle shorter 18

18. Lang's vesicle short and plump, about 1/2 the length of that of *rupicola* 19
18. Lang's vesicle small, reduced. 20

19. California; does not swim . . . . *Notoplana acticola*
19. Limited to Puget Sound; swims . . . . . . *Notoplana natans*

20. Penis papilla exceptionally long and slender; California . . . . *Notoplana saxicola*
20. Penis papilla short and conical; Puget Sound northward . . . . *Notoplana sanjuania*

21. Marginal tentacles formed of upturned folds; pharynx ruffled . . 22
21. Marginal tentacles not obvious folds; pharynx tubular. 24

22. Colored white and black 23
22. Not white and black; pale with brown flecks; Puget Sound northward . . . . *Pseudoceros canadensis*

23. White with narrow middorsal black stripe; California . . . . *Pseudoceros luteus*
23. White with black border and middorsal black markings; California . . . . *Pseudoceros montereyensis*

24. Orange-red. . . . . *Eurylepta aurantiaca*

List of Turbellaria

Order Acoela
- **Polychoerus carmelensis* Costello and Costello, 1938 (Often abundant on *Ulva* and stones in high tide pools of Monterey peninsula; excellent for class study)

Order Rhabdocoela
- **Dalyellia* sp. and similar spp. (Small fresh-water forms, brilliant green from contained or symbiotic algal cells or zoöchlorellae)
- **Mesostoma* sp. (Very transparent, with internal organs showing clearly in the flattened body; common in fresh-water ponds)
- **Syndesmis dendrastrorum* Stunkard and Corliss, 1951 (Reported from intestine of sand dollar *Dendraster* at La Jolla)
- **Syndesmis franciscana* (Lehman, 1946) (Common in intestines of sea urchins *Strongylocentrotus* on central Californian coast; good class material)

Order Tricladida
- **Bipalium adventitium* Hyman, 1943 (About greenhouses; single median stripe)
- **Bipalium kewense* Moseley, 1878 (Under pots in greenhouses; 5 dorsal stripes and a pair of dark spots on sides on "neck")
- **Dugesia tigrina* (Girard, 1850) (The most common "planarian" of ponds)
- **Geoplana mexicana* Hyman, 1939 (In lawns and gardens; dark color; body cylindrical)
- **Polycelis coronata* (Girard, 1850) (In streams and springs; eyes numerous)

Order Polycladida
- *Alloioplana californica* (Heath and McGregor, 1912)
- *Eurylepta aurantiaca* Heath and McGregor, 1912
- *Freemania litoricola* (Heath and McGregor, 1912)
- *Kaburakia excelsa* Bock, 1925
- *Plehnia caeca* Hyman, 1953
- *Pseudoceros canadensis* Hyman, 1953
- *Pseudoceros luteus* (Plehn, 1898)
- *Pseudoceros montereyensis* Hyman, 1953

Pseudostylochus burchami (Heath and McGregor, 1912)
Stylochoplana gracilis Heath and McGregor, 1912
Stylochus atentaculatus Hyman, 1953
Stylochus californicus Hyman, 1953
Stylochus exiguus Hyman, 1953
Stylochus franciscanus Hyman, 1953
Stylochus tripartitus Hyman, 1953
Notoplana acticola (Boone, 1929)
Notoplana inquieta (Heath and McGregor, 1912)
Notoplana longastyletta (Freeman, 1933)
Notoplana natans Freeman, 1933
Notoplana rupicola (Heath and McGregor, 1912)
Notoplana sanguinea Freeman, 1933
Notoplana sanjuania Freeman, 1933
Notoplana saxicola (Heath and McGregor, 1912)
Leptoplana chloranota (Boone, 1929)

Phylum Nemertea (Rhynchocoela)

Nemertean worms are readily recognized by their soft, elongated, narrow, highly contractile, unsegmented bodies, which are covered with cilia. The intestine is usually straight and opens at the posterior end of the body. A distinctive feature of the phylum is the long, slender, eversible proboscis, which is not usually connected with the digestive tract, but is an infolding of the anterior body walls. This proboscis may be caused to evert by adding alcohol, a few drops at a time, to a small amount of sea water containing the worms, or by putting them into fresh water. It is often everted like the finger of a glove when the worm is disturbed or injured. Like the polyclads, the nemerteans are predators with a wide range of prey. For this purpose the proboscis with its poisonous secretions or needlelike stylets is an efficient weapon.

Although the great majority of nemerteans are free-living, bottom-dwelling, marine forms, a few species are found in other situations or have another way of life. Thus *Prostoma rubrum* inhabits fresh water, a few species are found in damp situations on land, and certain forms such as *Nectonemertes* swim or float in mid-water, deep beneath the surface of the sea. *Malacobdella* lives as a commensal in the mantle cavity of bivalve molluscs, whereas *Carcinonemertes epialti* may be found parasitic (?) among the gills or egg masses of *Pugettia* and other crabs.

Nemerteans are handled with difficulty because of their extreme contractility, their secretion of large quantities of slimy mucus, and the tendency of many to break into fragments. They should be kept separated from other animals and must be anaesthetized carefully before preservation. For most purposes preservation in 10 per cent formalin is satisfactory, but since acids or formalin will destroy the stylets of hoplonemerteans, members of this group should be preserved in 80 per cent alcohol.

The field identification of nemerteans by the use of references such as Coe (1940) is made difficult by the fact that the division into orders is based in part upon the arrangement of muscle layers in the body wall. The general descriptive features below may be useful in deciding the probable order of an unknown specimen.

Class **Anopla.** Mouth posterior to brain. Proboscis not armed with stylets.

Order **Paleonemertea.** In general, paleonemerteans in this area are slender, soft, and extensile, heads somewhat blunt, bodies not much flattened, ocelli and longitudinal cephalic slits lacking.

Order **Heteronemertea.** Heads characteristically rather snakelike, with marked lateral slits (*Baseodiscus* is an exception). A small caudal cirrus is found in *Cerebratulus* and *Lineus*.

Class **Enopla.** Mouth anterior to brain. Proboscis armed with one or more stylets in all except the Bdellonemertea.

Order **Hoplonemertea.** The stylets are diagnostic (flatten animals cautiously between microscope slides and examine by transmitted light). This order includes most of the local species, including some of the most brightly and distinctively marked.

Order **Bdellonemertea.** A small group, commensal in mantle cavities of lamellibranchs. The leechlike form with a posterior sucking disc is distinctive.

Key to Nemertea

1. Commensal in mantle cavity of marine bivalve molluscs; body short, flattened, leechlike, with rounded sucker at rear . . . . *Malacobdella grossa*
1. Free-living; in fresh water; small; pink or reddish . . . . *Prostoma rubrum*
1. Free-living; marine; body usually elongate and never with a sucker 2

2. Entire body or dorsal surface solid color; no conspicuous lines or rings 3
2. With longitudinal lines, transverse rings, or spots of color contrasting with ground color 15

3. Of approximately same shade above and below 4
3. Darker above, distinctly lighter below 10

4. Small, slender; whitish, pink, or pale flesh-colored 5
4. Usually larger, with distinctive colorations (red, dark brown, brown, orange). 7

5. Whitish or pinkish, small; rather opaque; ocelli numerous, in 4 poorly defined rows on flattened head; with 2 or 3 pouches of accessory stylets . . . .*Amphiporus imparispinosus*

5. White or whitish; ocelli lacking 6

6. Minute, slender; somewhat transparent; in delicate tubes under stones, among algae, or in sand . . . . *Tubulanus pellucidus*
6. Elongate; body rounded anteriorly, with a long, very flattened posterior region in which the internal organs are visible as transverse darker lines in lateral areas . . . . *Carinoma mutabilis*

7. Lacking longitudinal cephalic grooves and caudal cirrus . . . 8
7. Possessing longitudinal cephalic grooves or slits on the sides of the head; small caudal cirrus (often missing) 9
7. Possessing cephalic grooves but lacking caudal cirrus; a row of 4-6 small ocelli on each side of head; color variable, olive to reddish; tolerates low salinities well . . . . . *Lineus ruber*

8. Color bright reddish-orange; soft-bodied and very extensile . . . . *Tubulanus polymorphus*
8. Color dark brownish-red, snout truncate and tipped with white, body may appear covered with whitish bloom; numerous small ocelli along sides of head; body flabby but not easily ruptured, tending to contract so that surface is thrown into transverse wrinkles . . . . *Baseodiscus punnetti*

9. Dark red to brown; deep cephalic grooves; anterior body firm and stout, posterior flattened and easily fragmented; in sand or mud. (The color of this species is extremely variable. Young individuals may be pale grayish, pinkish, or yellowish, whereas individuals from clean sand may be white with brown head) . . . . *Cerebratulus californiensis*
9. Anterior body flesh-colored or rosy, lighter on head, which tends to be pointed; shallow cephalic grooves; flattened posteriorly; in sand or mud . . . . *Micrura alaskensis*

10. Head either darker or lighter than rest of body 11
10. Color of head not markedly different from rest of body . . . 14

11. Head darker, or with dark patch 12
11. Head lighter, but with dark marking(s). 13

12. Very slender; head darker than rest of body; body light olive green to brown; a line of 2–8 ocelli on either side of anterolateral region of head. If placed in clean sea water and cut into many pieces, each piece will regenerate into a minute replica of the original worm within 3 or 4 weeks . . . . *Lineus vegetus*
12. Small, slender; head with 4 distinct ocelli, between which is a dark brown triangular or square patch; body color pale brown or flesh color. Young may lack dark patch, and color variations may include a brown body with central stripes, or a dark purplish body . . . . *Tetrastemma nigrifrons*

13. Small, slender, with 4 ocelli between which is a dark patch . . . . Dark variety of *T. nigrifrons*
13. Broad, fleshy; head light with 2 conspicuous dark brown or black oval or triangular spots; dorsal body color deep brownish orange (fig. 15, *b*) . . . . *Amphiporus bimaculatus*

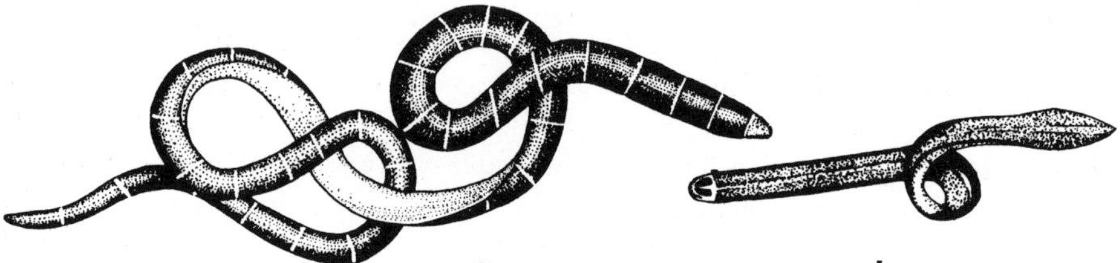

Fig. 15. Two common nemerteans (after Coe). *a, Micrura verrilli* Coe; its distinctive white caudal cirrus or tail has been lost as is often the case; *b, Amphiporus bimaculatus* Coe.

14. Very slender; uniform green above, pale below; very long curved stylets . . . . *Empleçtonema gracile*
14. Stouter; uniform brown, purplish, or orange-brown on dorsal surface and sides; yellowish beneath . . . . .*Paranemertes peregrina*

15. Small or very slender 16
15. Larger and stouter. 17

16. Dorsal surface pale brown or yellow, covered with small dark spots, sometimes elongated and arranged in irregular longitudinal lines, or grouped into rectangular areas separated by narrow transverse bands; body flattened; small caudal cirrus . . *Micrura pardalis*
16. Dorsal surface with 2 narrow longitudinal dark lines on light background; 2 large ocelli on each side . . . *Nemertopsis gracilis*
16. Extremely slender, a single row of 2–8 ocelli on either side of head; usually with fine longitudinal and circular lines (see 12 for those without lines) . . . . *. Lineus vegetus*

17. General body color deep brown; 5–6 slender, often broken, longitudinal white lines; numerous white narrow rings spaced at fairly regular intervals . . . . *Tubulanus sexlineatus*
17. General body color ivory white with a series of deep purple or wine-colored rectangular dorsal areas, separated by narrow white bands; dorsal side of head red or orange; small caudal cirrus (often lost) (fig. 15, *a*) . . . . *.Micrura verrilli*

List of Nemerteans

Order Paleonemertea

Carinoma mutabilis Griffin, 1898 (In mud flats)

Tubulanus pellucidus (Coe, 1895) (In mud, sands, gravel, and under stones)

Tubulanus polymorphus Renier, 1804 (Under heavy boulders bedded in gravel)

Tubulanus sexlineatus (Griffin, 1898)

Order Heteronemertea

Baseodiscus punnetti (Coe, 1904)

Cerebratulus californiensis Coe, 1905 (In sand and mud flats)

Lineus ruber (O. F. Müller, 1771) (Endures brackish water; common in Aquatic Park, Berkeley)

Lineus vegetus Coe, 1931

Micrura alaskensis Coe, 1901 (In mud flats)

Micrura pardalis Coe, 1905

Micrura verrilli Coe, 1901

Order Hoplonemertea

Amphiporus bimaculatus Coe, 1901

Amphiporus imparispinosus Griffin, 1898 (Among coralline algae on exposed rocks)

**Carcinonemertes epialti* Coe, 1902 (In egg masses of *Pugettia* and other crabs)

Emplectonema gracile (Johnston, 1837) (Occurs in masses among mussels and barnacles on pilings)

Nemertopsis gracilis Coe, 1904 (Under large boulders embedded in coarse sand)

Paranemertes peregrina Coe, 1901 (Among mussels and corallines on exposed rocks; often out in the open)

Prostoma rubrum (Leidy, 1850) (In fresh water, e.g., Lake Temescal)

Tetrastemma nigrifrons Coe, 1904

Order Bdellonemertea

Malacobdella grossa (O. F. Müller, 1776)

Aschelminth Complex

Under this heading are included several groups which have conventionally been considered separate phyla, but which Hyman (1951) places in a single phylum, Aschelminthes. They are characterized by being pseudocoelomate (i. e., having a body cavity that represents the persistent blastocoel or segmentation cavity), and by having a pronounced cuticle over the body, a marked tendency to cell-number constancy, a complete digestive tract with posterior anus, and other features for which the student is advised to consult Hyman. The Aschelminthes include the groups (classes or phyla, according to opinion) Rotifera, Gastrotricha, Kinorhyncha, Nematoda, Nematomorpha, and quite possibly the Priapuloidea. (The latter, in the past, have been placed near the sipunculids and echiuroids, which they superficially resemble; see p. 111).

Rotifera. These are among the smallest of metazoans, even smaller than many protozoans. They may be free-swimming or sessile, and are present in almost all fresh waters. Although said to be uncommon in the sea, they sometimes are present in great numbers in plankton hauls and might be confused with veliger larvae. Rotifers can be recognized by the corona or velum, edged with large cilia by which the animal swims and feeds. The corona is often produced into two or four lobes, which the active cilia appear to make revolve, hence the name "wheel animalcule" of the older microscopists. Internally (fig. 16) rotifers possess a set of clamping jaws known as the mastax, an intestine, gonads, and a flame cell or protonephridial excretory system. Certain rotifers provide the clearest demonstration of flame cells that we have encountered.

Gastrotricha. Gastrotrichs (fig. 17) are minute animals, of distinctive form, commonly encountered in fresh-water cultures and sometimes in marine collections. The lack of a corona, the spiny body, and a characteristic gliding motion distinguish them at a glance from rotifers.

Kinorhyncha (Echinoderida). Seldom encountered except after special search, echinoderids are very minute creatures with a body that appears segmented or covered with scales. See Hyman (1951) for a general account.

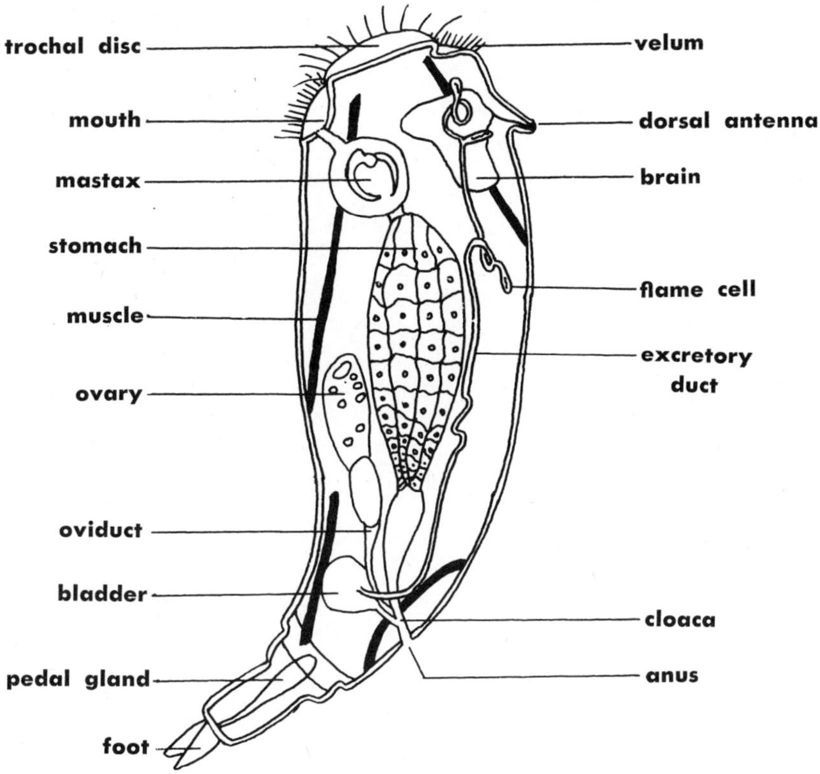

Fig. 16. Diagrammatic hemisection of a typical rotifer (after Shipley & MacBride).

Nematoda. Nematodes are an enormously abundant group, found in practically all ecological situations. Although the best-known ones are parasitic invertebrates (hookworm, *Ascaris*, and so on), and some are important parasites of crop plants, the majority of species are free-living in soil or water. Nematodes are easily recognized by the elongate body, usually of small size and pointed at the ends, by the unsegmented cuticle, and by a very characteristic, stiff, whipping or writhing motion. Large

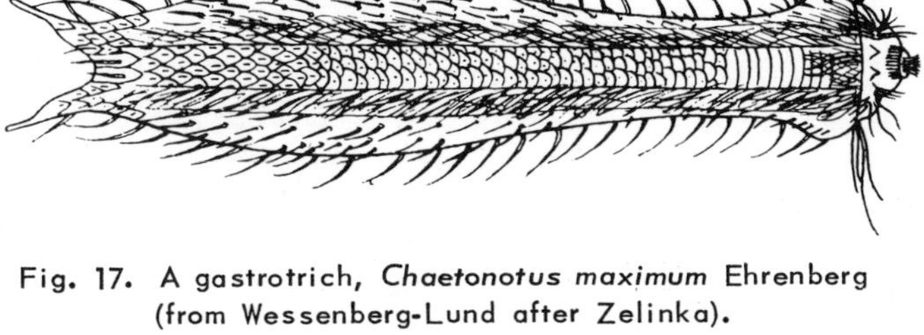

Fig. 17. A gastrotrich, *Chaetonotus maximum* Ehrenberg (from Wessenberg-Lund after Zelinka).

nematodes occur commonly in intertidal sponges, among bryozoans, and in detritus. They should be recognized as nematodes, but further identification is a specialized task.

Nematomorpha. These are the "horsehair snakes," whose larvae are parasitic in insects, crustaceans, and so forth.

Phylum Annelida

The annelids or segmented worms are characterized by an elongated body divided into segments and formed on the plan of a tubular jacket of muscle surrounding a large fluid-filled coelom. Although lacking any rigid internal skeleton, the annelids are able to utilize the hydrostatic pressure of this coelomic fluid, acted upon by the muscular body wall, as a sort of "fluid skeleton," aiding in extension of the body and in burrowing. Locomotion is also aided by numerous setae (or chaetae) which project from the sides of the body. The annelid body plan has proved to be extremely plastic and adaptable in an evolutionary sense, so that we find a very great diversity of form, habitat, and mode of life within the phylum.

Three classes are generally recognized among annelids proper: (1) Polychaeta (fig. 18, *a*), which have many setae (hence the name) and parapodia, as well as a diversity of tentacles and gill-like devices. They are almost entirely marine, exceedingly diverse in form, and are very numerous, comprising sixty families and some sixteen hundred genera. (The Archiannelida, a small and poorly known marine group, may be regarded as an appendage to the Polychaeta rather than as a distinct class. Archiannelids are small forms, showing a curious assortment of features including weak segmentation, poorly developed parapodia, and frequent absence of setae. Some are hermaphroditic. Some of the foregoing features may be primitive, others reduced. The group is probably an assortment of degenerate and rather specialized animals related to widely diverse, better-known families.) (2) Oligochaeta (figs. 18, *b*, *c*), which have no parapodia and few setae (as the name implies). They are predominately fresh-water or terrestrial, hermaphroditic annelids and comprise four orders with fifteen families. (3) Hirudinea or leeches (fig. 18, *d*), comprising three orders and six families, have no setae or parapodia, but are equipped with terminal suckers for attachment. Leeches are the most specialized of annelids in an anatomical sense and have developed a highly muscular, rather solid, body with a great reduction

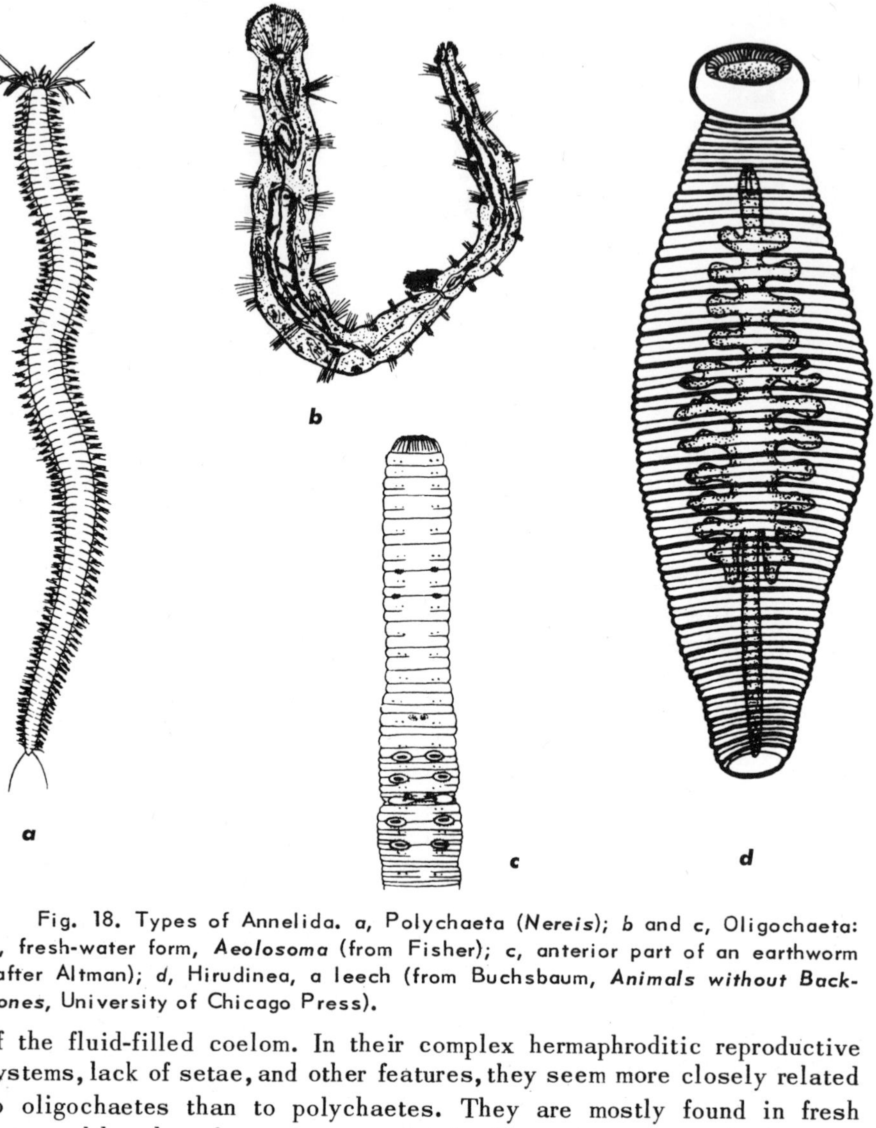

Fig. 18. Types of Annelida. *a*, Polychaeta (*Nereis*); *b* and *c*, Oligochaeta: *b*, fresh-water form, *Aeolosoma* (from Fisher); *c*, anterior part of an earthworm (after Altman); *d*, Hirudinea, a leech (from Buchsbaum, *Animals without Backbones*, University of Chicago Press).

of the fluid-filled coelom. In their complex hermaphroditic reproductive systems, lack of setae, and other features, they seem more closely related to oligochaetes than to polychaetes. They are mostly found in fresh water, although a few are terrestrial in damp situations and some are marine. All are predators, specialized for piercing their prey and feeding upon blood or soft tissues.

In addition to the above classes, there are certain annelidlike but unsegmented forms, the Echiuroidea and the Sipunculoidea, which are sometimes regarded as separate classes of annelids (or grouped into the single class Gephyrea). The current tendency is to consider them as separate phyla, somewhat distantly allied to the annelids (see pp. 108 and 109).

CLASS POLYCHAETA

There seems to be no generally acceptable scheme for dividing the class Polychaeta into orders. One method has been to recognize one order of more or less sedentary forms (Polychaeta Sedentaria) and another of free-living types (Polychaeta Errantia). Actually such a division is quite arbitrary, since the families of polychaetes present a veritable spectrum of adaptions, ranging from entirely free-swimming types, through crawlers, burrowers, temporary tube builders, to those which construct fixed tubes that they never leave. In practice it is far more useful to recognize the principal family types such as nereid, polynoid, spionid, syllid, sabellid, serpulid, and so on.

A typical free-living polychaete such as *Nereis* (which will serve as an example, although members of other free-living families differ greatly in detail) has paired locomotor appendages or parapodia, (singular, parapodium) (fig. 19), each composed of an upper lobe, the notopodium, and a lower, neuropodium. Each lobe typically contains a bundle of slender chitinous setae (singular, seta) which project from the parapodium, together with a larger dark spine known as the aciculum (plural, acicula). The shape, size, number, and position of the setae are of importance in classification. Arising from the base of the notopodium above and the neuropodium below, there are often slender flexible outgrowths, the dorsal cirrus (plural, cirri) and ventral cirrus, respectively. The notopodium consists of a dorsal lobe and a middle lobe, between which the setae and aciculum project; the neuropodium consists of a neuroacicular lobe provided with setae and an aciculum, and a ventral lobe. In many polychaetes, dorsal gills or branchiae, made conspicuous by the red or green blood within, arise from the bases of the parapodia in certain parts of the body.

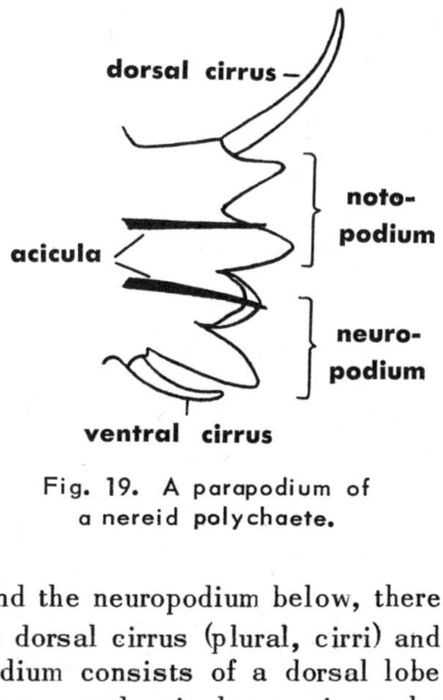

Fig. 19. A parapodium of a nereid polychaete.

The setae vary widely in form and furnish very precise characters for the determination of species. The use of such characters has been avoided where possible in the key but has been found necessary in many cases. Common types and their names should be recognized. Figure 20 gives some idea of the wide variety of types. First we may distinguish simple setae (fig. 20, nos. 1-20, 29-42) from composite or jointed setae (nos. 21-28). Long slender simple setae are spoken of as capillary setae

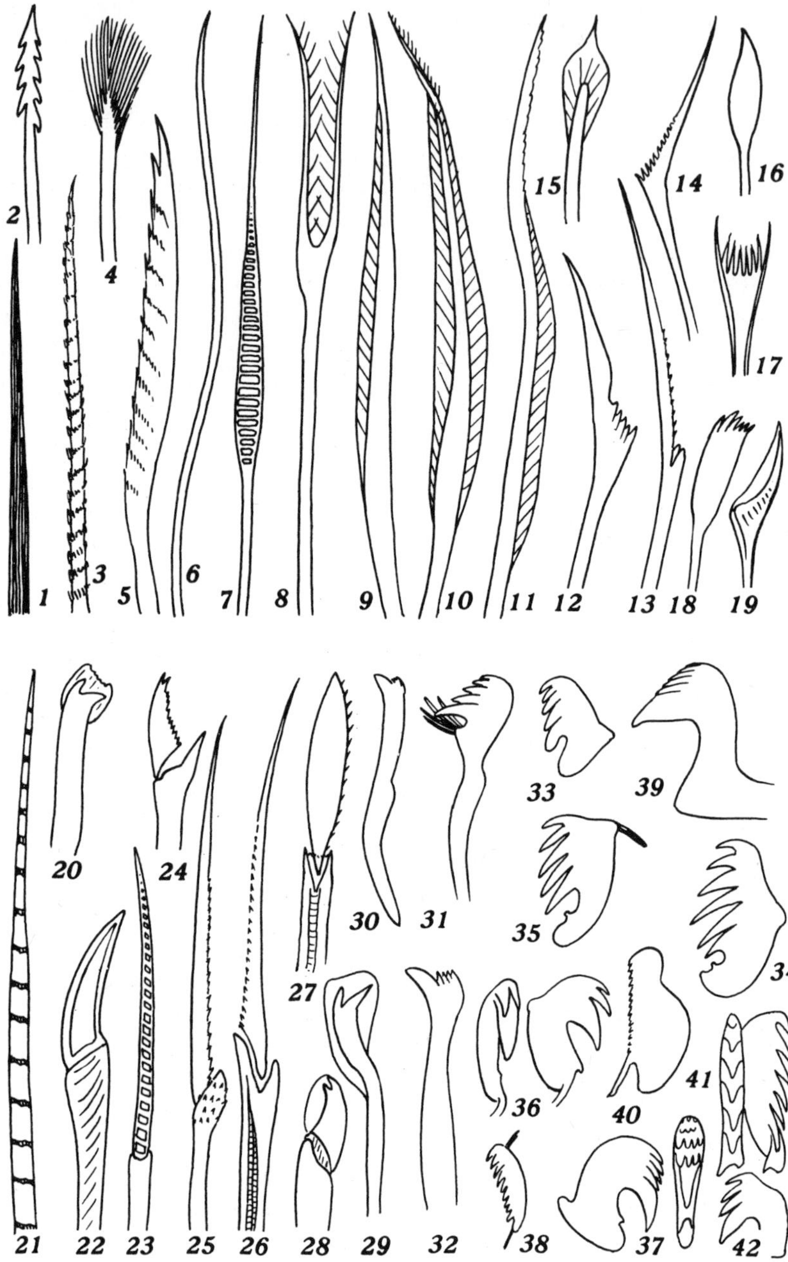

Fig. 20. Types of polychaete setae (after Fauvel). See discussion in text.

(no. 6). The tips of setae, whether simple or composite, may be entire, bifid (no. 8), or trifid. If bent like a sickle they are termed falcate (no. 11); if flattened like an oar blade, limbate (no. 10). Some simple setae have stubby, bent, usually bifid ends. These are spoken of as hooks or crotchets (nos. 29-32). These are usually relatively stout setae and graduate into short, broadened types known as uncini (nos. 33-42), usually set in close rows, which are especially characteristic of the Sedentaria.

Composite setae may be multiarticulate, as in the long bristles of some flabelligerids (fig. 20, no. 21), but are characteristically two-jointed, composed of a shaft and a blade (nos. 22-28). This blade in turn may have various shapes and may itself be a crotchet (no. 28). The blade rests in a notch in the end of the shaft. If the two sides of the notch are equal, it is spoken of as homogomph (no. 27); if unequal, as heterogomph (no. 26). Finally setae, either simple or compound, may be embedded at their tips in a clear matrix and are then spoken of as hooded. Thus, we may have simple hooded crotchets (nos. 20, 29) or composite hooded crotchets (no. 28).

The heads of polychaetes are of exceedingly diverse form. As an example of the head of an errant polychaete we may take the head of *Nereis* (fig. 21). This consists of a preoral prostomium, provided at its anterior margin with a pair of small antennae, and at its sides with paired, fleshy, biarticulated palps (palpi). The segment just behind the prostomium is the peristomium (in *Nereis* it may represent a fusion of two segments). It bears, in *Nereis*, four pairs of peristomial cirri (tentacular cirri) on short stalks at its anterior margin. The usage of the terms palp, antenna, tentacle, and cirrus varies greatly. Antennae, unless otherwise specified, are usually dorsal or marginal on the prostomium (the term tentacle may also be used). Palpi are usually associated with the mouth and tend to be lateral or ventral to the prostomium and bordering the anterior margin of the mouth. However, certain dorsal structures, especially if these are large, elongated, grooved, or prehensile (as in the spionids) are frequently called palps. The term cirrus is usually applied to structures arising dorsally or ventrally on the parapodia, whereas comparable structures on the anterior part of the body, if elongated, may be designated tentacular cirri (or peristomial tentacles). Tentacle is a very general term and is used to signify any of a variety of elongated sensory or feeding structures, usually on the head.

The prostomium of *Nereis* bears two pairs of eyes; other polychaetes may have one pair or none, and in some there are numerous eyespots scattered on the peristomium or even on the tentacles or sides of the body.

The first parapodia of *Nereis* are borne on the segment behind the peristomium. The first two pairs of parapodia are uniramous, that is,

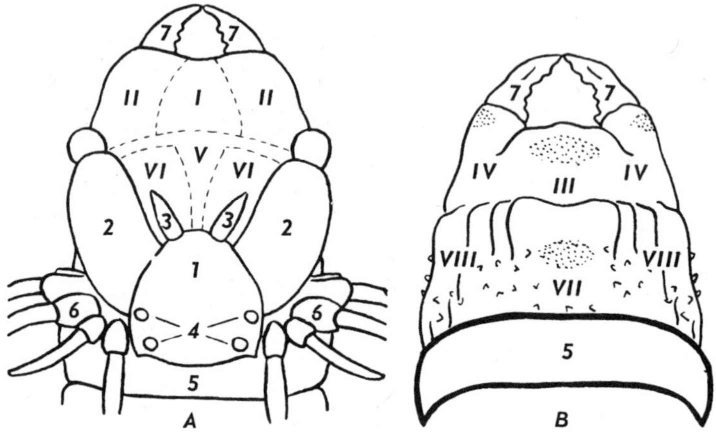

Fig. 21. Head of *Nereis* with everted proboscis: *a*, dorsal view; *b*, ventral view. (1) prostomium; (2) palpi; (3) prostomial tentacles; (4) eyes; (5) peristomium; (6) peristomial or tentacular cirri; (7) jaws; I-VIII, areas of proboscis; V-VIII, on oral ring; I-IV, on maxillary ring.

each contains a single aciculum and setal fascicle (bundle). All succeeding parapodia are biramous. In other polychaetes the parapodia vary, some having only uniramous, others only biramous parapodia, some with both types, some with parapodia greatly reduced.

In most free-living polychaetes the pharyngeal region may be everted (pushed forward and turned inside out) to form a proboscis, which often bears stout jaws and small horny teeth (paragnaths). The proboscis of *Nereis* is divisible into two external regions (fig. 21): (1) an **oral ring** which is just external to the mouth (internal when proboscis is retracted), bounded by the mouth on one end and the maxillary ring on the other, and divisible into six areas, numbered from V to VIII: and (2) beyond this a second ring, the **maxillary** ring, bounded on one end by the oral ring, provided with stout, horny jaws at its other end, and divisible into six areas numbered from I to IV. Even-numbered areas are paired; odd-numbered areas lie in the dorsal and ventral mid-line. A study of the arrangement of paragnaths is a necessary step in the identification of nereids; in other families the pattern is less complex, but may be referred to.

The sedentary polychaetes depart widely from the body form which characterizes free-living types. Prostomium and eyes are often reduced, proboscis and jaws may be absent, and the anterior end, especially in types dwelling in fixed tubes, greatly elaborated for feeding and respiration. In sabellids and serpulids the peristomial cirri (or tentacles) form a great "branchial crown" of featherlike "gills," which serves both for feeding and respiration. Cilia pass water between the branches of the plumes and transport food, entangled in mucus, down to the mouth. In other forms such as terebellids, the peristomial tentacles are long, filamentous, and extensile, serving to bring in food by ciliary action in a groove running along each filament. Just behind the head there arise tufted, blood-filled branchiae which serve in respiration. Parapodia in tubicolous (tube-dwelling) polychaetes tend to be small and are provided with rows of hooklike setae (uncini) for gripping the sides of the tube. Special glands may secrete tube-forming material. One or more peristomial cirri may, in serpulids, which form a rigid calcareous tube, be modified to a pluglike operculum that can block the tube entrance. The body in tube-dwelling forms is often divisible into a more anterior and specialized thorax and a less specialized posterior abdomen. The thorax may bear anteriorly a collar, and this may be extended rearward to the posterior end of the thorax as a pair of folds, the thoracic membranes. The preceding accounts are of extreme types, free-living and tubicolous respectively, and can convey but a poor idea of the actual diversity of pattern that characterizes the numerous annelid families.

The key that follows begins as a key to separate families, and a study of it will give an impression of the characteristics of each family type. Since the ability to recognize families in the field is of great practical value to the zoölogist, the student should early learn the family types commonly encountered. Generic and specific identification usually requires more careful study. Fortunately the West Coast polychaetes are fairly well known, thanks chiefly to the work of Dr. Olga Hartman in California and of Cyril and Edith Berkeley in British Columbia.

Preservation of annelids for future identification and study is best done in formalin or in a fixative such as Bouin's, after anaesthetization (isotonic magnesium chloride, 73 gm. of $MgCl_2 \cdot 6H_2O$ per liter, and sea water in equal parts is recommended). Free-living polychaetes should be killed with the proboscis everted. Eversion generally occurs spontaneously if the animals are preserved before anaesthetization is complete. In some cases it is necessary to stretch the worm out on a flat surface and, by pressing back of the head, to force out the proboscis while formalin is applied with a dropper. Worms should generally be laid out straight at fixation, or may be draped limply over a needle and kept straightened out by their own weight as fixative is applied. Such worms are more satisfactory for study than if fixed curled up, highly contracted, or with proboscis uneverted. Alcohol leaves most worms rather flabby and is not recommended unless preceded by formalin or a suitable fixing agent.

Key to the Families of Polychaeta

by Olga Hartman
Allan Hancock Foundation, University of Southern California

1. Dorsal surface more or less completely covered by overlapping scales (elytra) (fig. 22, *b*), by paleae (modified flattened setae, fig. 22, *d*), or overlain by a thick, felty layer (fig. 22, *a*) . . . 2
1. Dorsal surface not covered with elytra, paleae, or felt . . . 4

2. Dorsal surface more or less concealed by felty layer that covers elytra; body short and oval; sea mice . . . . APHRODITIDAE (Genus *Aphrodita,* subintertidal, dredged or occasionally washed in; may be large and stout, with gleaming metallic hairs. See Hartman, 1939*a*:19)
2. Dorsal surface more or less concealed by paleae; usually minute . . . . CHRYSOPETALIDAE (p. 80)
2. Dorsal surface more or less concealed by elytra; medium-sized to large 3

3. Elytra and dorsal cirri alternate regularly from 5th segment to posterior end of body . . . . POLYODONTIDAE (Represented by *Peisidice aspera* Johnson, 1897 in rocky intertidal, especially in mussel beds; neuropodia with composite setae, prostomium without median antenna; body short, subrectangular, to 12 mm. long. See Johnson, 1897:184)
3. Elytra and dorsal cirri alternate in anterior segments, but on the posterior segments elytra are present on all segments and entirely replace cirri . . . . SIGALIONIDAE (Represented by *Sthenelais fusca* Johnson, 1897; median prostomial antenna (fig. 22, *c*) with broad base and lateral wings; elongate, dusky brown; among roots of *Phyllospadix* or under rocks. See Hartman, 1939*a*:61)
3. Elytra and dorsal cirri alternate regularly from the 4th to about the 23d segment; thereafter each 2 elytra are followed by a dorsal cirrus . . . . POLYNOIDAE (p. 95)

4. Anterior end with long, strong spinous setae, not forming operculum, but projecting forward and concealing prostomium (fig. 32) . . . . FLABELLIGERIDAE (p. 84)
4. Anterior end completely concealed by branching tentacles (figs. 43, 44) 5
4. Anterior end completely concealed by chitinous spines forming operculum 6
4. Anterior end more or less completely covered by many long filamentous outgrowths 7
4. Anterior end otherwise; prostomium generally not concealed . . 9

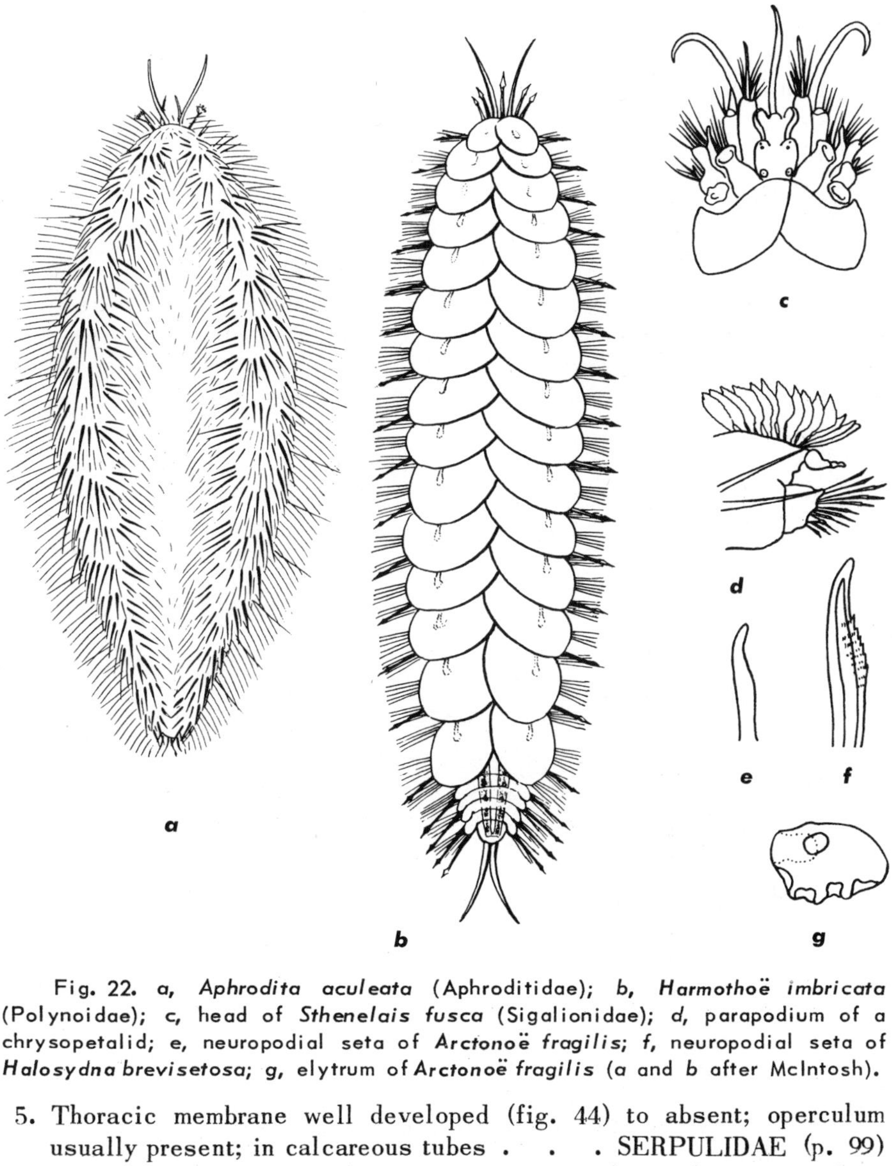

Fig. 22. *a, Aphrodita aculeata* (Aphroditidae); *b, Harmothoë imbricata* (Polynoidae); *c,* head of *Sthenelais fusca* (Sigalionidae); *d,* parapodium of a chrysopetalid; *e,* neuropodial seta of *Arctonoë fragilis; f,* neuropodial seta of *Halosydna brevisetosa; g,* elytrum of *Arctonoë fragilis* (*a* and *b* after McIntosh).

5. Thoracic membrane well developed (fig. 44) to absent; operculum usually present; in calcareous tubes . . . SERPULIDAE (p. 99)
5. Thoracic membrane little developed or absent (fig. 43); no operculum; in sandy chitinized tubes . . . . . . SABELLIDAE (p. 97)

6. Thorax with 2 pairs of pectinate (comblike) branchiae; caudal appendage (tail) annulated and with setae; operculum formed of single series of a few large spines; construct solitary, conical tubes open at both ends (fig. 40) . . . . . . PECTINARIIDAE (p. 93)
6. Thorax without pectinate branchiae; with many filiform branchiae in longitudinal rows above oral aperture; caudal appendage smooth,

slender; operculum formed of 3 whorls of numerous closely spaced spines; constructs masses of tubes more or less firmly cemented together . . . . SABELLARIIDAE (p. 97)

7. Filamentous outgrowths present anteriorly and continued along sides of body (fig. 29) except in *Dodecaceria*; body not divided into 2 distinct regions . . . . CIRRATULIDAE (p. 80)
7. Filamentous outgrowths concentrated on anterior region (figs. 23, 47); body usually with 2 distinct regions 8
8. Tentacles retractile into mouth, leaving branchiae exposed and extending over anterior end; prostomium often clearly visible (fig. 23); body usually short, linear . . . . . AMPHARETIDAE (Represented by *Schistocomus hiltoni* Chamberlin, 1919, in rocky intertidal. See Okuda, 1947) . . . .
8. Tentacles not retractile into mouth; prostomium and branchiae usually not visible except by lifting tentacles away; branchiae, when present, dorsal (fig. 47); body longer and irregularly coiled . . . . TEREBELLIDAE (p. 105)

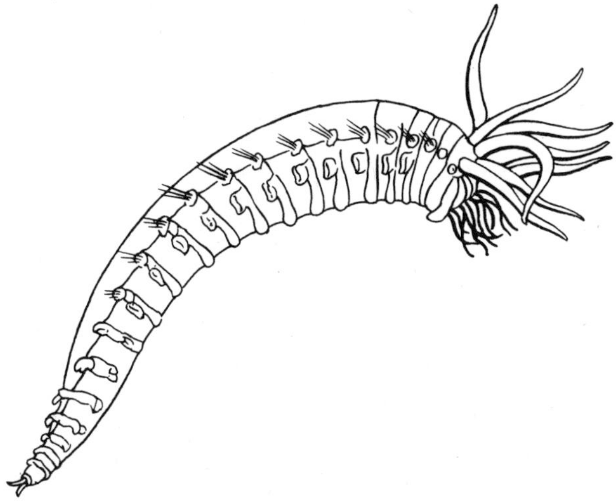

Fig. 23. *Amage auricula* (Ampharetidae)

9. Prostomial palpi long, prehensile, and tentacular, sometimes accompanied by pair of minute anterior antennae (figs. 28, 45) . . . 10
9. Prostomial palpi short, usually fleshy, or absent; anterior tentacles various. 12
10. Body divisible into 2 or more clearly defined regions (fig. 28) . 11
10. Body not divisible into regions, although sometimes single anterior segment is specially modified (fig. 45) . . SPIONIDAE (p. 100)

11. Prostomial palpi greatly elongated, heavily papillated; body consists of anterior region of about 8 segments (note parapodia), separated from longer posterior region by 9th segment, which is quite different . . . . MAGELONIDAE
(Represented by *Magelona pitelkai* Hartman, 1944, in intertidal sand flats. See Hartman, 1944*b*:260, pl. 19)
11. Palpi not greatly elongated, without papillations; body consists of 2 or more regions, each markedly different from the others (fig. 28) . . . . CHAETOPTERIDAE (p. 79)

12. Prostomium and peristomium ciliated; body short; usually of 9–11 segments . . . . CTENODRILIDAE
(Represented by the minute *Ctenodrilus serratus* (Schmidt, 1857) formerly placed with Archiannelida or in Cirratulidae. See Hartman, 1944*c*:323)
12. Prostomium not ciliated, without tentacles, although occasionally with lobed, flaring membrane; body with many more than 15 segments 13
12. Prostomium with tentacles, which may be minute 21

13. Posterior end of body provided with conspicuous sternal shields (fig. 24, *e*) bordered with stiff bristles; anal branchiae present; body short, distended, grublike . . . . STERNASPIDAE
(Represented by *Sternaspis fossor* Stimpson, 1854; subintertidal, rarely washed in. See Moore, 1923:218)
13. Posterior end without such anal plates 14

14. Anterior end, or both ends, flat; at least anterior segments elongate 15
14. Anterior and posterior ends pointed or rounded; segments usually not elongate 16

15. Anterior end forms a flat plaque (fig. 24, *a*); posterior end terminates in plaque with or without cirri; segments very long, segmental grooves nodelike . . . . MALDANIDAE
(Represented by *Axiothella rubrocincta* (Johnson, 1901); in protected, sandy intertidal. Tube of sand grains, often shared by pinnotherid crab, *Pinnixa longipes*. See Hartman, 1944*b*:265)
15. Anterior end with flat, lobed membrane (fig. 24, *b*, *c*, *d*) . . . . OWENIIDAE
(Represented by *Owenia fusiformis* (della Chiaje, 1841) in rocky intertidal. See *Ammochares occidentalis* in Johnson, 1901:420)

16. Proboscis provided with dark, chitinous jaw pieces (fig. 26, *b*, *c*); body smooth, elongate, cylindrical, resembling an earthworm; parapodia weakly developed or, at most, simple lobes; dorsal and ventral cirri often tiny (fig. 26, *a*) 17

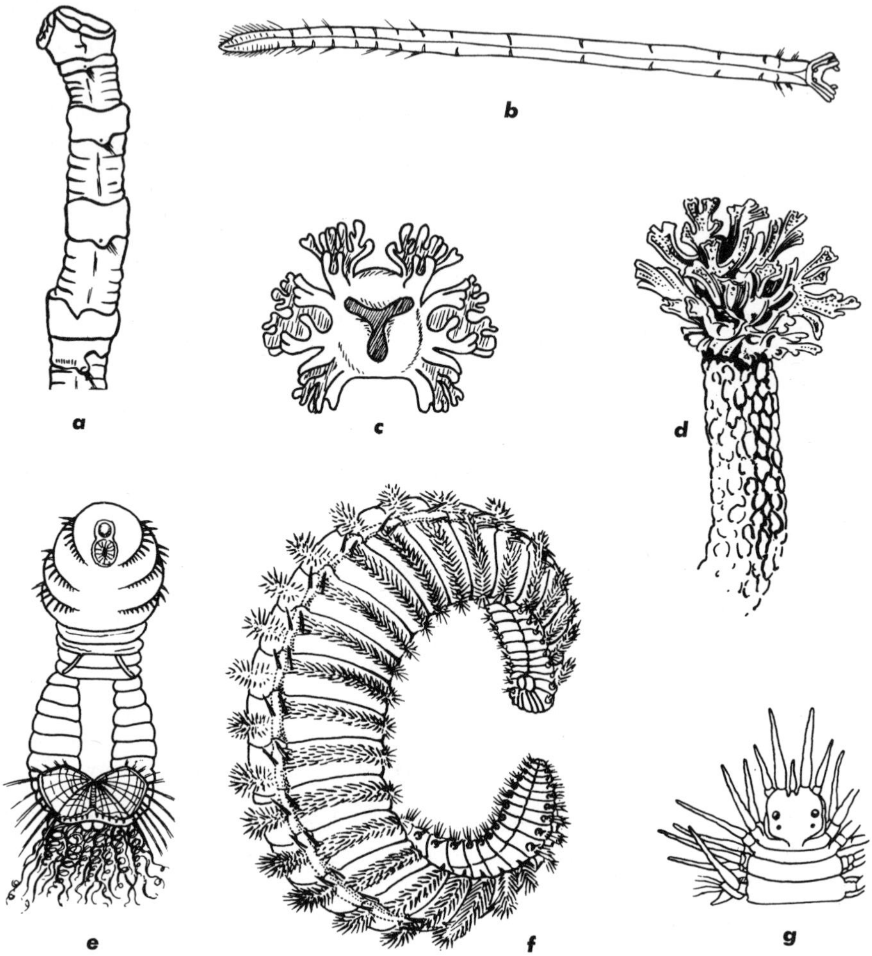

Fig. 24. *a, Axiothella* (Maldanidae), anterior part; *b, Owenia* (Oweniidae); *c*, head expansion of *Owenia*; *d, Owenia fusiformis*, in tube; *e, Sternaspis scutata* (Sternaspidae); *f, Euphrosine foliosa* (Euphrosinidae); *g, Podarke pallida* (Hesionidae) (*d, e, f*, and *g* after McIntosh).

16. Proboscis without chitinous jaws; form variable; body more or less divisible into regions 18

17. Parapodia provided with hooded hooks (fig. 26, *d*) and pointed setae; prostomium without eyespots . . . . LUMBRINERIDAE (p. 86)

17. Parapodia with only pointed setae; prostomium with or without eyespots . . . . ARABELLIDAE (p. 77)

18. Thorax with palisaded ranks of neurosetae; dorsum sometimes more or less completely concealed by overlapping, cirriform branchiae . . . . ORBINIIDAE (p. 91)

18. Thoracic neuropodia without palisaded ranks of setae . . . 19

19. Body consisting of 3 regions, an anterior and a posterior without branchiae, and a median region with conspicuous branchiae; sometimes of great size (up to 50 cm.) . . . ARENICOLIDAE (p. 78)

19. Body consisting of 2 regions, which are not easily distinguishable from each other 20

20. No uncini; all setae slender, pointed; body eel-like, broad, or short and grublike; some with lateral eyespots; segments usually closely multiannulate (fig. 38) . . . . OPHELIIDAE (p. 90)

20. Some parapodia with uncini, body slender, sometimes very long and fragile; without lateral eyespots (fig. 27) . CAPITELLIDAE (p. 78)

21. Prostomium with frontal horns; body short; prostomium with eyespots . . . . SCALIBREGMIDAE
(Represented by *Oncoscolex pacificus* (Moore, 1909). See Moore, 1909:282)

21. Prostomium with minute tentacles, which are usually inconspicuous unless rendered conspicuous because of reduced size of head; prostomium usually reduced in size, or, if large, partly concealed by first few segments of body 22

21. Prostomium with tentacles more or less conspicuous; these accompanied by a pair of palpi (except in PHYLLODOCIDAE) . . . 26

22. Prostomium quadrate or subcircular (fig. 35) 23

22. Prostomium pointed, provided with 4 tiny tentacles in a cross at tip; proboscis large, stout, cylindrical to barrel-like, terminating distally in dark chitinous jaw pieces (figs. 33, *a*; 34) 25

23. With branchia-like structures in the form of a recurved cirrus between two rami of parapodium (fig. 35, *c*); prostomium trapezoidal, anterior part with two pairs of small antennae (fig. 35, *b*) . . . . NEPHTYIDAE (p. 86)

23. With branchia-like structures in tufts above notopodia (fig. 25, *b*); a peculiar dorsal appendage (caruncle) extending back from the prostomium (fig. 25, *a*) 24

23. Lacking branchia-like structures; prostomium deeply cleft; body ribbonlike, resembling a 2-edged saw . . . . . PILARGIIDAE
(Represented by *Pilargis berkeleyi* Monro, 1933, a dead white, very elongate worm found in muddy sand flats, and by *P. maculata* Hartman, 1947, deep red, in rock crevices. See Hartman, 1947*b*)

24. Setal tufts form transverse series, segmentally arranged (fig. 24, *f*), broad, almost reaching across dorsum; branchiae not conspicuous . . . . EUPHROSINIDAE
(Genus *Euphrosine*, represented by several species; bright red to orange in life, crawling over surfaces or in crevices; typically sub-

intertidal, but occasionally washed in. See Hartman, 1940*a*:208)

24. Setal tufts limited to sides of body, not forming transverse series (fig. 25, *b*); branchial tufts conspicuous, usually arborescent. . . . . AMPHINOMIDAE (p. 77)

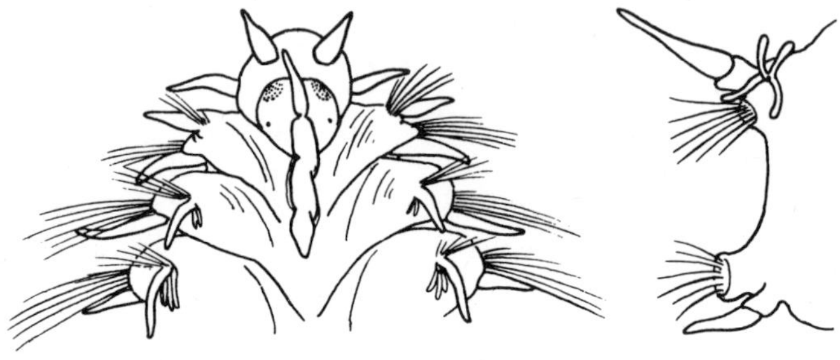

Fig. 25. *Pareurythoe californica* (Amphinomidae). *a*, anterior part showing short caruncle; *b*, parapodium.

25. Parapodia similar throughout, either all uniramous or all biramous; distal end of proboscis with four jaws . . GLYCERIDAE (p. 84)
25. Parapodia uniramous in anterior part, biramous and larger in posterior part of body; distal end of proboscis with many small black jaw pieces . . . . GONIADIDAE (p. 85)

26. Palpi present, sometimes globular and obscure 27
26. Palpi absent; parapodia uniramous, dorsal cirri often large, foliaceous, glandular, sometimes resembling the elytra of a polynoid; proboscis cylindrical, with soft papillae in specific patterns . . . . PHYLLODOCIDAE (p. 93)

27. Palpi prominent, elongate, curving, and laterally directed; one pair of cirriform or annulated prostomial tentacles; black jaws usually visible through body wall (fig. 30). . . . DORVILLEIDAE (p. 82)
27. Palpi slender, pointed, biarticulate (fig. 24, *g*); prostomium with 3 tentacles . . . . HESIONIDAE (Represented by *Podarke pugettensis* Johnson, 1901; short, with long dorsal cirri, purple to brown in life; occurs in ambulacral grooves of asteroids or under rocks. See Johnson, 1901)
27. Palpi fleshy and forwardly directed; usually biarticulated and prominent 28
27. Palpi short and globular, not especially prominent (fig. 31, *a*, *c*) . 29

28. With an unpaired median prostomial antenna, with or without paired peristomial tentacles; parapodia uniramous; dorsal cirri often moniliform (beaded) (fig. 46); barrelet of proboscis usually visible through

body wall . . . . SYLLIDAE (p. 103)

28. With paired prostomial tentacles and peristomial cirri (fig. 21); parapodia biramous; almost never with moniliform cirri; proboscis with paragnaths and jaws . . . . NEREIDAE (p. 86)

29. Palpi globular, obscure, giving prostomium a bilobed appearance; either 1, 3, or 5 prostomial tentacles; pharynx with chitinized, *paired* mandibles and maxillae . . . . EUNICIDAE (p. 82)
29. With 7 prostomial tentacles, of which anterior pair are small and simple, and the other 5 large with thickened annulated bases; in addition usually a pair of small dorsolateral peristomial cirri; pharyngeal maxillae not completely paired . . . ONUPHIDAE (p. 90)

AMPHINOMIDAE (Fig. 25)
(Reference: Hartman, 1940*a*:201)

1. Caruncle reduced, extends posteriorly to middle of 2d setigerous segment; smaller, to 50 mm. long . . . . *Pareurythoë californica* (Johnson, 1897)
1. Caruncle larger, conceals much of the prostomium and extends posteriorly behind the 3d setigerous segment; larger, to about 350 mm. long . . . . . . *Eurythoë complanata* (Pallas, 1766) (The amphinomids or "fire worms" possess tufts of glistening whitish setae, which may penetrate and irritate the hands.)

ARABELLIDAE (Fig. 26, *a-c*)
(Reference: Hartman, 1944*a*:170)

1. Parapodia with heavy acicular seta projecting from the lobe . . . . *Drilonereis* 2
1. Parapodia without heavy projecting acicular seta. . *Arabella* 3

2. Mandibles (ventral jaw pieces) absent. . . . *Drilonereis nuda*
2. Mandibles present . . . . *Drilonereis falcata*

3. Parapodial lobes short throughout; prostomium with 4 eyespots in a transverse row; smaller and more slender . . . *Arabella iricolor*
3. Postsetal lobe in posterior parapodia elongate, directed obliquely upward; prostomium usually lacks eyespots in adult; larger, more robust . . . . *Arabella semimaculata*

List of Species

Arabella iricolor (Montagu, 1804)
Arabella semimaculata (Moore, 1911)
Drilonereis falcata Moore, 1911
Drilonereis nuda Moore, 1909

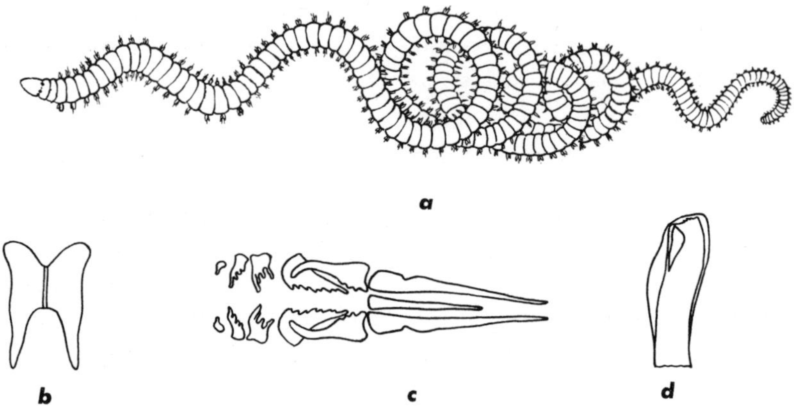

Fig. 26. *a*, *Arabella iricolor* (Arabellidae) after McIntosh); *b*, mandible of *Arabella*; c, maxillary apparatus of *Arabella*; *d*, hooded crotchet of *Lumbrineris*.

ARENICOLIDAE

(Reference: Ashworth, 1912)

1. With 11 pairs of branchiae; prostomium without lateral lobes . . . . *Arenicola cristata* Stimpson, 1856
1. With 13 pairs of branchiae; prostomium with lateral lobes . . . . *Arenicola pusilla* Quatrefages, 1866

CAPITELLIDAE (Fig. 27)

(Reference: Hartman 1947*a*)

1. Thorax consists of 9 segments. . . . . *Capitella* 4
1. Thorax consists of 11 segments . . . . *Mediomastus*
1. Thorax consists of 12 segments 2
1. Thorax consists of 13 segments . . . . *Leiochrides*
1. Thorax consists of 14 segments . . . . *Dasybranchus*
1. Thorax consists of more than 16 segments . . . . *Anotomastus*

2. Hooks only in last 6 thoracic segments . . . . . *Heteromastus*
2. Hooks lacking from thorax . . . . *Notomastus* 3

3. Nephridial apertures limited to some posterior thoracic segments, or absent; body long, very slender . . . . . . *Notomastus tenuis*
3. Nephridial apertures on abdomen or also thoracic segments; body thicker and larger . . . . *Notomastus magnus*

4. Segments 6 and 7 with only pointed setae . . . . *Capitella capitata*
4. Segments 6 and 7 with some hooks . . . . . *Capitella ovincola*

List of Species

Anotomastus gordiodes (Moore, 1909)

Capitella capitata (Fabricius, 1780)
Capitella ovincola Hartman, 1947 (In squid egg jelly)
**Dasybranchus lumbricoides* Grube, 1878
**Leiochrides pallidior* (Chamberlin, 1918)
**Heteromastus filiformis* (Claparède, 1864)
**Mediomastus californiensis* Hartman, 1944
Notomastus (*Notomastus*) *magnus* Hartman, 1947 (Occurs among surf grass, *Phyllospadix*, along rocky coasts)
Notomastus (*Clistomastus*) *tenuis* Moore, 1909 (Common in mud flats, Tomales Bay)

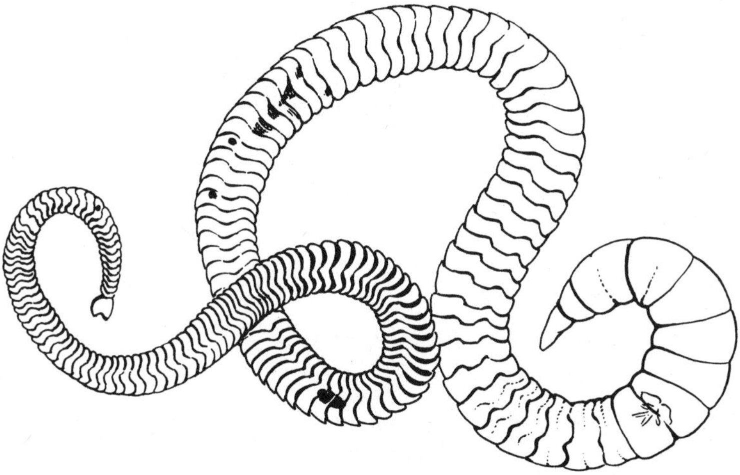

Fig. 27. *Capitella capitata* (Capitellidae) (after McIntosh).

CHAETOPTERIDAE

(References: Fauvel, 1927:77; Hartman 1944*b*:262; Potts, 1914:958)

1. Inhabiting large, coarse, leathery U-shaped tubes, both ends of which are exposed (fig. 28); in sand . . . . *Chaetopterus variopedatus* (Renier, 1804)
1. Inhabiting large, coarse, somewhat curved, but not U-shaped tubes; with one end exposed; in sand or eelgrass beds . . . . *Mesochaetopterus taylori* Potts, 1914
1. Inhabiting smaller (1 mm. diameter), annulated, more or less translucent tubes; usually in clusters; common under ledges in low intertidal (in Monterey Harbor forms enormous masses on pilings of Municipal Wharf) . . . *Phyllochaetopterus prolifica* Potts, 1914

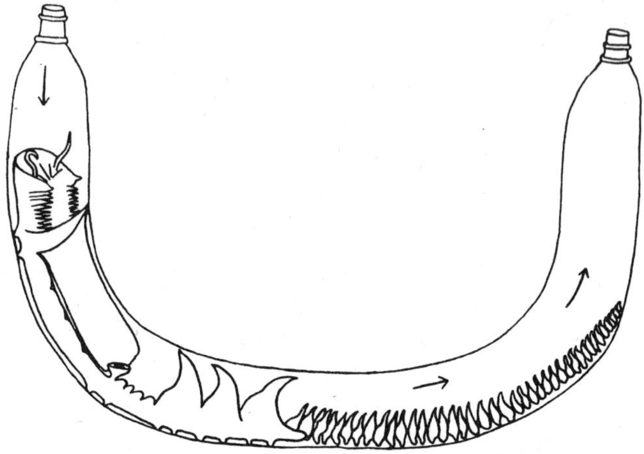

Fig. 28. *Chaetopterus variopedatus* in tube (after MacGinitie). Arrows show direction of water flow.

CHRYSOPETALIDAE (Fig. 22, *d*)

(References: Hartman, 1940*a*:201; 1944*b*:246; Johnson, 1897:161)

1. Paleae of 1 kind; color in life pale rust to yellowish . . . . *Chrysopetalum occidentale* Johnson, 1897
1. Paleae of 2 kinds; color in life glistening white or greenish . . . . *Paleanotus chrysolepis* Schmarda, 1861

CIRRATULIDAE

(References: Berkeley, 1929:307; Fauvel, 1927:93; Hartman, 1944*b*:263)

1. Body with few pairs of long, tentacular cirri near front; palpi posterior to prostomium grooved; constructing calcareous tube masses; dark to black in life . . . . *Dodecaceria*
1. Body with more numerous tentacular cirri (fig. 29, *a*, *d*), not constructing calcareous masses; not black 2

2. Anterior end with pair of thicker, grooved palpi (fig. 29, *c*) in addition to tentacles 5
2. Anterior end without palpal structures (fig. 29, *a*, *b*, *d*) . . . 3

3. Dorsal transverse group of tentacles arise from same segment as first setae; prostomium with eyespots on either side . . *Cirratulus*
3. Dorsal transverse group of tentacles arise from 3d to 5th setigerous segment (laterally situated tentacles start on 1st setiger); no eyespots . . . . *Cirriformia* 4

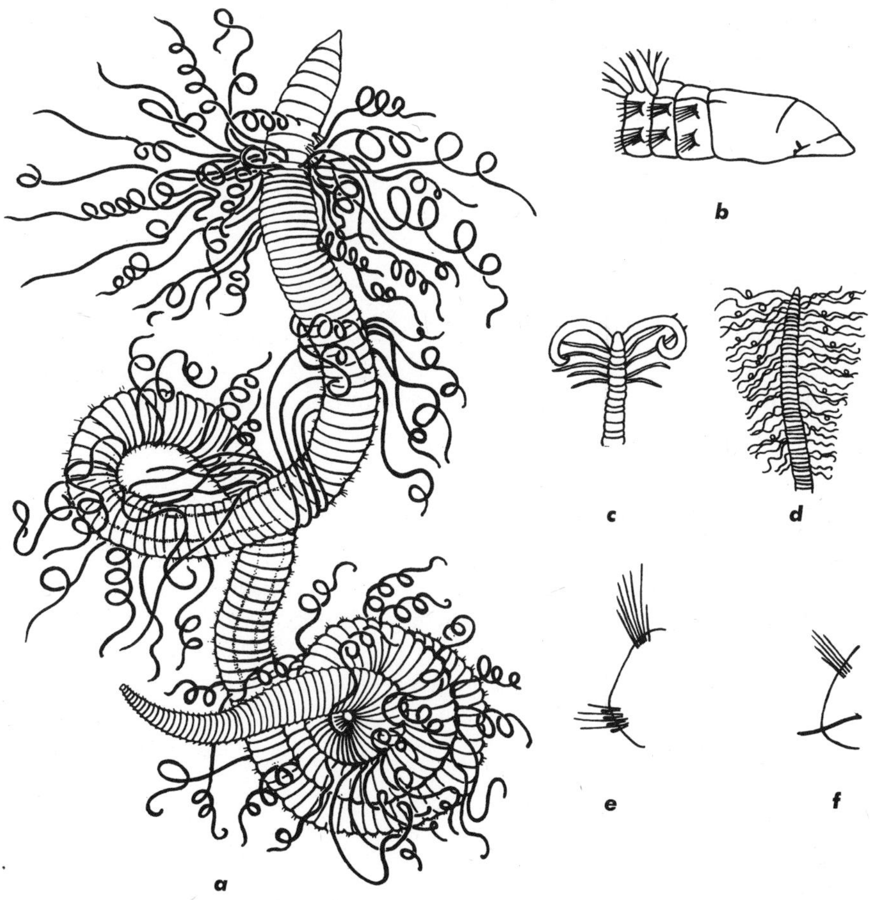

Fig. 29. Cirratulidae. *a, Cirriforma tentaculata* (after McIntosh); *b,* anterior end of *Cirriformia; c,* anterior end of *Tharyx; d,* anterior end of *Cirratulus; e,* setae of posterior segments of *Cirriformia spirabrancha; f,* setae of posterior segments of *C. luxuriosa.*

4. Posterior segments with few to single thick, dark spines (fig. 29, *f*); inhabits burrows or crevices in rocks . . . *Cirriformia luxuriosa*
4. Posterior segments with more numerous, finer, pale spines (fig. 29, *e*); inhabits sand among rocks or pilings *Cirriformia spirabrancha*

5. Setae entirely simple and slender . . . . *Tharyx*
5. Setae include acicular bifid spines in some anterior neuropodia . . . . *Caulleriella*

List of Species

**Cirratulus cirratus* (O. F. Müller, 1865)

Cirriformia luxuriosa (Moore, 1904) (In crevices in rocky intertidal)
Cirriformia spirabrancha (Moore, 1904) (Abundant in protected bays, burrowing in sand [Monterey] or mud [Tomales Bay] leaving only greenish tentacles exposed)
**Dodecaceria fistulicola* Ehlers, 1901
**Tharyx parvus* Berkeley, 1929 (In San Francisco Bay mud flats)
**Tharyx* sp. (An undescribed species occurring at Moss Beach, San Mateo County)
**Caulleriella pacifica* (Berkeley, 1929)

DORVILLEIDAE (=STAURONEREIDAE) (Fig. 30)

(References: Hartman, 1938*b*:99; 1944*a*:187; Fauvel, 1923)

1. Minute; antennae and palpi hardly visible . . *Ophryotrocha puerilis*
1. Larger; antennae and palpi conspicuous . . . . *Dorvillea* 2

2. Parapodia with dorsal acicula and dorsal cirrophores . . . 3
2. Parapodia without dorsal acicula or dorsal cirrophores . . . . *Dorvillea gracilis*

3. Large, brilliantly banded with coral and cream; prostomium thick, blunt; median neuropodia longest in their dorsal regions . . . . *Dorvillea moniloceras*
3. Small, pale or colorless; prostomial antennae long, slender, articulated; median neuropodia longest in their ventral regions . . . . *Dorvillea articulata*

List of Species

Dorvillea articulata (Hartman, 1938)
Dorvillea gracilis (Hartman, 1938)
Dorvillea moniloceras (Moore, 1909)
Ophryotrocha puerilis Claparède and Metschnikow, 1864 (Common in intertidal debris, especially among bryozoans; a contaminant in culture tanks)

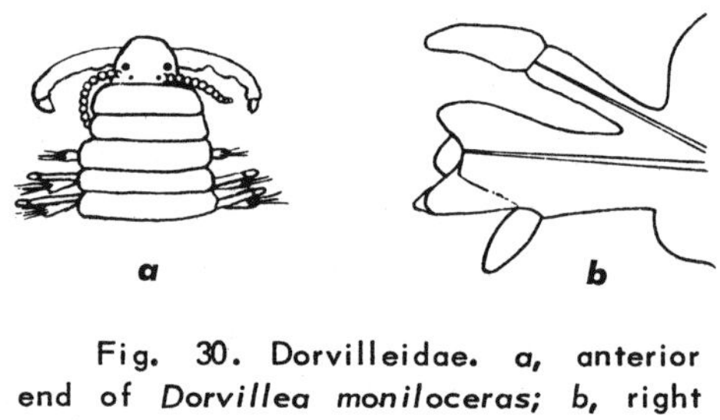

Fig. 30. Dorvilleidae. *a*, anterior end of *Dorvillea moniloceras*; *b*, right median parapodium of *Dorvillea articulata*.

EUNICIDAE (Fig. 31)

(References: Hartman, 1938*b*:94 (genus *Eunice*); 1944*a*:96)

1. Adult with 1 prostomial antenna . . . . *Nematonereis*
1. Adult with 5 prostomial antennae 2

2. Peristomium with a pair of tentacular cirri dorsally 3

2. Peristomium without cirri; acicula black . . . . . *Marphysa* 6
3. Acicular hooks and pectinate setae absent; mandibles massive, much larger than maxillary plates . . . . *Palola*
3. Acicular hooks and pectinate setae present; mandibles not greatly enlarged . . . . *Eunice* 4
4. Subacicular hooks distally bidentate; branchiae present from 3d setiger, continued through anterior 1/3 of body or even to the middle . . . . *.Eunice longicirrata*
4. Subacicular hooks distally tridentate 5
5. Prostomial antennae smooth, long . . . . . *Eunice americana*
5. Prostomial antennae more or less articulate or moniliform . . . . *Eunice antennata*
6. Branchiae with 4–7 filaments where best developed; composite setae distally pointed . . . . *Marphysa sanguinea*
6. Branchiae simple, cirriform throughout; composite setae distally falcate, with bidentate tip . . . . . *Marphysa stylobranchiata*

List of Species

Eunice antennata (Savigny, 1818)
Eunice americana Hartman, 1944
Eunice longicirrata Webster, 1884
Marphysa sanguinea (Montagu, 1804)
Marphysa stylobranchiata Moore, 1909
**Nematonereis unicornis* Schmarda, 1861
**Palola paloloides* (Moore, 1909)

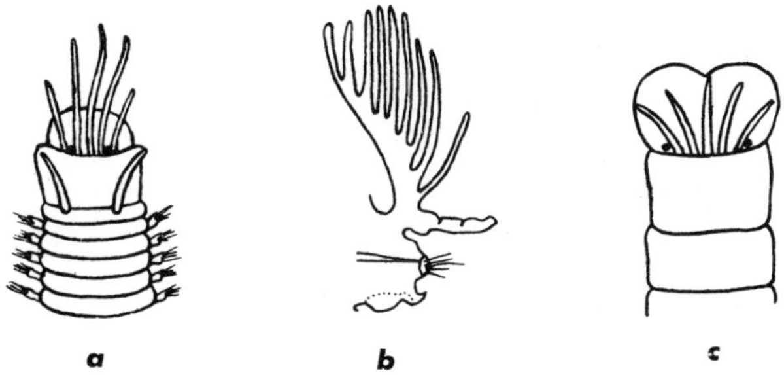

Fig. 31. Eunicidae (after Hartman). *a*, anterior end of *Eunice; b*, right parapodium of *Eunice longicirrata* Webster; c, anterior end of *Marphysa.*

FLABELLIGERIDAE (Fig. 32)

(References: Fauvel, 1927-115; Johnson, 1901:417; Moore, 1909:286)

1. With only simple setae; body surface may be jacketed in sand grains adherent to cuticle . . . . *Stylarioides* 3
1. With some composite (jointed) setae; inhabits gelatinous transparent tube . . . . *Flabelligera* 2

2. Among spines of *Strongylocentrotus*; purple in life . . . . *Flabelligera commensalis*
2. Free-living; dark gray in life . . . . *Flabelligera infundibularis*

3. Anterior end abruptly oblique; in U-shaped burrows in shales . . . . *Stylarioides inflata*
3. Anterior end not abruptly oblique; habitat otherwise 4

4. Ventral hooks distally entire . . . . . . *Stylarioides papillata*
4. Ventral hooks distally bifid . . . . . . *Stylarioides eruca*

List of Species

Flabelligera commensalis Moore, 1909
Flabelligera infundibularis Johnson, 1901
Stylarioides eruca (Claparède, 1868)
Stylarioides inflata (Treadwell, 1914) (See Hartman, 1952)
Stylarioides papillata (Johnson, 1901) [This is very close to, and possibly identical with, the European *S. plumosa* (O. F. Müller)]

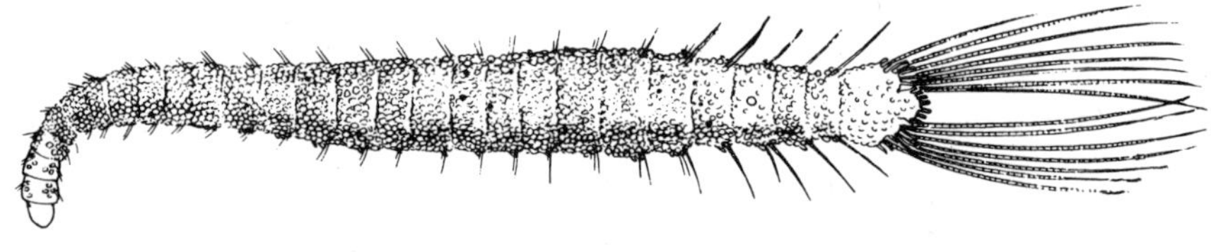

Fig. 32. *Stylarioides flabellata* (Sars) (Flabelligeridae) (after McIntosh).

GLYCERIDAE

(Reference: Hartman, 1950:57)

1. Parapodia uniramous, with single aciculum (fig. 33, *b*); setae composite . . . . *Hemipodus*
1. Parapodia biramous, with 2 acicula (fig. 33, *c*); setae both simple and composite . . . . *Glycera* 2

2. Parapodial lobes broadest subdistally; segmental retractile branchiae emerge from posterior faces of parapodia . . . *Glycera americana*
2. Without retractile branchiae; parapodial lobes otherwise . . . 3

3. Parapodia with 2 presetal and 2 postsetal lobes 4

3. Parapodia with 2 presetal and 1 postsetal lobe . *Glycera capitata*
4. Parapodial rami long, with blisterlike branchia on dorsal surface . . . . *Glycera robusta*
4. Parapodia without such branchiae 5
5. The 2 posterior lobes of parapodium unequal, dorsal the longer . . . . *Glycera alba*
5. The 2 posterior lobes of about same length . . . *Glycera tesselata*

List of Species

Glycera alba (Müller, 1788)
Glycera americana Leidy, 1855
Glycera capitata Oersted, 1843
**Glycera convoluta* Keferstein, 1862
Glycera robusta Ehlers, 1868
Glycera tesselata Grube, 1863
**Hemipodus borealis* (Johnson, 1901)

(Glycerids living in a soft substrate burrow by protruding the proboscis with great speed and force into the mud or sand, anchoring it, and pulling the body ahead as proboscis is retracted)

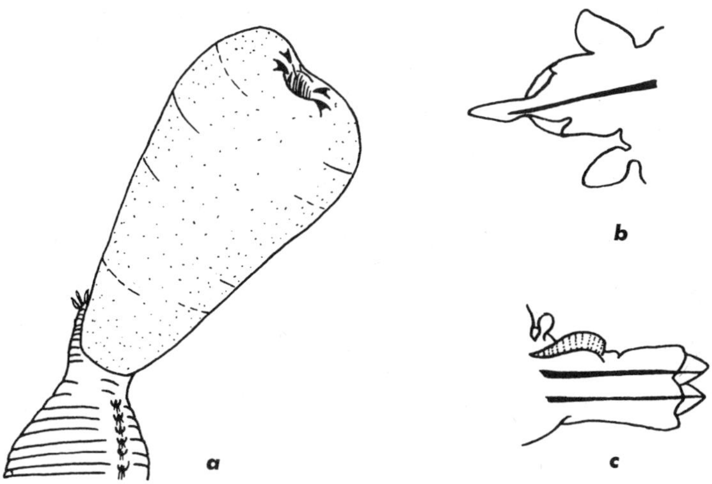

Fig. 33. Glyceridae. *a*, anterior end of a glycerid to show everted proboscis; *b*, left parapodium of *Hemipodus* showing single aciculum; c, right parapodium of *Glycera robusta* showing two acicula.

GONIADIDAE (Fig. 34)

(Reference: Hartman, 1950:4)

1. Parapodia change abruptly from uniramous in anterior region to biramous in posterior region; proboscis with chevron at base (consists of 10 or more V-shaped pieces) . *Goniada brunnea* Treadwell, 1906

1. Parapodia change gradually; proboscis without chevron at base, but with hard yellow organs resembling those on radula of snail . . . . *Glycinde armigera* Moore, 1911

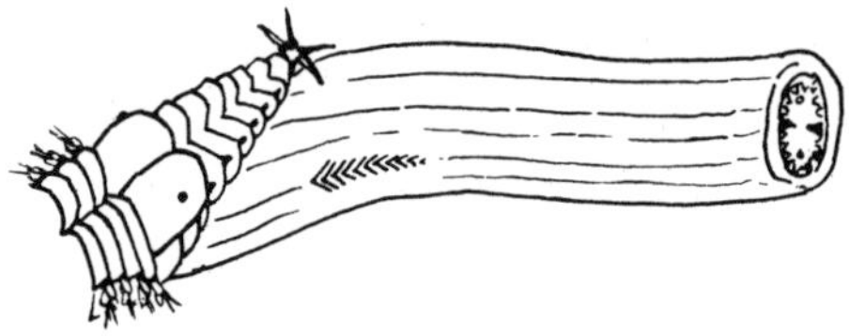

Fig. 34. Anterior part of *Goniada maculata* (*Goniadidae*) showing everted proboscis with chevron at base (after McIntosh).

LUMBRINERIDAE (Fig. 26, *d*)
(Reference: Hartman, 1944*a*:133)

1. Parapodial lobes elongate, erect in posterior parapodia . . . . *Lumbrineris erecta* (Moore, 1904)
1. Parapodial lobes short throughout 2
2. Anterior parapodia with some composite hooded crotchets in addition to pointed setae; prostomial lobe spatulate . . . . *Lumbrineris latreilli* Audouin and Milne-Edwards, 1833
2. Anterior parapodia without composite setae; prostomial lobe bluntly conical. . . . . *Lumbrineris zonata* (Johnson, 1901)

NEPHTYIDAE (Fig. 35)
(Reference: Hartman, 1938*c*)

1. Smaller; prostomium and first few segments with broad, dark pattern on the dorsal side; setae rather stiff; inhabits muddy sand flats . . . . *Nephtys caecoides* Hartman, 1938
1. Larger; prostomium with limited spread-eagle pattern; first few segments usually pale dorsally; setae soft, flowing; inhabits clean sandy outer beaches . . . *Nephtys californiensis* Hartman, 1938

NEREIDAE
(References: Hartman, 1938*a*; 1944*b*:251)

1. Peristomium forms a conspicuous ventral flap that projects forward under the prostomium (fig. 36, *a*); commensal with pagurids . . . . *Cheilonereis*
1. Peristomium does not form such a flap 2
2. Posterior notopodia have simple, dark, uncinate setae (fig. 36, *e*); peristomial cirri elongate, extend posteriorly to about segment 8; tube-building, gregarious (fig. 36, *b*) . . . . *Platynereis agassizi* (see note in species list)

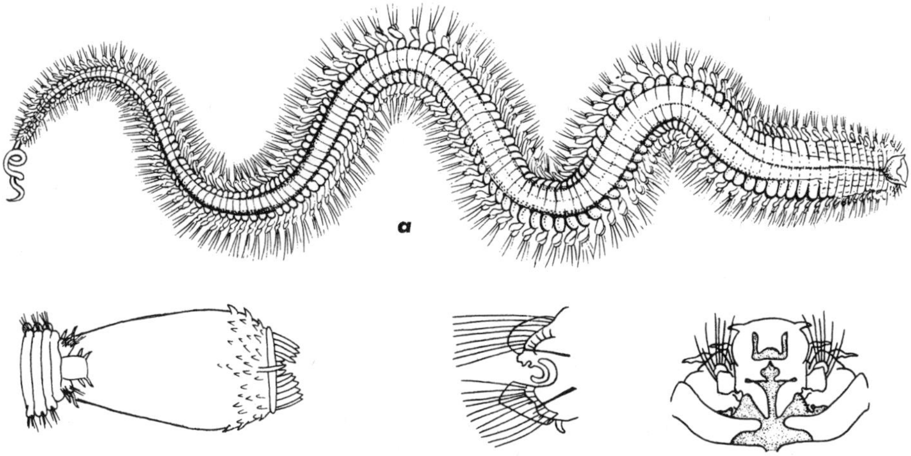

Fig. 35. Nephtyidae. *a, Nephtys caeca* (after McIntosh); *b, Nephtys* with everted proboscis; *c*, left parapodium of *Nephtys* showing recurved cirrus; *d, Nephtys caecoides*, anterior dorsal part showing dark pattern on prostomium and first few segments.

2. Posterior notopodia without simple, dark, uncinate setae . . . 3

3. Paragnaths absent from one or the other of the proboscideal rings 4
3. Paragnaths present on both rings 5

4. Paragnaths present only on oral ring (fig. 36, *c*) . . . *Eunereis*
4. Paragnaths present only on maxillary ring (fig. 36, *p*) *Ceratonereis*

5. Posterior notopodia with a few homogomph falcigerous setae (fig. 36, *f*) . . . . *Nereis* 6
5. Posterior notopodia without homogomph falcigerous setae . . 12

6. Posterior parapodial dorsal lobes greatly elongate, straplike (fig. 36, *q*) . . . . *Nereis vexillosa*
6. Posterior parapodial lobes not greatly elongate, never straplike . 7

7. Dorsum with broken transverse pattern; small, usually less than 30 mm. long (fig. 36, *k*) . . . . *Nereis latescens*
7. Dorsum without broken transverse pattern; usually larger . . 8

8. Proboscis provided with many tiny paragnaths over both oral and maxillary rings . . . . *Nereis eakini*
8. Proboscis otherwise 9

9. Paragnaths of proboscis all unusually tiny, inconspicuous; body greatly prolonged, inhabits mud flats . . . . . . *Nereis procera*
9. Paragnaths of proboscis include some larger ones; body not nearly so prolonged 10

10. Parapodial lobes typically dark; dorsal lobes in posterior region do not become noticeably longer than those in median region . . . . *Nereis pelagica neonigripes*
10. Parapodial lobes not dark 11
11. Area VI of proboscis (fig. 21) with 4–5 large cones disposed more or less in a diamond . . . . *Nereis grubei*
11. Area VI of proboscis with an oval mass of 6–7 small cones . . . . *Nereis zonata*
12. Area VI with transverse paragnaths (fig. 36, *d*) . . . *Perinereis*
12. Area VI with only conical paragnaths . . . . . . *Neanthes* 13
13. Marine; posterior parapodial lobes broad, foliaceous (fig. 36, *h*) . . . . *Neanthes brandti*
13. In brackish water 14
14. Posterior parapodial lobes elongate (fig. 36, *i*); Lake Merritt and San Francisco Bay . . . . *Neanthes succinea*
14. Posterior parapodial lobes short (fig. 36, *o*); in coastal streams and lagoons . . . . . . *Neanthes lighti* (see note in species list)

List of Species

**Ceratonereis* spp.

**Cheilonereis cyclurus* (Harrington, 1897)

**Eunereis longipes* Hartman, 1936

Neanthes brandti (Malmgren, 1866) (May reach a length of 3 feet or more)

Neanthes lighti Hartman, 1938 [Found in estuaries, extending into nearly fresh water. Viviparous. This may prove to be *Neanthes limnicola* (Johnson), originally described from Lake Merced, San Francisco]

Neanthes succinea (Frey and Leuckart, 1847) (The "heteronereid" stage occasionally swarms in immense numbers in sloughs bordering San Francisco Bay)

Nereis eakini Hartman, 1936

Nereis latescens Chamberlin, 1919

Nereis grubei (Kinberg, 1866) (Includes *N. mediator* Chamberlin, 1918, according to D. J. Reish, unpublished Ph. D. dissertation, University of Southern California)

Nereis pelagica neonigripes Hartman, 1936 (A subspecies of the widespread *N. pelagica* Linnaeus)

Nereis procera Ehlers, 1868

Nereis vexillosa Grube, 1851

Nereis zonata Malmgren, 1867

**Perinereis monterea* (Chamberlin, 1918)

Platynereis agassizi (Ehlers, 1868) [This has recently been found very widely distributed in the north and south Pacific; the oldest name appears to be *P. bicanaliculata* (Baird).]

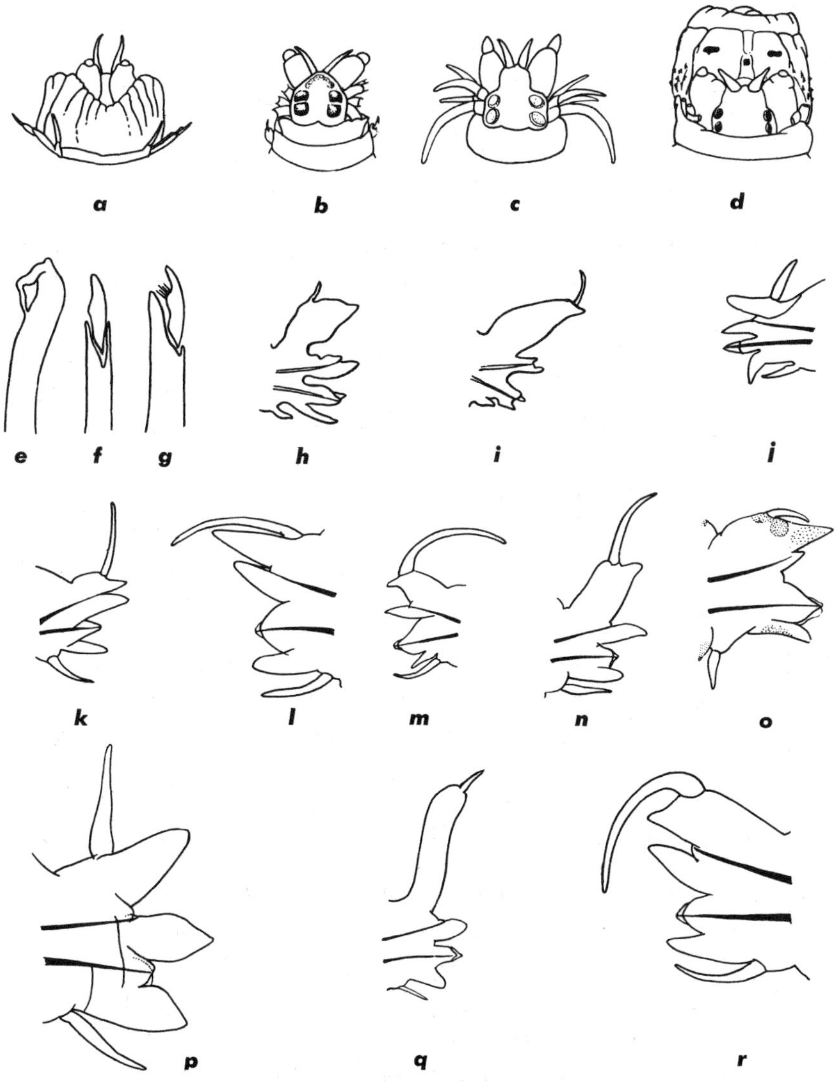

Fig. 36. Nereidae. Prostomium: *a, Cheilonereis cyclurus* ventral view; *b, Platynereis agassizi; c, Eunereis longipes; d, Perinereis monterea.* Setae: *e,* simple seta; *f,* homogomph falcigerous (composite) seta; *g,* heterogomph composite seta. Parapodia: *h, Neanthes brandti,* posterior segments; *i, Neanthes succinea; j, Eunereis longipes; k, Nereis latescens; l, Nereis procera; m, Nereis zonata; n, Nereis mediator; o, Neanthes lighti; p, Ceratonereis tunicatae; q, Nereis vexillosa; r, Nereis pelagica.*

ONUPHIDAE (Fig. 37)

(Reference: Hartman, 1944*a*:42)

1. Without dorsal tentacular cirri on 1st segment; subintertidal; in slender quill-like tubes . . . . *Hyalinoecia*
1. With dorsal tentacular cirri on 1st segment; tubes not quill-like . 2

2. Branchial filaments spiraled on main stalk . . . . *Diopatra* 3
2. Branchial filaments pinnately arranged on main stalk . . *Onuphis*
2. Branchial filaments simple, cirriform . . . . *Nothria* 4

3. Pectinate setae with relatively few large teeth, ventral cirri padlike from 5th segment . . . . *Diopatra splendidissima*
3. Pectinate setae with relatively numerous fine teeth; ventral cirri padlike from 6th segment . . . . *Diopatra ornata*

4. Branchiae present from 1st parapodium 5
4. Branchiae lacking from a long anterior region (about 19 segments or more) . . . . *Nothria stigmatis*

5. Ventral cirri padlike from segments 5–6; intersegmental spaces dark . . . . *Nothria elegans*
5. Ventral cirri padlike from segment 8; intersegmental spaces pale . . . . *Nothria iridescens*

List of Species

Dioptatra ornata Moore, 1911
Diopatra splendidissima Kinberg, 1857
**Hyalinoecia juvenalis* Moore, 1911 (Subtidal; rarely washed in)
Nothria elegans (Johnson, 1901)
Nothria iridescens (Johnson, 1901)
Nothria stigmatis (Treadwell, 1922)
**Onuphis eremita* Audouin and Milne-Edwards, 1834 (On exposed sand beaches)

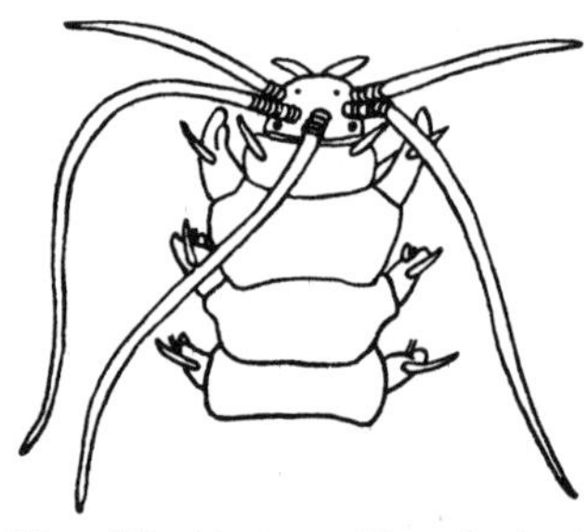

Fig. 37. *Nothria* (Onuphidae).

OPHELIIDAE (Fig. 38)

(References: Fauvel, 1927:129; Hartman, 1938*b*:102)

1. Ventral groove present, well marked 2
1. Ventral groove absent; a pair of setal fascicles anterior to mouth . . . . *Travisia*

2. Ventral longitudinal groove extending throughout length . . . 3
2. Ventral longitudinal groove in posterior region only 6

3. Cirriform branchiae present on some parapodia 4

3. Parapodia without branchiae . . . . *Polyophthalmus*

4. Lateral eyespots between some successive parapodia *Armandia* 5
4. Without lateral eyespots . . . . *Ammotrypane*

5. Lateral eyespots crescentic; lateral lobes of anus obliquely truncated above and slightly indented at end . . . . *Armandia brevis*
5. Lateral eyespots circular; lateral lobes of anus terminating in 5 dorsal and 2 sublateral papillae . . . . *Armandia bioculata*

6. Anterior region including 1st 2 setigerous segments set off from thorax by median constricted ring; anal cirri consisting of larger median ventral cirrus and numerous dorsolateral cirri disposed in inverted V; 5 pairs of nephridial pores, situated about midway between parapodial ridges. 7
6. Anterior region not set off from thoracic region by constriction; anal cirri consisting of a pair of larger ventral papillae and several smaller dorsal papillae disposed in transverse series; 6 pairs of nephridial pores, situated on or near parapodial ridges . . *Ophelia*

7. Branchial cirri forked, consisting of 2 simple cirriform branches . . . . *Thoracophelia*
7. Branchial cirri pectinately or compoundly branched *Pectinophelia* 8

8. Usually with 15 pairs of pectinately branched branchiae . . . . *Pectinophelia dillonensis*
8. Usually with 16 pairs of irregularly pinnately branched branchiae . . . . *Pectinophelia williamsi*

List of Species

**Ammotrypane aulogaster* Rathke, 1843
Armandia bioculata Hartman, 1938
Armandia brevis (Moore, 1906)
**Ophelia limacina* (Rathke, 1843)
Pectinophelia dillonensis Hartman, 1938
Pectinophelia williamsi Hartman, 1938
**Polyophthalmus pictus* (Dujardin, 1839)
**Thoracophelia mucronata* (Treadwell, 1914)
**Travisia gigas* Hartman, 1938

ORBINIIDAE

(References: Fauvel, 1927:10; Hartman 1944*b*:258; 1948:30)

1. Prostomium terminating in broad, spatulate lobe (fig. 39, *a*) . . . . *Nainereis*
1. Prostomium terminating in pointed cone (fig. 39, *b*) 2

2. Anterior thoracic region with ventral papillae. . . . . *Orbinia*

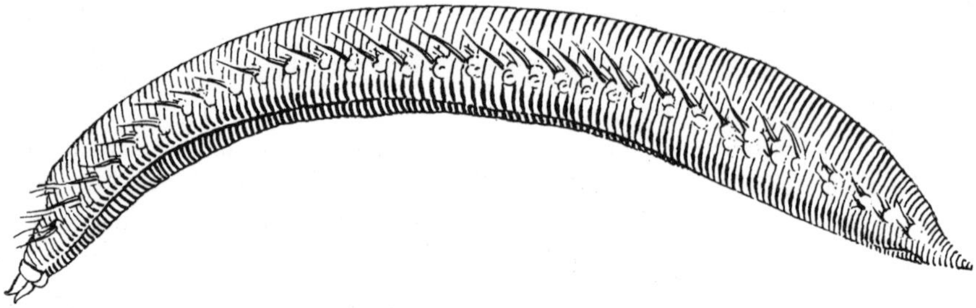

Fig. 38. *Ophelia limacina* (H. Rathke), Opheliidae (after McIntosh).

2. Anterior thoracic region without ventral papillae 3

3. Thoracic neuropodia with pointed setae only . . *Haploscoloplos*
3. Thoracic neuropodia with blunt spines and pointed setae . . . . *Scoloplos* 4

4. Thoracic neuropodial ridge with 2 postsetal lobes . . . . *Scoloplos armiger*
4. Thoracic neuropodial ridge with single postsetal lobe . . . . *Scoloplos acmeceps*

List of Species

Haploscoloplos elongata (Johnson, 1901)
Nainereis dendritica (Kinberg, 1867)
Scoloplos acmeceps Chamberlin, 1919
Scoloplos armiger (O. F. Müller, 1776)
Orbinia johnsoni (Moore, 1909)

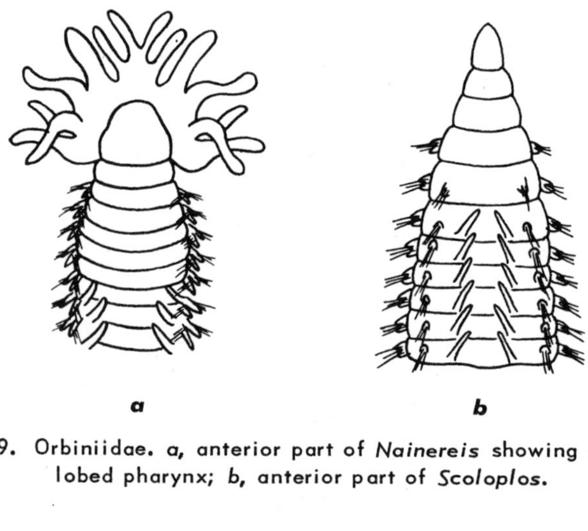

a b

Fig. 39. Orbiniidae. *a*, anterior part of *Nainereis* showing everted, lobed pharynx; *b*, anterior part of *Scoloplos*.

PECTINARIIDAE (Fig. 40)

(Reference: Hartman, 1941*b*)

1. Uncini with major teeth in single series of 4 only; tubes curved, constructed of moderately coarse black and white sand grains; cephalic spines blunt, short . *Cistenides brevicoma* (Johnson, 1901)
1. Uncini with major teeth in 2 series, numbering 4 or 5 in each longitudinal row; tubes nearly straight, constructed of fine reddish sand; cephalic spines long, tapering . . . . *Pectinaria californiensis* Hartman, 1941

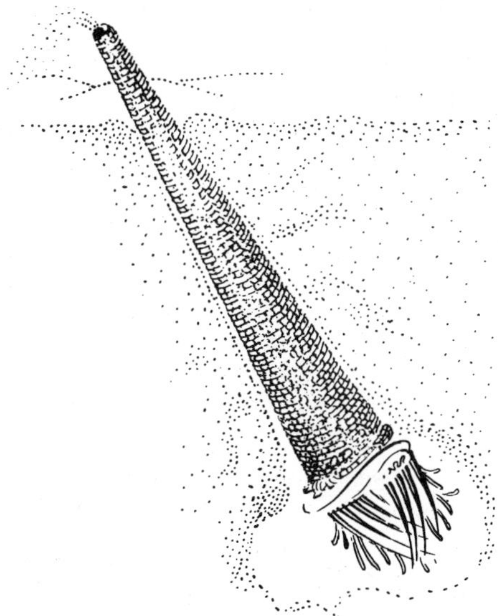

Fig. 40. *Lagis kareni* (Pectinariidae) excavating in sand (after McIntosh).

PHYLLODOCIDAE

(References: Hartman 1936*d*; 1944*b*:247)

1. With 4 pairs of tentacular cirri 2
1. With 2 pairs of tentacular cirri . . . . *Eteone* 9
2. First segment fused with prostomium, 2d and 3d segments free . . . . *Hypoeulalia*
2. Segments 1 and 2 more or less fused but free from prostomium . . . . *Genetyllis*
2. Segments 1 and 2 free but sometimes more or less reduced dorsally 3
3. Tentacular segments complete rings; prostomium with median antenna; proboscis diffusely papillated 4
3. First segment dorsally reduced; prostomium with or without median antenna 5

3. First segment dorsally and ventrally reduced, 2d segment dorsally reduced; prostomium with median antenna . . . . *Clavadoce*

4. Tentacular cirri cirriform (fig. 41, *a*) . . . . *Eulalia*
4. Ventral cirrus of segment 2 foliaceous, asymmetrical; other tentacular cirri cirriform . . . . *Steggoa*

5. Prostomium with median antenna . . . . *Eumida*
5. Prostomium without median antenna, with or without nuchal papilla 6

6. Proboscis proximally set with longitudinal rows of papillae . . . . *Anaitides* 7
6. Proboscis proximally set with diffuse papillae . . *Phyllodoce*

7. Prostomium broadest in anterior 1/2, hexagonal in outline; median notocirri distally pointed; larger . . . *Anaitides medipapillata*
7. Prostomium broadest in posterior half, cordate; smaller . . . 8

8. Prostomium longer than broad; neurocirri bluntly rounded distally; with conspicuous pigmentation consisting of 3 longitudinal lines on dorsal and ventral sides (fig. 41, *b*) . . *Anaitides williamsi*
8. Prostomium as long as broad; neurocirri pointed distally; without conspicuous pigmentation pattern . . . . . *Anaitides mucosa*

9. Larger, more than 50 mm. long; dorsal cirri broadly rounded, subcordate; ornamented with irregularly spaced black spots . . . . *Eteone pacifica*
9. Smaller, less than 50 mm. long, dorsal cirri not cordate; not ornamented with black spots 10

10. Prostomium broader than long; 1st segment dorsally reduced; dorsal cirri distally pointed, approximately triangular in shape . . . . *Eteone lighti*
10. Prostomium as long as broad or longer; 1st segment fully developed dorsally 11

11. Parapodia anteriorly tiny, appearing dwarfed because of inflated nature of anterior body segments; prostomium long, slender, trapezoidal (fig. 41, *c*) . . . . *Eteone dilatae*
11. Parapodia well developed, provided with thickened cirri and stout glandular parapodial bases; prostomium as long as broad or slightly longer, tending to be conical . . . . *Eteone longa*

List of Species

Anaitides mediapapillata Moore, 1909
Anaitides mucosa (Oersted, 1843)
Anaitides williamsi Hartman, 1936

**Clavadoce splendida* Hartman, 1936
**Eteone californica* Hartman, 1936
Eteone dilatae Hartman, 1936
Eteone lighti Hartman, 1936
Eteone longa (Fabricius, 1780)
Eteone pacifica Hartman, 1936
**Eulalia aviculiseta* Hartman, 1936
**Eumida sanguinea* (Oersted, 1843)
**Genetyllis castanea* (Marenzeller, 1879)
**Hypoeulalia bilineata* (Johnston, 1840)
**Phyllodoce ferruginea* Moore, 1909
**Steggoa californiensis* Hartman, 1936

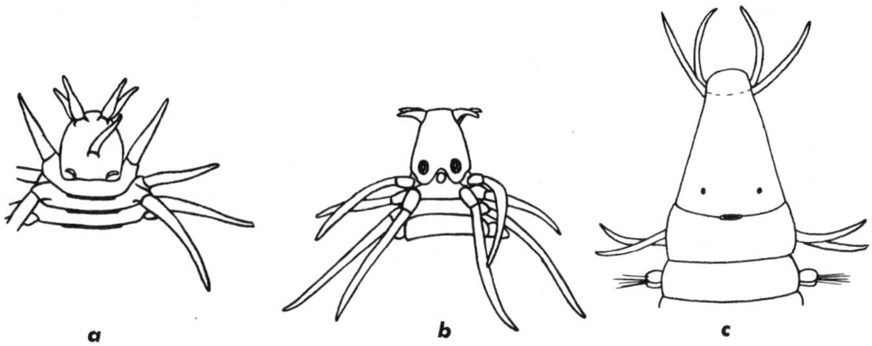

Fig. 41. Phyllodocidae. *a*, *Eulalia*; *b*, *Anaitides williamsi*; *c*, *Eteone dilatae*.

POLYNOIDAE

(References: Hartman, 1939*a*:27; 1944*b*:243; Pettibone, 1953)

1. With 12 pairs of elytra, their surface more or less tuberculated . . . . *Lepidonotus*
1. With 15 pairs of elytra. 5
1. With 18 pairs of elytra. . . . . *Halosydna* 2
1. With more than 20 pairs of elytra 3

2. Elytra more or less firmly attached; neuropodial setae entire at end (fig. 22, *f*); common in mussel beds and under rocks; also commensal with *Thelepus* . . . . *Halosydna brevisetosa*
2. Elytra easily shed; neuropodial setae bifid at end; free-living, under stones . . . . *Halosydna johnsoni*

3. Elytra more or less completely covering dorsum in anterior region; dorsal cirri and antennae club or pear-shaped; neuropodial setae falcate (sickle-shaped, fig. 22, *e*), sides smooth . . . *Arctonoë* 4
3. Elytra reduced in size throughout, those in posterior 1/2 not noticeably smaller than those in anterior 1/2; dorsal cirri and antennae cirriform; neuropodial setae serrated . . . . . *Lepidasthenia*

4. Elytra strongly frilled at external margins (fig. 22, *g*); in ambulacral grooves of asteroids . . . . *Arctonoë fragilis*
4. Elytral margin smooth externally; broad, transverse dark band across anterior segments; in branchial grooves of molluscs, echinoderms, and other organisms, rarely free-living . . . *Arctonoë vittata*
4. Elytra slightly undulate at margins; commensal with holothurians . . . . *Arctonoë pulchra*

5. Notopodial and neuropodial setae each of 1 kind 7
5. Notopodial and neuropodial setae each of 2 kinds, a stout and a slender form . . . . *Hesperonoë* 6

6. Elytral surface smooth or with very few minute, low papillae; commensal with *Urechis* . . . . *Hesperonoë adventor*
6. Elytra with numerous scattered low papillae; commensal with *Callianassa* . . . . *Hesperonoë complanata*

7. Neuropodial setae entire at distal end 9
7. Some neuropodial setae with tips bifid . . . . . *Harmothoë* 8

8. Larger, typically more than 20 mm. long; free-living; first pair of elytra usually pale, others dark . . . . . *Harmothoë imbricata*
8. Usually smaller, typically less than 20 mm. long; commensal with leptosynaptids; elytra pale except for a dark spot near the center . . . . *Harmothoë lunulata*

9. Neuropodial setae long, slender, distal ends hairlike. . . *Antinoë*
9. Neuropodial setae with distal ends thicker, not hairlike . . *Eunoë*

List of Species

Antinoë sp.

Arctonoë fragilis (Baird, 1863) (Commonly commensal on *Evasterias*, rarely on *Pisaster giganteus*)

Arctonoë pulchra (Johnson, 1897) (Commensal on sea cucumber *Stichopus*)

Arctonoë vittata (Grube, 1855) (Commensal on keyhole limpet *Diodora aspera*)

**Eunoë senta* (Moore, 1902)

Halosydna brevisetosa Kinberg, 1855 (Cosmopolitan in distribution)

Halosydna johnsoni (Darboux, 1899)

**Harmothoë hirsuta* Johnson, 1897

Harmothoë imbricata (Linnaeus, 1767) (Cosmopolitan in distribution)

Harmothoë lunulata (delle Chiaje, 1841)

Hesperonoë adventor (Skogsberg, 1928) (Commensal in burrows of the echiuroid *Urechis caupo*)

Hesperonoë complanata (Johnson, 1901) (Commensal in burrows of the mud shrimp *Callianassa*)

**Lepidasthenia gigas* (Johnson, 1897) (Commensal with terebellids)
**Lepidonotus caelorus* Moore, 1903 (Free-living, under rocks)

SABELLARIIDAE (Fig. 42)

(References: Hartman, 1944:269; 1944*c*:323)

1. Opercular spines form neat, black cone . . . . . *Phragmatopoma californica* (Fewkes, 1889)
1. Opercular spines form open pattern, amber-colored. . . . . *Sabellaria cementarium* Moore, 1906

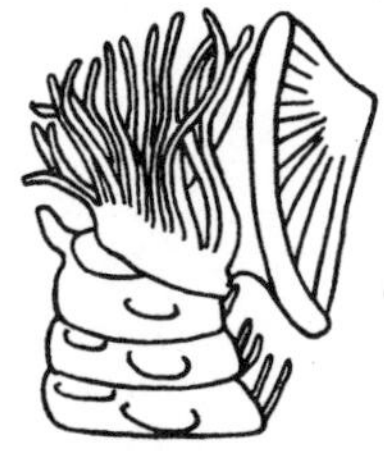

Fig. 42. *Phragmatopoma* (Sabellariidae).

SABELLIDAE

(References: Fauvel, 1927:296; Hartman, 1944*a*:278; 1951*a*)

1. With thick gelatinous cover; thoracic hooks long-handled (Myxicolinae) . . . . *Myxicola*
1. Without gelatinous cover; thoracic hooks with or without handles 2

2. Thoracic hooks long-handled (fig. 43, *d*) . . . . . Fabriciinae 3
2. Thoracic hooks avicular (no handle) . . . . . . Sabellinae 4

3. With a thoracic collar . . . . *Chone* 12
3. Without a thoracic collar . . . . *Fabricia*

4. Tentacular radioles divided several times. . . . *Schizobranchia*
4. Tentacular radioles not divided 5

5. Composite eyes at distal end of some radioles, especially the dorsalmost pairs (fig. 43, *f*) . . . . *Megalomma*
5. Without composite eyes in such position 6

6. Thoracic notosetae of 2 abruptly different kinds; thoracic collar 4-lobed 7
6. Thoracic notosetae differing gradually among themselves; thoracic collar 2-lobed . . . . *Sabella* 11

7. Each base of tentacular membrane conspicuously spiraled . . . . *Eudistylia* 8
7. Each base of tentacular membrane forms a semicircle. . . . . *Pseudopotamilla* 9

8. Tentacular radioles crossed by alternating bars of red and white; dorsal edge of tentacular base not cleft; in sand or mud flats. . . . . *Eudistylia vancouveri*
8. Tentacular radioles deep maroon, with pale or orange tips; dorsal edge of tentacular base deeply cleft; in rocky crevices, especially on reefs . . . . *Eudistylia polymorpha*

9. Dorsal tentacular membrane with notched flaps (fig. 43, *a*, see 1); dorsal collar lobes short or obscure; tentacular eyespots conspicuous . . . . *Pseudopotamilla ocellata*
9. Dorsal tentacular membrane without flaps (fig. 43, *b*, see 1); dorsal collar lobes present (see 2); eyespots present or absent . . . 10

10. Larger; collar membrane with high dorsal lobes (fig. 43, *b*, see 2); eyespots absent . . . . *Pseudopotamilla intermedia*
10. Smaller; collar membrane with low, oblique dorsal lobes, widely separated from lateral lobes by a deep dorsolateral cleft; eyespots variable . . . . *Pseudopotamilla socialis*

11. Tentacular radioles with paired eyespots (fig. 43, *e*); radioles angular in cross section . . . . *Sabella crassicornis*
11. Tentacular radioles without paired eyespots; radioles not angular in cross section . . . . *Sabella media*

12. Minute; thoracic spatulate setae with distal mucron (tooth); inhabits shell-covered tube . . . . *Chone minuta*
12. Large, robust; thoracic spatulate setae lack distal mucron; inhabits thin, sand-covered tube . . . . *Chone mollis*

List of Species

Chone minuta Hartman, 1944
Chone mollis (Bush, 1904)
Eudistylia polymorpha (Johnson, 1901) (The large and showy "feather-duster" worm; a rock-crevice dweller)
Eudistylia vancouveri (Kinberg, 1867) (A mud-flat dweller)
**Fabricia dubia* Wesenberg-Lund, 1941 (See Hartman 1951*a*. Members of this genus are minute worms, found in marine, brackish, and even fresh water. Some can rebuild tubes after removal, and may move about either backward or forward, being conveniently equipped with a pair of eyespots at each end of the body)
Megalomma sp.
**Myxicola infundibulum* (Montagu, 1808)
Pseudopotamilla intermedia Moore, 1905
Pseudopotamilla ocellata Moore, 1905 (The tiny 2-tentacled hydroid "*Lar*" occurs on the tube of this species in the Monterey area)
Pseudopotamilla socialis Hartman, 1944
Sabella crassicornis Sars, 1851
Sabella media (Bush, 1904)
**Schizobranchia insignis* Bush, 1904

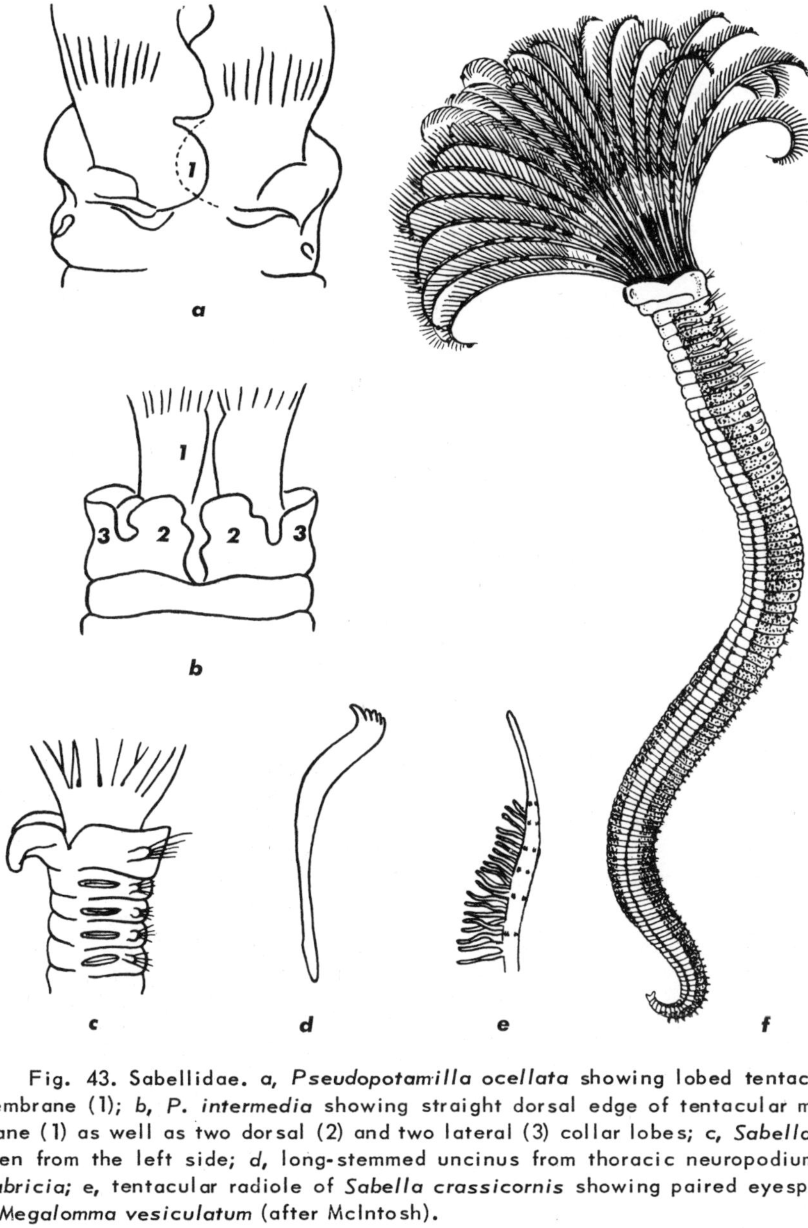

Fig. 43. Sabellidae. *a, Pseudopotamilla ocellata* showing lobed tentacular membrane (1); *b, P. intermedia* showing straight dorsal edge of tentacular membrane (1) as well as two dorsal (2) and two lateral (3) collar lobes; *c, Sabella,* as seen from the left side; *d,* long-stemmed uncinus from thoracic neuropodium of *Fabricia; e,* tentacular radiole of *Sabella crassicornis* showing paired eyespots; *f, Megalomma vesiculatum* (after McIntosh).

SERPULIDAE (Fig. 44)

(References: Fauvel, 1927:348; Hartman, 1944*b*:285; 1948)

1. Tube minute, neatly coiled; body asymetrical. 2
1. Tube larger, straight or somewhat twisted, but not coiled; body not asymetrical. 3

2. Tube sinistral . . . . *Laeospira*
2. Tube dextral . . . . *Dexiospira*

3. Without operculum; tentacles few; forming small fragile tube masses; red in life . . . . *Salmacina*
3. With operculum; tentacles few to numerous 4

4. Operculum bladderlike with smooth stalk; tubes usually solitary, fragile, sometimes longitudinally ribbed . . . *Vermiliopsis*
4. Operculum funnel-shaped 5

5. Stalk of operculum smooth; tubes solitary, fairly large, cylindrical to somewhat twisted . . . . *Serpula*

5. Stalk of operculum with lateral wings; tubes strongly ridged longitudinally . . . . *Spirobranchus*
5. Operculum with spines but without lateral wings on stalk . . . 6

6. Spines of operculum produced from a single terminal plaque; in brackish water . . . . *Mercierella enigmatica*
6. Spines produced in 2-storied structure; marine 7

7. Distal series of spines with lateral projections . . . . *Hydroides*
7. Distal series of spines with smooth sides . . . . *Eupomatus*

List of Species

**Crucigera zygophora* Johnson, 1901
**Dexiospira spirillum* (Linnaeus, 1758)
**Eupomatus gracilis* Bush, 1904
**Hydroides uncinatus* (Philippi, 1844)
Mercierella enigmatica Fauvel, 1923
(Abundant in Lake Merritt, Oakland, and Aquatic Park, Berkeley)
**Laeospira borealis* (Daudin, 1800)
**Salmacina tribranchiata* (Moore, 1923)
(Reproduces by fission while in calcareous tubes. Sometimes considered a variety of *S. dysteri* Huxley)
**Serpula vermicularis* Linnaeus, 1767
**Spirobranchus spinosus* Moore, 1923
Vermiliopsis sp.

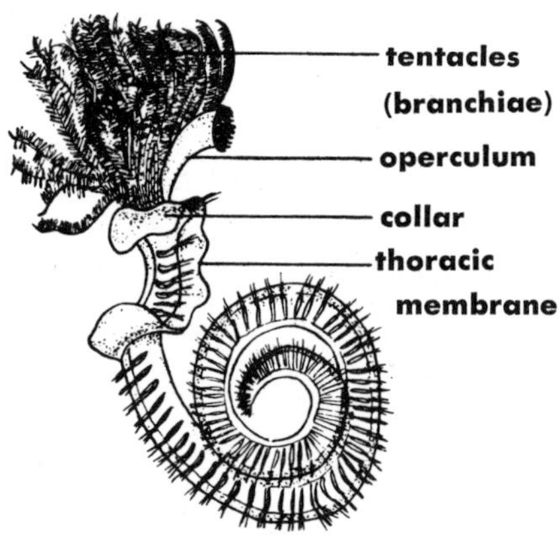

Fig. 44. *Mercierella enigmatica* (Serpulidae), removed from its tube and viewed from the left side.

SPIONIDAE

(References: Fauvel, 1927:27; Hartman, 1936*c*; 1940*b*; 1941*a*; 1944*b*:258)

1. With 5th segment modified and provided with heavy hooks (fig. 45, *b*) 2

1. Without such a modified segment 3
2. Branchiae present in front of and behind the modified segment . . . . *Boccardia* 9
2. Branchiae not present until after modified segment (fig. 45, *a*) . . . . *Polydora* 11

3. With only 1 pair of branchiae, these simple and inserted just behind paired palpi (fig. 45, *c*); 2d segment with raised dorsal membrane; pygidium a simple ring . . . . *Streblospio*
3. With more than 1 pair of branchiae; without raised dorsal membrane behind palpi; pygidium disclike or with long processes . . . 4

4. Prostomium with pair of lateral horns at front end; branchiae present from 2d setiger and present on most body segments . *Rhynchospio*
4. Prostomium without lateral horns; branchiae present from 2d segment or later 5

5. Successive neuropodia in long posterior region connected by interramal pouches . . . . *Laonice*
5. Successive neuropodia without interramal pouches 6

6. Branchiae present from 1st setiger and continued throughout; pygidium surrounded by cirriform processes . . . . *Spio*
6. Branchiae first present after 1st setiger; pygidium with cirriform processes or disclike 7

7. Pygidium disclike; branchiae present from 2d setiger and continued back through a long region 8
7. Pygidium with cirriform processes; branchiae not present until about segment 19, and continued through a median region . . *Pygospio*

8. Hooded hooks (fig. 45, *d*) present in notopodia and neuropodia *Nerine*
8. Hooded hooks present only in neuropodia . . . . . *Nerinides*

9. Posterior segments with modified heavy spines . *Boccardia uncata*
9. Posterior segments without heavy spines 10

10. Branchiae present on anterior segments 2, 3, and 4 . . . . *Boccardia proboscidea*
10. Branchiae present on anterior segments 2 and 3 . *Boccardia truncata*

11. Posterior segments with heavy spines; in coralline algae . . . . *Polydora armata*
11. Posterior segments without heavy spines 12

12. Prostomial caruncle with median antenna . . . . *Polydora ligni*
12. Prostomial caruncle without median antenna 13

13. Branchiae present from 7th segment and continued on many segments . . . . *Polydora brachycephala*
13. Branchiae present from 10th segment and continued to about 25th segment . . . . *Polydora giardi*

List of Species

Boccardia proboscidea Hartman, 1940 (In rock crevices in high tide pools; usually only the 2 tentacles are visible.)
Boccardia truncata Hartman, 1936
Boccardia uncata Berkeley, 1927
**Laonice cirrata* (Sars, 1861)
**Nerine cirratulus* (delle Chiaje, 1828)
**Nerinides acuta* (Treadwell, 1914)
Polydora armata Langerhans, 1880
Polydora ligni Webster, 1879 (= *P. amarincola* Hartman)
Polydora giardi Mesnil, 1896
Polydora brachycephala Hartman, 1936

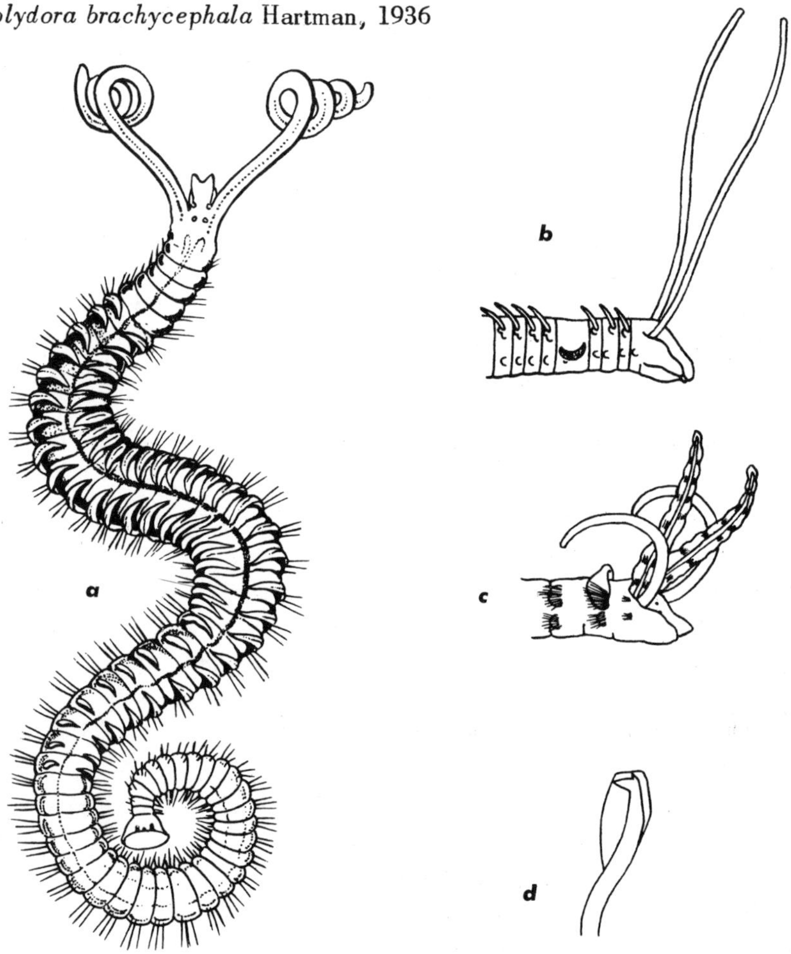

Fig. 45. Spionidae. *a, Polydora ciliata* (after McIntosh); *b, Boccardia; c, Streblospio; d,* hooded hook of *Nerine.*

**Pygospio californica* Hartman, 1936
**Rhynchospio arenincola* Hartman, 1936
**Spio filicornis* (O. F. Müller, 1776)
**Streblospio benedicti* Webster, 1879 (=*S. lutincola* Hartman)

SYLLIDAE

(References: Berkeley and Berkeley, 1948; Fauvel, 1923:254; Hartman, 1944*b*:248)

1. Without ventral cirri . . . . *Autolytus*
1. With ventral cirri 2

2. Short, blunt, with few segments; dorsal cirri very long throughout; with conspicuous nuchal lappets on dorsal side of peristomium . . . . *Pterosyllis*
2. Longer, often threadlike; dorsal cirri otherwise 3

3. Prostomium covered over its posterior 1/2 with a nuchal hood . 4
3. Prostomium without nuchal hood 6

4. Pharnyx unarmed; dorsum purple; in algal clumps . . . *Hesperalia*
4. Pharnyx armed with a semicircle of papillae (fig. 46, *f*); dorsal surface pale, with black transverse stripes; in sponges and bryozoans . . . . *Odontosyllis* 5

5. Dorsum marked with dark pigment bands alternating with yellow bands at irregular intervals . . . . *Odontosyllis phosphorea*
5. Red in life; tiny . . . . *Odontosyllis parva*

6. Parapodia provided with only, or also, composite hooks . . . 7
6. Parapodia provided with only simple, coarse, distally prolonged setae . . . . *Haplosyllis*

7. Minute; ventral and dorsal cirri reduced, inconspicuous; palpi fused throughout (fig. 46, *c*); median antenna of prostomium usually longer than paired ones . . . . *Exogone*
7. Larger; ventral and dorsal cirri more or less conspicuous; palpi free from each other for some distance (fig. 46, *d*) 8

8. Body more or less depressed; proboscis with a saw-toothed edge (trepan) . . . . *Trypanosyllis* 9
8. Body not greatly flattened; proboscis without a trepan . . . 10

9. Dorsum with bold red transverse stripes; body greatly depressed . . . . *Trypanosyllis gemmipara*
9. Dorsum with brownish diamond-shaped spots, segmentally arranged, body only slightly depressed . . . . *Trypanosyllis adamanteus*
9. Dorsum uniformly purple; body slightly depressed *Trypanosyllis* sp.

10. Antennae and dorsal cirri smooth (fig. 46, *e*) . . . *Pionosyllis*
10. Antennae and dorsal cirri articulated 11

11. Setae include simple and composite ones . . . . . . *Syllis* 15
11. Setae all composite 12

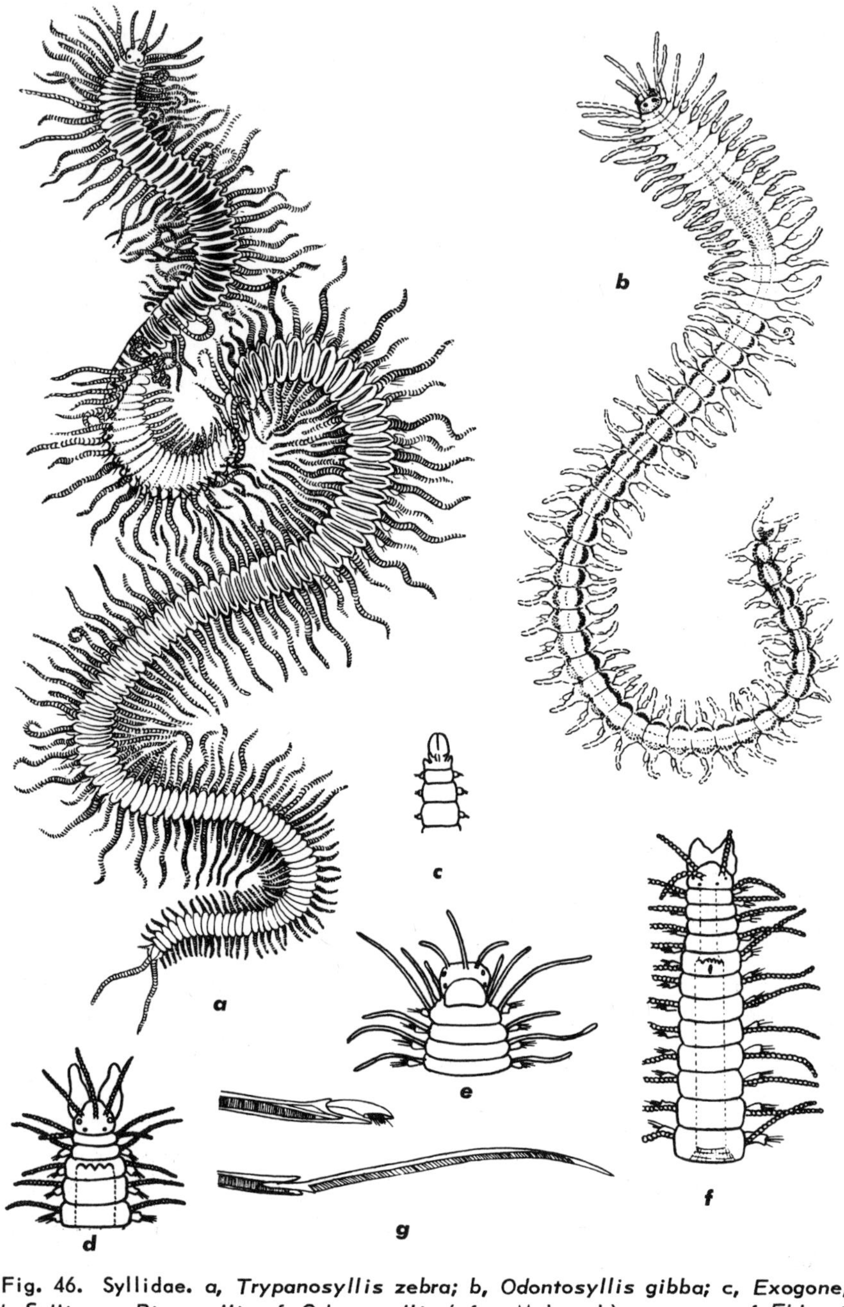

Fig. 46. Syllidae. *a, Trypanosyllis zebra; b, Odontosyllis gibba;* c, *Exogone;* d, *Syllis;* e, *Pionosyllis;* f, *Odontosyllis* (after McIntosh); g, setae of *Ehlersia.*

12. Composite setae all alike . . . . *Typosyllis* 13
12. Composite setae of two kinds, most with a short appendage, but some posterior ones with a long appendage (fig. 46, *g*) . . . *Ehlersia*

13. Dorsum chocolate brown; tentacles pale . . . *Typosyllis pulchra*
13. Dorsum not dark 14

14. Dorsal cirri alternately long and short . . . *Typosyllis hyalina*
14. Dorsal cirri all short and fusiform in shape . *Typosyllis armillaris*

15. Dorsalmost seta heavy, often lacking its appendage; anterior region without long appendaged setae . . . . *Syllis elongata*
15. Dorsalmost seta not conspicuously heavy; those in anterior segments with prolonged appendage . . . . *Syllis alternata*

List of Species

**Autolytus varius* Treadwell, 1914
**Ehlersia cornuta* (Rathke, 1843)
Exogone spp.
Haplosyllis sp.
**Hesperalia californiensis* Chamberlin, 1919
Odontosyllis parva Berkeley, 1923
Odontosyllis phosphorea Moore, 1909
**Pionosyllis gigantea* Moore, 1908
Pterosyllis sp.
Syllis alternata Moore, 1908
Syllis elongata (Johnson, 1901)
Trypanosyllis adamanteus Treadwell, 1914
Trypanosyllis gemmipara Johnson, 1901
Trypanosyllis sp. (From Carmel, California)
Typosyllis armillaris (O. F. Müller, 1776)
Typosyllis hyalina (Grube, 1863)
Typosyllis pulchra (Berkeley, 1938)

TEREBELLIDAE (Fig. 47)

(References: Berkeley and Berkeley, 1952; Hartman, 1944*b*:270)

1. Uncini absent from thorax, or, if present, arranged in a single row[1] 2
1. Uncini in thorax in double rows from the 11th segment[2] . . . 4

2. With 3 pairs of branchiae, each consisting of numerous filaments . . . . *Thelepus*
2. Without branchiae 3

3. Uncini absent; prostomial lobe greatly prolonged as a long flap with 3 lobes . . . . *Amaea*

[1] Resembling one side of a zipper fastener, sometimes in form of a nearly closed loop.
[2] Resembling a closed zipper.

3. Uncini present; prostomium not so prolonged . . . . *Polycirrus*

4. Without branchiae; peristomium with transverse rows of minute eyespots . . . . *Spinosphaera*
4. With single pair of branchiae, these somewhat branched; with thirteen thoracic setigers . . . . *Ramex*
4. With 2 pairs of branchiae, these richly branched; with 24 or more thoracic setigers . . . . *Terebella*
4. With 3 pairs of branchiae 5

5. Anterior end provided with conspicuous lateral lappets . . . 6
5. Anterior end without lateral lappets; with 17 thoracic setigers . 8

6. Thoracic uncini avicular . . . . *Pista* 7
6. Thoracic uncini pectiniform; anterior end with 2 pairs of large lateral lappets . . . . *Loimia*

7. Large lappets on 2d branchial segment; tube with spongelike, reticulated top . . . . *Pista elongata*
7. Large lappets on 2d and 3d branchial segments; tube with large hoodlike overlapping membrane . . . . *Pista pacifica*

8. Chocolate brown in life; tentacles white, crowded; branchiae red; constructs soft mud tubes . . . . *Amphitrite*
8. Greenish; tentacles light tan; branchiae deep reddish-brown; in sandy tubes . . . . *Eupolymnia*

List of Species

**Amaea occidentalis* Hartman, 1944
**Amphitrite robusta* Johnson, 1901
**Eupolymnia crescentis* Chamberlin, 1919
**Loimia montagui* (Grube, 1878)
Pista elongata Moore, 1909 (In rocky intertidal)
Pista pacifica Berkeley, 1942 (In sandy mud flats)
Polycirrus spp. (The species on this coast are not sufficiently well known to be distinguished)
**Ramex californiensis* Hartman, 1944
**Spinosphaera oculata* Hartman, 1944
**Terebella californica* Moore, 1904
**Thelepus crispus* Johnson, 1901

Fig. 47. *Amphitrite gracilis,* Terebellidae (after McIntosh).

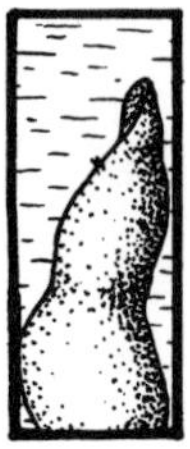

Phylum Echiuroidea

Echiuroids are worms which in a general way resemble annelids but are entirely unsegmented. The body is usually sausage-shaped, consisting of a muscular wall about a spacious coelom, and contains a very long, looped intestine opening by a posterior anus. A characteristic solid, extensible proboscis just anterior to the mouth has given the group the name of "spoon worms," from its shape when contracted. There is usually a pair of setae placed ventrally, behind the mouth, and one or two posterior rings of setae.

Despite the resemblances of echiuroids to annelids and to sipunculids, the embryological studies of Newby (1940) on *Urechis*, together with other evidence, seem to indicate that echiuroids are actually no closer to annelids than they are to molluscs.

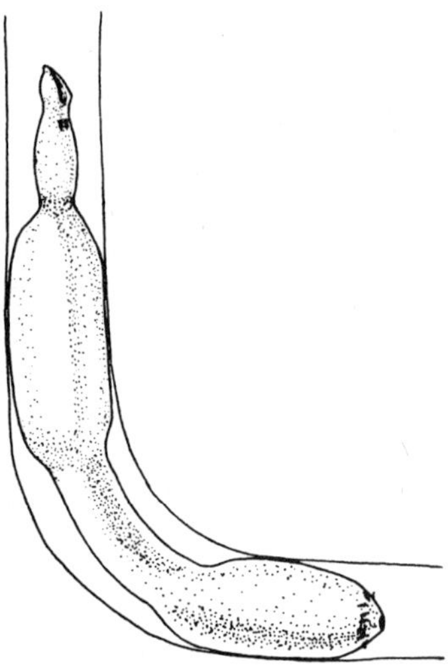

Fig. 48. *Urechis* (from Ricketts and Calvin after MacGinitie).

The group is entirely marine. Most echiuroids burrow in sand or mud, or inhabit rock crevices, empty shells, sand dollar tests, pholad holes, and so on, and swallow large quantities of bottom material or lighter detritus gathered by the long proboscis. However, the most common echiuroid of this area, *Urechis caupo* (fig. 48), has the very specialized habit of collecting fine particulate material, including bacteria, in a secreted net of mucus through which it pumps a flow of water, periodically consuming both net and collected food. The natural history of *Urechis* is interestingly described by Fisher and MacGinitie (1928*b*), and a good general account of the group will be found in Fisher's monograph (1946).

Key to Local Echiuroids

1. Size large; flesh-colored; a circle of conspicuous setae at posterior end of body; preoral proboscis very short, never deciduous; forms permanent burrows in muddy sand; locally common . . . . *Urechis caupo* Fisher and MacGinitie, 1928
1. Size small to medium; no posterior setae; proboscis elongate, soft, sometimes deciduous; color green to gray-violet, proboscis yellow; rare . . . . *Listriolobus pelodes* Fisher, 1946

Phylum Sipunculoidea

The common local sipunculids are known as "peanut worms" because of their color, size, and shape when contracted, although the term would not be descriptive for many other species. The body is more or less cylindrical and elongate when relaxed, with a slender anterior part. Upon contraction this anterior "introvert" is pulled back into itself, and the worm becomes more ovoid, with the body very firm and turgid. The muscular body wall encases a spacious coelum, containing a long, coiled intestine. The anus opens dorsally at the base of the introvert (in contrast to its posterior position in the echiuroids). Although sipunculids resemble echiuroids sufficiently to have been placed with them for many years in the class Gephyrea of the annelids, it is now felt that they represent a distinct phylum, differing from annelids in the lack of segmentation and setae, and from echiuroids in the position of the anus and in other characteristics. The introverted anterior body wall, which can be extended for feeding, bears tentacles about the mouth at its tip and is quite unlike the proboscides of either annelids or echiuroids.

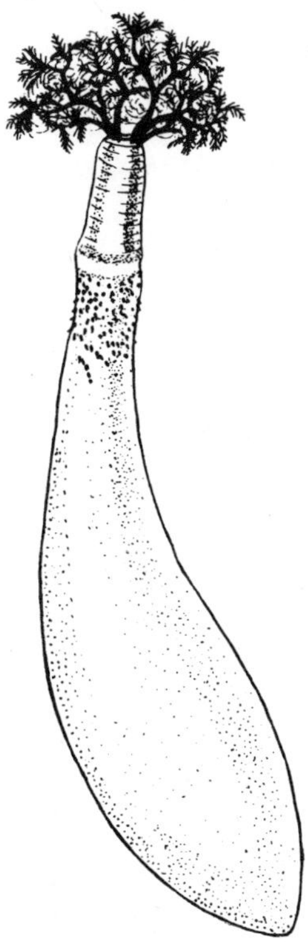

Fig. 49. A Sipunculid (after Fisher).

Sipunculids are entirely marine or estuarine. They commonly, in this area, inhabit rock crevices, empty shells, algal holdfasts, or other protected situations along the open coast. Some may be found in burrows in firm sands or in muds, or among eelgrass roots, although they are absent from the shifting sands of exposed beaches. The rocky intertidal forms apparently collect detritus by means of the tentacles;

burrowing forms may swallow large quantities of mud and sand.

In order to identify sipunculids it is best to let them relax naturally until the tentacles are exposed, and if possible to anaesthetize them in the extended condition. Because of their tendency to pull in the introvert and tentacles when placed in the preservative, even after heavy anaesthetization, it is well to grasp the animals with forceps just behind the tentacles, and to hold them thus until killing is completed. Fisher's monograph (1952) covers the local sipunculids thoroughly.

Key to Sipunculids

1. Tentacles conspicuous when extended, bushy (fig. 49) . . . 2
1. Tentacles inconspicuous, fingerlike 4

2. Introvert armed with small black or brown hooks; tentacles arising on 4 stems. . . . . *Dendrostomum pyroides*
2. Introvert devoid of hooks 3

3. Body very long, cylindrical; tentacles arising on 6 stems, all about equal; living in mud flats . . . . . *Dendrostomum perimeces*
3. Body fusiform (spindle-shaped) to pyriform (pear-shaped), not extremely long; resembling *D. pyroides*; tentacles on 6 stems, dorsal pair smaller; living in sand among rocks . . . . *Dendrostomum dyscritum*

4. Muscles of body wall divided into separate longitudinal bands . 5
4. Muscles of body wall without trace of bands; tentacles very inconspicuous; body small, slender, threadlike. . . *Golfingia hespera*

5. Skin longitudinally ribbed; introvert short, with scalelike papillae, sharply marked off from body . . . . . . *Sipunculus nudus*
5. Skin not ribbed; introvert without scalelike papillae, not sharply set off from body 6

6. Adult size very large; tentacles arranged in numerous longitudinal series, forming a sort of head; no spots . . *Siphonosoma ingens*
6. Medium-sized; tentacles short, in a single crescentic series (open dorsally) dorsal to mouth; introvert and body variably spotted with black, brown, or purple. . . . . . *Phascolosoma agassizii*

List of Species

Dendrostomum pyroides Chamberlin, 1919 (Generally in crevices of granite. *Dendrostoma petraeum* Fisher, 1928 is a synonym)

Dendrostomum perimeces Fisher, 1928 (Found in mud flats)
Dendrostomum dyscritum Fisher, 1952 (In sand among rocks)
Golfingia hespera (Chamberlin, 1919) (E. Ray Lankester named this genus to commemorate a holiday spent golfing at St. Andrews, Scotland, in 1885)
Phascolosoma agassizii Keferstein, 1866 (The commonest sipunculid in central California intertidal; in rock crevices, shells, holdfasts, and so on)
Siphonosoma ingens (Fisher, 1947) (Most northern record, Elkhorn Slough)
Sipunculus nudus Linnaeus, 1766 (Subtidal; occasionally washed ashore)

Phylum Priapuloidea

The members of this phylum bear a superficial resemblance to sipunculids or echiuroids, but the current tendency is to place them among or near that phylum or complex of forms known as Aschelminthes (Hyman, 1951), together with nematodes, rotifers, and other groups. The priapuloid body is muscular and cylindrical, with a bulbous spiny proboscis anterior and a hollow lobed structure posterior, whose cavity is continuous with the perivisceral space of the body. There are only three species in the phylum, all marine, of which *Priapulus caudatus* Lamarck, 1816 occurs in sticky mud bottom in Tomales Bay. It apparently does not construct a permanent burrow, but moves about through the substrate. According to Lang (1948*a*), *Priapulus* is an active predator on polychaetes and even other priapuloids, which it swallows whole. The species is known also from Puget Sound, northern Japan, and the North Atlantic. Its early embryology, which might shed light on its affinities, is not known, but its rotiferlike larval form has been described (Lang, 1948*b*).

Phylum Arthropoda

The diverse and successful arthropods are segmented animals which have developed a firm jointed body exoskeleton and jointed appendages. The parts of this body armor are movable upon each other by discrete strandlike muscles, which are always striated. Mechanically, the body plan is radically different from that of the segmented annelids, in which the body wall is a soft "dermo-muscular tunic" of smooth muscle, surrounding a fluid that contributes rigidity by being under pressure from the soft body wall. The arthropod, quite unlike the average annelid, has a greatly reduced coelom, but has opened out its vascular system into a system of blood-filled spaces, the haemocoel. The hearts of arthropods (and of the arthropodlike Onychophora) possess inlet valves (ostia) by which the blood enters. The eggs are yolky, and the cleavage and embryology are of a highly modified type. Only in the case of the crustacean "nauplius" do we find anything comparable to a simple larval form, and even this shows marked structural complication. Probably as a consequence of the stiff and relatively impermeable exoskeleton, the arthropods have been the most successful phylum in adapting to a terrestrial existence, several great classes occurring only on land (or secondarily reëntering the water).

Since the arthropods constitute by far the largest animal phylum it is necessary that the student have a clear over-all grasp of the major groups, not only those which are well represented in the intertidal, but others which will not be encountered there. The phylum may be divided, for our purposes, into two main lines or subphyla (not including Trilobita, Tardigrada, and certain other groups of doubtful affinities).

1. Subphylum CHELICERATA. The first pair of appendages are usually pincerlike (chelicerate); this subphylum includes the arachnids and their allies. Although predominately a terrestrial group, the Chelicerata include the ancient marine "horseshoe crabs," *Limulus* (class Xiphosura), not found on this coast, and the weird "sea spiders" (class Pycnogonida) which are discussed on page 201 and following.

2. Subphylum MANDIBULATA. The first pair of appendages are antennae (antennules), the second are also antennae or are missing, the third pair form "jaws" (mandibles) which give the group its name.
   a.) Superclass Crustacea. The first two pairs of appendages are antennae. Crustaceans are predominately aquatic, and although extremely diverse, preserve to a large extent the more primitive features of arthropods in general. The group will be treated in some detail on succeeding pages.
   b.) Superclass Labiata. Only the first pair of antennae are present. The bases of the second maxillae are fused to form a "lower lip" (labium) which gives the group its name. Here occur the great class of the Insecta and several smaller classes collectively referred to as "myriapods." The Labiata are a terrestrial group, although numerous insects have reëntered fresh water, and a few occur intertidally. We shall not cover the group in detail in this manual.

SUPERCLASS CRUSTACEA

A general classification of the Crustacea indicates the diverse nature of the group. The student of invertebrate zoölogy should be able to place crustaceans "on sight" into the classes and orders listed below. For the beginning student, the summary of Crustacea in Borradaile and Potts, *The Invertebrata,* is excellent; the student wishing a more complete and well-rounded account will find that Calman in Part VII of Lankester, *Treatise on Zoology,* provides a fine, only slightly out-of-date discussion.

KEY TO THE MAJOR GROUPS OF CRUSTACEA

1. Firmly attached, either to solid substrate (fig. 50, *a*) or rarely as a parasite that is partly internal (fig. 50, *b*) . . . . Class CIRRIPEDIA (p. 128)
1. Free-living, or, if parasitic, generally external 2
2. Body appendages leaflike and numerous (except in Cladocera) . . . . Class BRANCHIOPODA 3
2. Body appendages not leaflike, generally differentiated into several types, relatively few in number 6

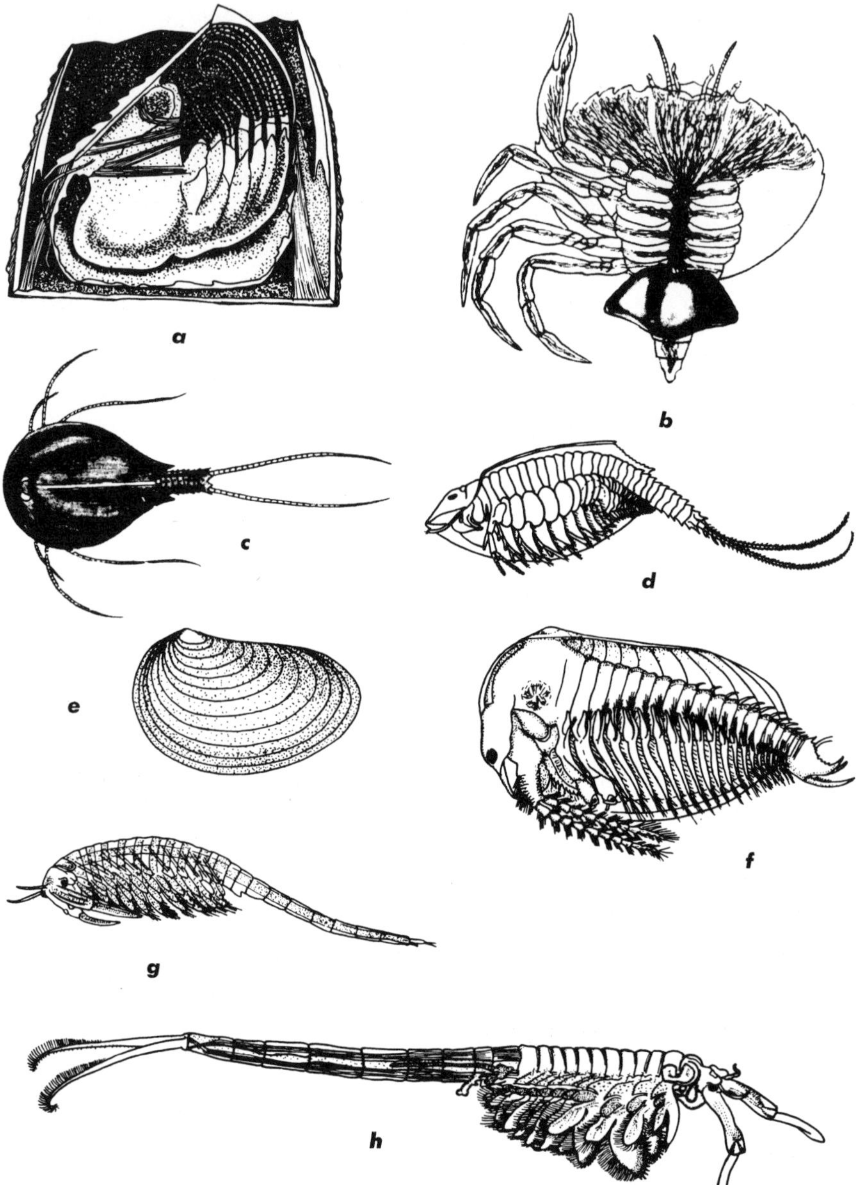

Fig. 50. Types of Crustacea. *a*, Acorn barnacle (Cirripedia, Thoracica), half of shell removed (after Darwin); *b*, *Sacculina* (Rhizocephala) on a crab (after Delage); *c*, *Apus* (Notostraca), dorsal view (after Bonn); *d*, *Apus*, appendages from the left side, carapace removed; *e*, *Cyzicus* (Conchostraca), exterior view of carapace (after Lankester); *f*, *Cyzicus* (=*Estheria*), left valve removed to show appendages (after Lankester); *g*, *Artemia* (Anostraca) (after Bronn); *h*, *Branchinecta gigas* Lynch (Anostraca) (after Lynch).

3. Without a carapace (fig. 50, *g*, *h*) . . . Order Anostraca. (p. 115)
3. With a carapace 4

4. Carapace broad, shieldlike (fig. 50, *c*) . Order Notostraca (p. 117)
4. Carapace conspicuously folded; body laterally compressed . . 5

5. Carapace bivalved, hinged dorsally, enveloping entire body and head; appendages very numerous (fig. 50, *f*) . . . . Order Conchostraca (p. 117)
5. Carapace apparently bivalved, actually not hinged; head not covered by carapace; appendages few (fig. 51, *d*) . Order Cladocera (p. 117)

6. With a bivalved carapace enveloping body and head, externally resembling the Conchostraca, but differing in smaller size (not more than 2–3 mm. in length) and in small number of appendages (fig. 51, *a*, *b*) . . . . Class OSTRACODA (p. 120)
6. Carapace absent or present, but covering at most the thoracic region. 7

7. Flattened fish parasites with compound eyes, suctorial mouth, and limbless abdomen; often placed with next class (fig. 58, *a*) . . . . Genus *Argulus*, Class BRANCHIURA (p. 127)
7. With a median simple eye; consisting of very small swimming forms and of parasites which may be somewhat larger; abdominal region without appendages (fig. 51, *c*) . . .Class COPEPODA (p. 120)
7. With paired compound eyes; free-swimming or crawling forms, generally much larger than copepods; abdomen with appendages except where abdomen is greatly reduced. . .Class MALACOSTRACA 8

8. Carapace, if present, does not fuse with more than four thoracic segments; oöstegites present (plates on ventral side of body of female which form a brood pouch for the hatching of the eggs and the carrying of the young) (fig. 51, *e*-*h*) . . . . Subclass PERACARIDA (p. 135)
8. Carapace fusing with all the thoracic segments; no oöstegites . . . .Subclass EUCARIDA, including Order Decapoda (p. 171)

CLASS BRANCHIOPODA

Of the four orders of the BRANCHIOPODA, only one, the Cladocera, is represented in the sea. The others, chiefly to be found in seasonal ponds, may be briefly discussed as follows. The Anostraca are represented by the brine shrimp, *Artemia salina* (fig. 50, *g*), found in this region in the highly concentrated sea water of pools for the commercial production of salt by evaporation. The brine shrimp is much used as

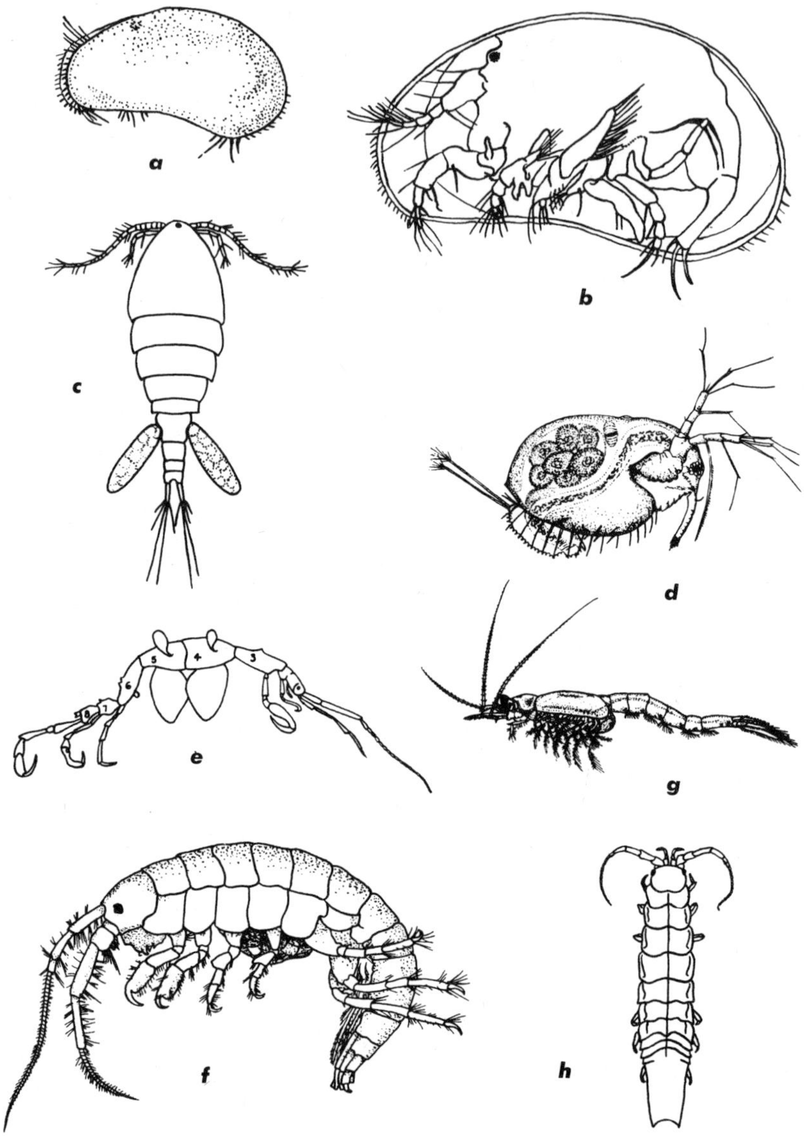

Fig. 51. Types of Crustacea (cont.). *a, Ostracoda* (after Sars); *b, Asterope* (Ostracoda) with left valve removed to show appendages (after Sars); *c, Cyclops* (Copepoda) (after Hartog); *d, Macrothrix rosea* (Cladocera) (after Birge); *e, Caprella* (Amphipoda) (after Eales); *f, Ampithoë humeralis* (Amphipoda); *g, Mysis* (Mysidacea) (after Gerstacker); *h, Idothea (Pentidotea) resecata* (Isopoda) (after Richardson).

food for small fish in aquaria, and can often be obtained cheaply from aquarium stores, or raised from dried "eggs." Other genera, commonly known as fairy shrimps (fig. 50, *h*) occur in enormous numbers at times in seasonal fresh-water ponds. The Notostraca are represented by *Lepidurus* and *Apus* (fig. 50, *c*, *d*), also occurring in seasonal fresh-water pools or rice fields, and probably will not be encountered in the coastal area. A member of the Conchostraca, *Cyzicus* (fig. 50, *e*, *f*), is also widely found in seasonal fresh-water ponds and is often mistaken for a small bivalved mollusc, which its external appearance strongly suggests. *C. californicus* (Packard), formerly called *Estheria californica*, is a common local species.

Order Cladocera

Members of this order are all planktonic, most of them being found in fresh-water ponds, although two genera, *Evadne* and *Podon*, are commonly taken in marine plankton tows. For class use, it is usually most practical to collect cladocerans from fresh-water ponds. For study of fresh specimens it is convenient to anaesthetize them with nicotine or to stick

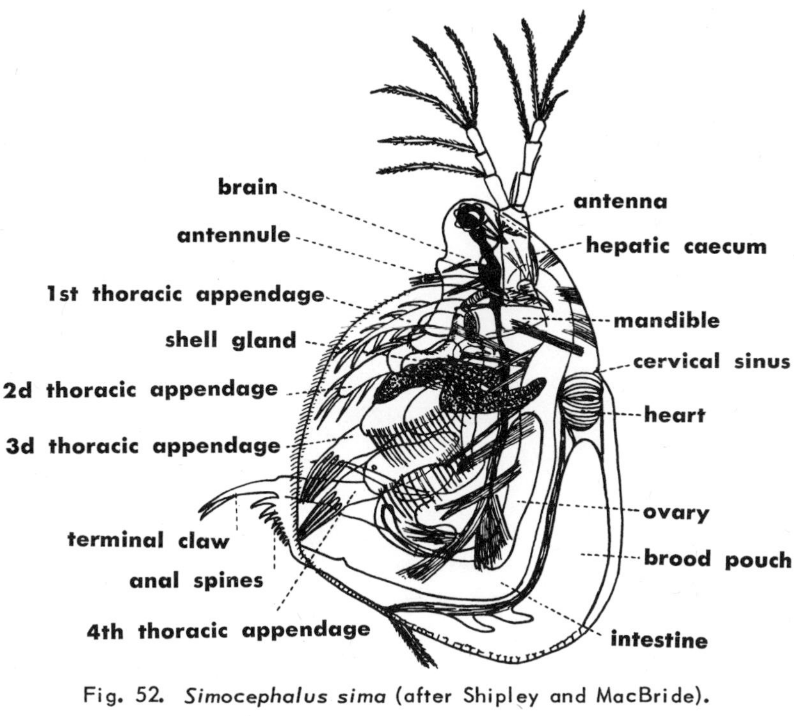

Fig. 52. *Simocephalus sima* (after Shipley and MacBride).

larger ones to a dab of vaseline placed in the bottom of a dish before adding water. The body (fig. 52) is completely encased in a carapace, which appears, but actually is not, bivalved, and has neither hinge nor adductor muscle. The head bears the conspicuous antennae (second antennae) used in swimming, and terminates as a depressed rostrum near which are the vestigial antennules (first antennae). Above the origin of the antennae is a ridge known as the fornix (see *Pleuroxus*, fig. 53, *c*). A notch or transverse groove, the cervical sinus, marks the dorsal junction of head and body in some species. There is a single prominent movable compound eye anterior to the brain, and below and behind the compound eye there is usually a small dark eyespot, the ocellus, closely associated with the brain.

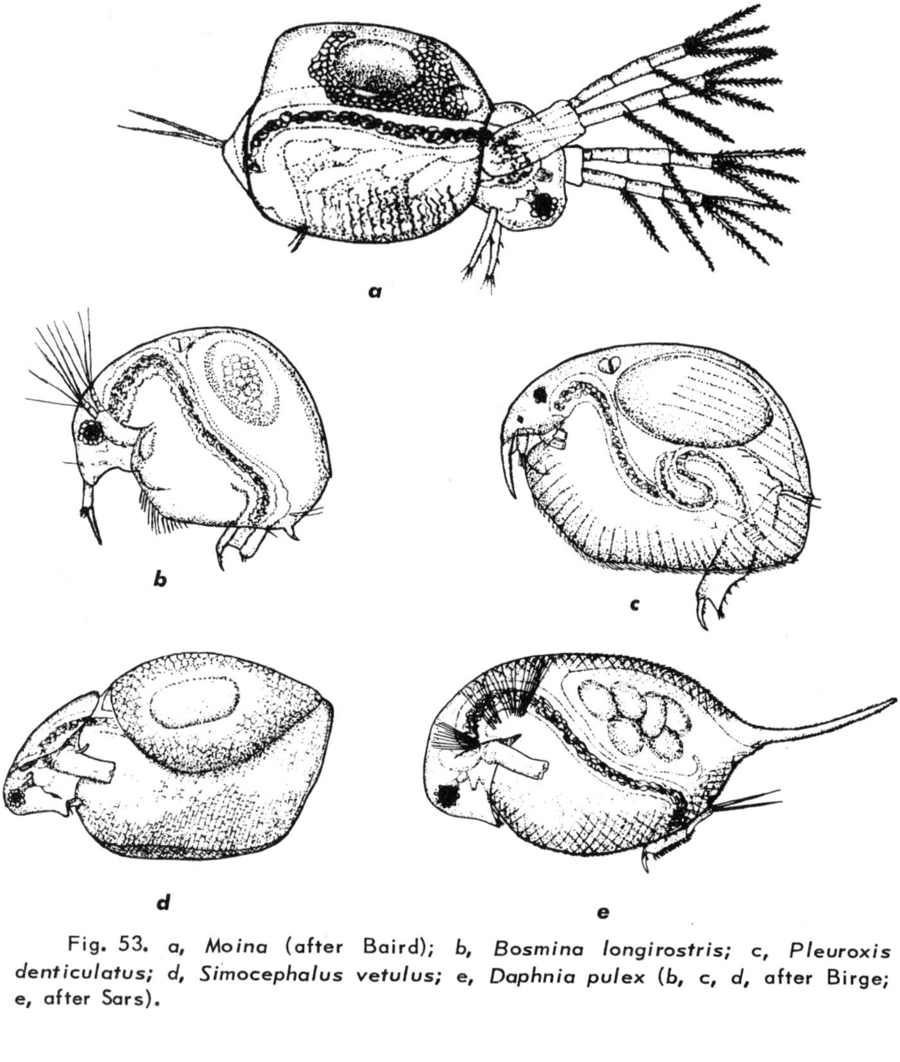

Fig. 53. *a, Moina* (after Baird); *b, Bosmina longirostris; c, Pleuroxis denticulatus; d, Simocephalus vetulus; e, Daphnia pulex* (*b, c, d,* after Birge; *e,* after Sars).

The head may be in line with the body or depressed. The posterior end of the shell or carapace may bear a spine as in *Daphnia* (fig. 53, *e*). The spine may lie at the posterodorsal angle or at the posteroventral (= inferoposteal) angle, when the posterior end of the shell is truncated. The posteroventral angle may be acute, rounded, smooth, or toothed. The nature of the shell surface is often of taxonomic importance. The number of setae on the various joints of the two branches of the antennae beginning at the proximal end is written in a formula of systematic value. The formula for *Daphnia* is $\frac{0\text{-}0\text{-}1\text{-}3}{1\text{-}1\text{-}3}$

For a good general account of, and an extensive key to, the Cladocera, consult Ward and Whipple, *Fresh-water Biology*.

KEY TO THE COMMON GENERA OF CLADOCERA

1. In fresh water 3
1. Marine 2

2. Head not distinctly set off from body; body oval to pyriform, pointed posteriorly; abdomen not protruding . . . . *Evadne*
2. Head distinctly set off from rest of body; shell forming a dorsal hump; abdomen protruding . . . . *Podon*

3. Antennules (fig. 52) of female very small (except in *Moina*, where large), freely movable, and inserted near or behind middle of ventral surface of head (fig. 53, *a*) . . . . . Family DAPHNIIDAE 4
3. Antennules of female long, immovable, inserted near anterior end of ventral surface of head and recurved, forming a beak (fig. 53, *b*) Family BOSMINIDAE . . . . *Bosmina*
3. Antennules of female long, freely movable, inserted at anterior surface of head (fig. 51, *d*) . . . Family MACROTHRICIDAE
3. Antennules of female inserted on ventral surface of head, partly covered by fornices that form a beak with the rostrum extending downward in front of antennules (fig. 53, *c*). Family CHYDORIDAE 3a
   3a. Rostrum long and slender, often upcurved at tip; posterior margin short, often toothed . . . . *Pleuroxus*

4. Cervical sinus present (fig. 52) 5
4. Cervical sinus absent (fig. 53, *e*). . . . . *Daphnia*

5. Antennules very small, hardly movable if at all; head small, depressed, with a rostrum 6
5. Antennules large, freely movable; head large, extended, thick in front; without a rostrum; valves small, not covering body; long

ciliated spines from dorsal side of postabdomen (fig. 53, *a*) . . . . *Moina*

6. Head moderately depressed; valves more or less quadrate . . 7
6. Head greatly depressed; eye almost directly below bases of antennae; valves oval or round . . . . *Ceriodaphnia*
7. Large; valves subquadrate with rounded angles; body large and heavy; yellow to yellow-brown (fig. 52, 53, *d*) . . *Simocephalus*
7. Smaller; valves almost rectangular, posteroventral angle of each produced into a spine; color variable . . . . . *Scapholeberis*

CLASS OSTRACODA

Ostracods present a most characteristic appearance: they are encased in a bean-shaped bivalved shell, from which project a few appendages that are not leaflike in form (fig. 51, *a*, *b*). The carapace entirely encloses both head and body; there is no visible body segmentation, and the body is extremely shortened, having at most seven pairs of appendages. The scurrying motion of these small creatures identifies them almost at a glance. They will be found in almost all collections of fresh-water and marine plants, of bottom material, or of such sessile animals as hydroids and bryozoans. Recognition of ostracods is relatively simple, but identification is a very difficult task. There is a good general account in Ward and Whipple.

CLASS COPEPODA

The Copepoda are the largest class of the entomostracan crustaceans, abundant in all marine and fresh-water situations, and in some groups showing extreme modifications for a parasitic life. For most class use it is easier to obtain copepods from fresh-water pools than from the sea, hence the following discussion of copepod features is based upon the large and common fresh-water genus *Diaptomus*. Living specimens may be anaesthetized for study with a drop of dilute nicotine solution. The copepod body is divided into an anterior part known as the metasome in front of a movable articulation, and a smaller posterior urosome (fig. 54, *a*). These terms are safer than thorax and abdomen, about the limits of which opinions differ. The last or anal segment bears a pair of caudal or furcal rami (or furcae). The uniramous first antennae (antennules) are large and conspicuous, used in swimming; in addition there are smaller, often biramous, second antennae, and five pairs of thoracic (swimming) legs. These legs, particularly the last, are often of very great systematic importance.

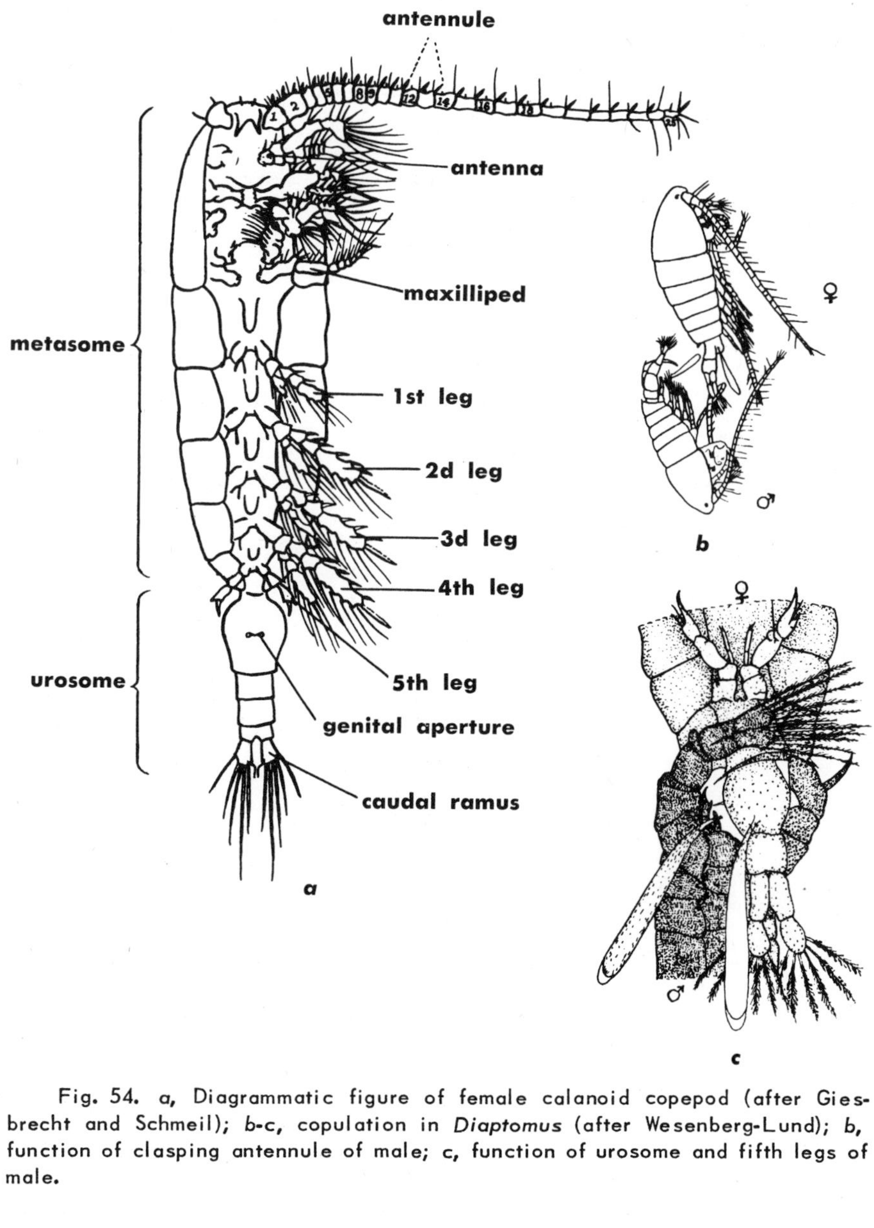

Fig. 54. *a*, Diagrammatic figure of female calanoid copepod (after Giesbrecht and Schmeil); *b-c*, copulation in *Diaptomus* (after Wesenberg-Lund); *b*, function of clasping antennule of male; *c*, function of urosome and fifth legs of male.

The fifth legs of the female calanoid (fig. 54) are greatly modified but symmetrical. In contrast, those of the male are very asymmetrical and have a special function in copulation. Figure 54, *b*, *c*, illustrates the functions in copulation of the peculiarly modified right antennule of the male, the male fifth legs, and the usually asymmetrical urosome of the male. Only in male calanoids are the fifth legs thus elaborately modified. In other orders these appendages do not show such exaggerated dimorphism.

There is a current trend of critical reëvaluation and revision of the higher categories in copepod classification. This may presage the

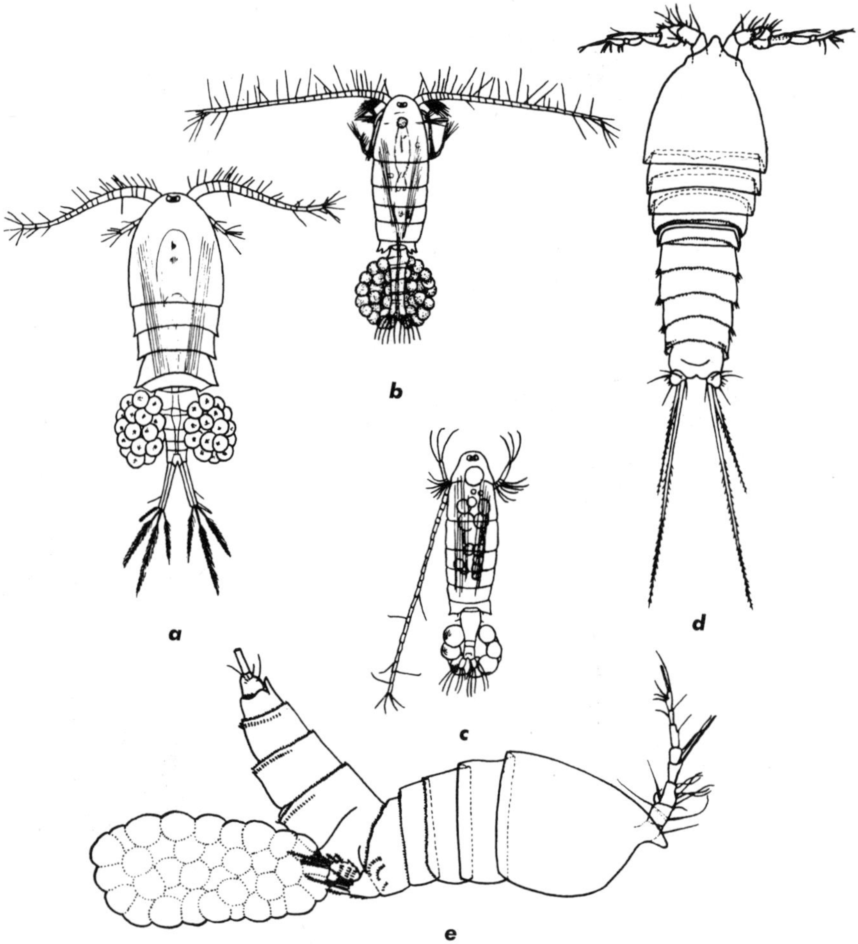

Fig. 55. Types of copepods (variously magnified): *a*, *Cyclops strenuus* Fischer (after Wesenberg-Lund); *b*, *Diaptomus wierzejskii* Richard (after Wesenberg-Lund); *c*, *Diaptomus gracilis* Sars (after Gurney); *d*, *e*, *Canthocamptus illinoisensis* (Forbes) (after Coker).

abandonment, in part at least, of the earlier widely used ordinal names. However, more than 90 percent of all copepods would fall into the modernized counterparts of the old orders Calanoida, Cyclopoida, Harpacticoida, and Caligoida. As a practical expedient, we may continue to use these names, but remembering that changes in diagnosis and perhaps substitution of different names may be expected in the not-too-distant future. The first three of these groups include the commonly encountered free-living copepods of fresh water and many marine forms, both free-living and commensal. Calanoida are usually larger (the examples in fig. 55, *b*, *c* are entirely out of scale), the antennules very long (with twenty-five segments), the urosome sharply distinct from the metasome, which is much longer and broader than the urosome (furcal rami included in measurement, but not setae). Cyclopoida (fig. 55, *a*) are smaller, the antennules shorter, usually with less than seventeen segments, the urosome and metasome distinct and of about the same length. Harpacticoida (fig. 55, *d*, *e*) are usually very small, the antennules short with few segments, and the urosome is not conspicuously set off from the metasome. Caligoida (fig. 58, *b*), include several hundred species, predominately parasitic upon marine fishes.

Because of the enormous number of copepod genera and species, the following key to a few of the more common ones in the San Francisco area should be used with caution, and not relied upon for a positive identification. In the forthcoming revision of Ward and Whipple, edited by W. T. Edmondson, to appear in 1954 or earlier, extensive keys to fresh-water copepods will be included. These embody major revisionary work which will probably not be supplanted for a long time to come.

Key to the Most Common Copepoda of the San Francisco Bay Region

1. Body not abruptly narrowed behind; urosome nearly as wide as metasome (fig. 55, *d*, *e*); first antennae short, rarely of more than 8 or 9 segments (both prehensile in male); largely bottom-living . . . . Order Harpacticoida 2
1. Body abruptly narrowed behind, urosome narrow; antennules long, usually with numerous segments; chiefly free-swimming . . . 3
2. Fresh water (fig. 55, *d*, *e*) . . . . *Canthocamptus* (= *Atthyella*)
2. Marine; orange-red present in high-tide pools *Tigriopus californicus*
3. Antennules very long, usually with more than 20 segments (fig. 55, *b*, *c*); a single, ventrally located egg sac, or egg sacs paired but unequal . . . . Order Calanoida 5

3. Antennules moderately long, with 8–20 segments, usually 10–16; 2 lateral or dorsolateral egg sacs (fig. 55, *a*) Order Cyclopoida 4

4. Fresh-water; free-living; with relatively narrow body . . . . *Cyclops* and related genera
4. Marine, on surface of *Callianassa*; red; female broad; male much smaller (fig. 56) . . . . *Clausidium vancouverense*

5. Marine; urosome of female 4-segmented; left 5th leg of males enlarged . . . . *Calanus* spp.
5. In brackish to hypersaline bays and lagoons; urosome of female 2-segmented; right 5th leg of male the larger . . . . *Pseudodiaptomus euryhalinus*
5. Urosome of female 3-segmented; right 5th leg of male with a long curved claw 6

6. Fresh-water, 4th and 5th segments imperfectly separated . . . . *Diaptomus* 7
6. Brackish-water; 4th and 5th segments distinct; 5th segment in female produced into flattened processes . . . . *Eurytemora*

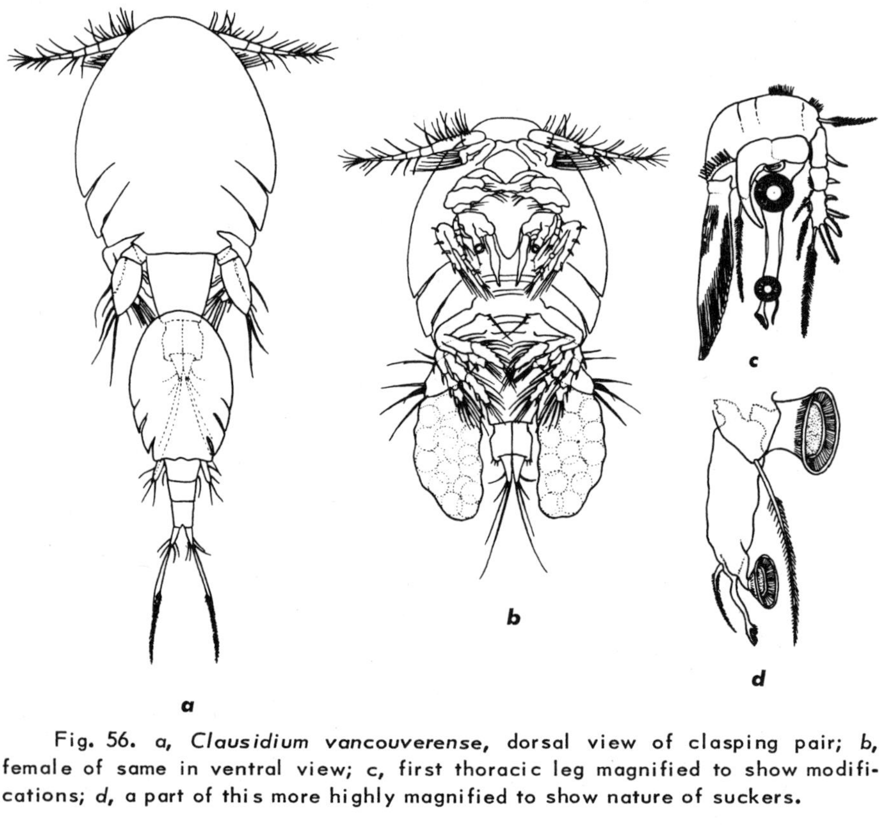

Fig. 56. *a*, *Clausidium vancouverense*, dorsal view of clasping pair; *b*, female of same in ventral view; *c*, first thoracic leg magnified to show modifications; *d*, a part of this more highly magnified to show nature of suckers.

7. Spine of right 5th leg of male nearly straight, situated near distal end of 2d segment of exopodite (fig. 57, *a*); medial process of left exopodite of male 5th leg a slender curving, flat spine (fig. 57, *a*); 3d segment of exopod of female 5th leg distinct (fig. 57, *b*) . . 8
7. Spine of right 5th leg of male curved, much nearer middle of segment (fig. 57, *c*); medial apical process of left exopod digitiform; 3d segment of exopod of female 5th leg vestigial, represented only by seta and spine 10

8. Antennules with setae more numerous than usual, 2 each on segments 13–19, and 4 on segment 2 (fig. 57, *a*) . . . *Diaptomus caducus*

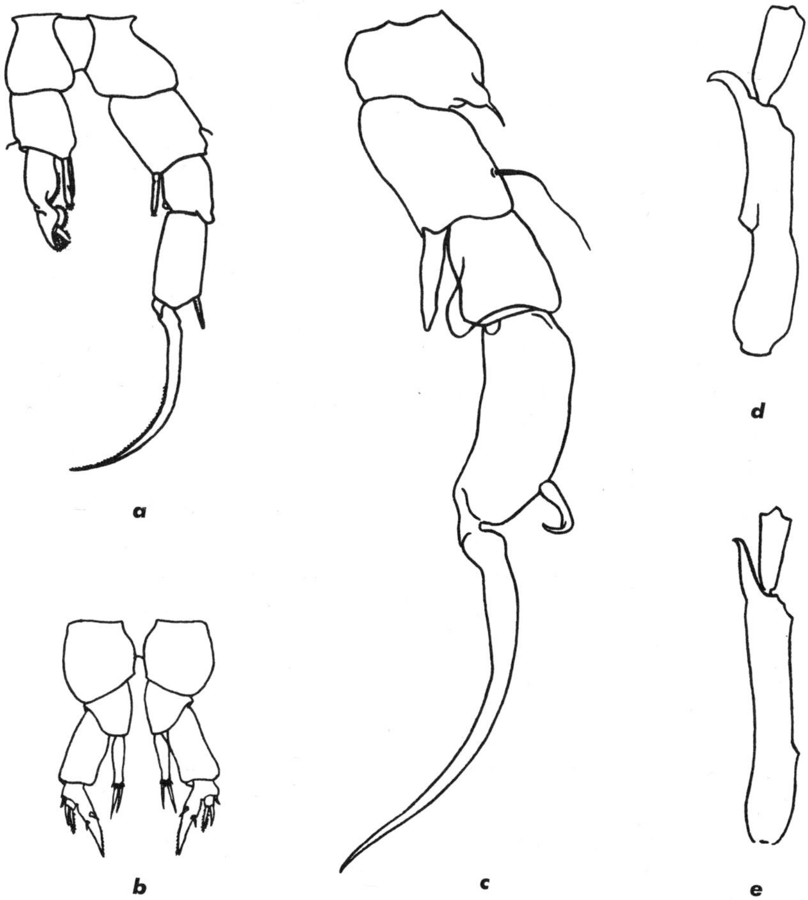

Fig. 57. Fifth legs and right antennules of local species of *Diaptomus*, variously magnified; *a*, fifth legs of male of *Diaptomus caducus* Light; *b*, fifth legs of female of same species; c, right fifth leg of male *Diaptomus siciloides* Lilljeborg, much enlarged; *d*, antepenultimate segment of right antennule of male of *Diaptomus novamexicanus* Herrick; e, same of *D. siciloides*.

8. Antennules with but 1 seta on segments 13–19 and 3 on 2d segment 9

9. Larger; process on antepenultimate segment of right male antennule much longer than penultimate segment, slender distally; right 5th leg of male with a prominent rugose swelling on median margin of 2d basal segment . . . . *Diaptomus eiseni*
9. Smaller; process on antepenultimate segment of right male antennule rarely as long as penultimate segment, thicker distally; no such swelling on right 5th foot of male . . . *Diaptomus franciscanus*

10. Antepenultimate segment of right male antennule without hyaline lamella but with a short process abruptly outcurved near tip (fig. 57, *e*); 5th leg of male with prominent hyaline lamella on distomedial margin of basal segment of right exopod (fig. 57, *c*); this segment nearly twice as long as wide . . . . . *Diaptomus siciloides*
10. Antepenultimate segment of right male antennule with both hyaline lamella and a process; process with gentle distal curvature (fig. 57, *d*); basal segment of right exopodite of male 5th leg wider than long, without hyaline lamella . . . . *Diaptomus novamexicanus*

Partial List of Local Copepoda

(M, marine; B, brackish; F, fresh water; MC, marine, commensal)

Order Harpacticoida

F *Canthocamptus* (= *Atthyella*) sp.

M *Tigriopus californicus* (Baker, 1912) (= *T. triangulus* Campbell) Many other genera and species, fresh-water and marine. (See Monk, 1941, for marine species.)

Order Calanoida

M **Calanus finmarchicus* (Gunner, 1765)

M *Calanus* spp.

F *Diaptomus caducus* Light, 1939

F *Diaptomus eiseni* Lilljeborg, 1889

F *Diaptomus franciscanus* Lilljeborg, 1889

F *Diaptomus novamexicanus* Herrick, 1895 (= *D. washingtonensis* Marsh)

F *Diaptomus siciloides* Lilljeborg, 1889

B **Eurytemora hirundoides* (Nordquist, 1888) (In brackish waters of San Francisco Bay area)

B *Pseudodiaptomus euryhalinus* Johnson, 1939 (Reported from San Francisco Bay, especially in hypersaline waters)

Order Cyclopoida

MC *Clausidium vancouverense* (Haddon, 1912) (On *Callianassa* and *Upogebia*)

F *Cyclops* spp. (Many other fresh-water genera and species)
MC **Paranthessius columbiae* (Thompson, 1897) (On gills of clams *Schizothaerus* and *Protothaca*)
MC **Paranthessius panopeae* Illg, 1949 (From geoduck, *Panope*)
MC **Paranthessius tivelae* Illg, 1949 (From Pismo clam, *Tivela*)
MC **Paranthessius saxidomi* Illg, 1949 } (From Washington clam, *Saxidomus*)
MC **Paranthessius perplexus* Illg, 1949 }

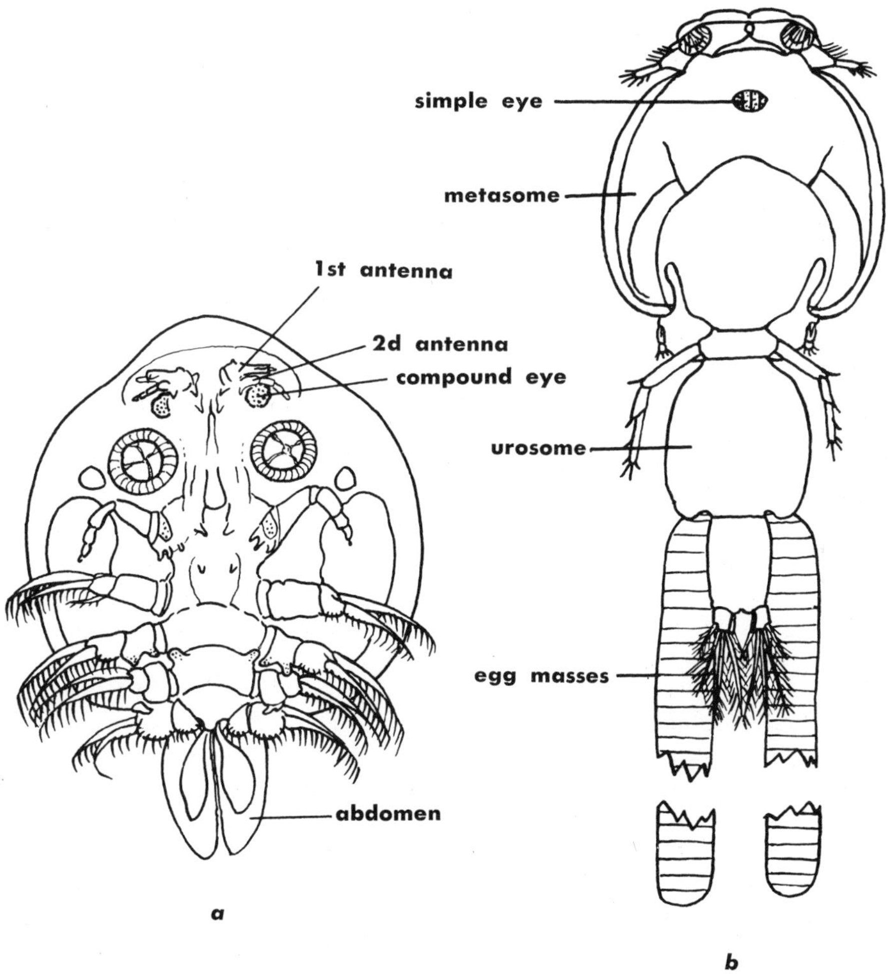

Fig. 58. "Fish lice." *a*, class Branchiura, ventral view of *Argulus*, (after Meehean). *b*, *Caligus rapax*, dorsal view of female with egg strings, class Copepoda, order Caligoida (redrawn from Scott and Scott).

CLASS BRANCHIURA OR ARGULOIDEA

These animals are ectoparasites of fish, fresh-water and marine. Two species are commonly encountered in this region, *Argulus japonicus* Thiele (fig. 58, *a*) common on goldfish, and the much larger *A. pugettensis* Dana common on the various types of marine perch. The group is variously considered to represent an order or subclass of the Copepoda or a separate class, since they differ in the presence of compound eyes and in other characteristics from copepods proper. The papers by Meehean (1940) and Wilson (1944) provide keys to this group.

CLASS CIRRIPEDIA

Of all the crustaceans, the cirripedes may be considered the most specialized, since as adults they are all attached forms, some of them parasitic and extremely modified. Indeed, some could not be recognized as crustaceans except for their larval stages. Three orders of cirripedes occur locally. Of these, the Thoracica or typical stalked and sessile barnacles are familiar to all, and will be discussed in more detail below.

A second order, Acrothoracica, is made up of small forms that burrow into molluscan shells or corals. The tiny *Trypetesa* (see Tomlinson, 1953) may be found in as many as 20 per cent of the "dead" *Tegula* shells inhabited by hermit crabs in the Monterey area and occasionally in other snail shells similarly occupied. It inhabits the body whorl, each animal excavating a small shallow burrow with a slitlike opening on the inside of the host's shell.

The order Rhizocephala is the most specialized of all. Starting life as a typical nauplius, then cypris larva, the rhizocephalan attaches itself to the abdomen of some crustacean, usually an anomuran or brachyuran decapod. Effecting a penetration of the host integument, the larva casts off its limbs and migrates as a mass of cells into the host's body. Eventually it develops into a mass of nutritive rootlets and a globose body, which pushes through to the exterior at some point on the abdomen of its host. The rounded *Sacculina* is fairly common on older *Pugettia producta*, the kelp crab; a small percentage of hermit crabs bear the more elongated *Peltogaster* (Reinhard, 1944). The order, in this area, is not well known.

In a typical stalked or goose barnacle such as *Lepas* (fig. 59, *c*) the armor is made up of five plates, a pair of large scuta (singular, scutum) at the stalk end, a pair of small terga at the free end, and a single narrow median carina at the hinge side. The homologues of these five plates are present in the sessile barnacles as well as several other plates. In *Mitella*, the leaf barnacle, many accessory platelets are found, some of which may be homologous with the extra plates in the sessile barnacles.

In a sessile or acorn barnacle such as *Balanus* (fig. 59, *a*) there are three calcareous regions: (1) the basis by which it is attached, (2) the parapet or wall, and (3) the movable portion, the valves, shutting the aperture at the free end of the wall. The wall of a sessile barnacle consists primitively of eight plates (fig. 59, *b*), each known as a compartment. The one farthest from the head (dorsal or posterodorsal) is the carina, homologous with that in *Lepas*; that at the opposite end is the rostrum. Those on either side of the carina are the carinolaterals; the pair next to the rostrum are the rostrolaterals; the middle compartments

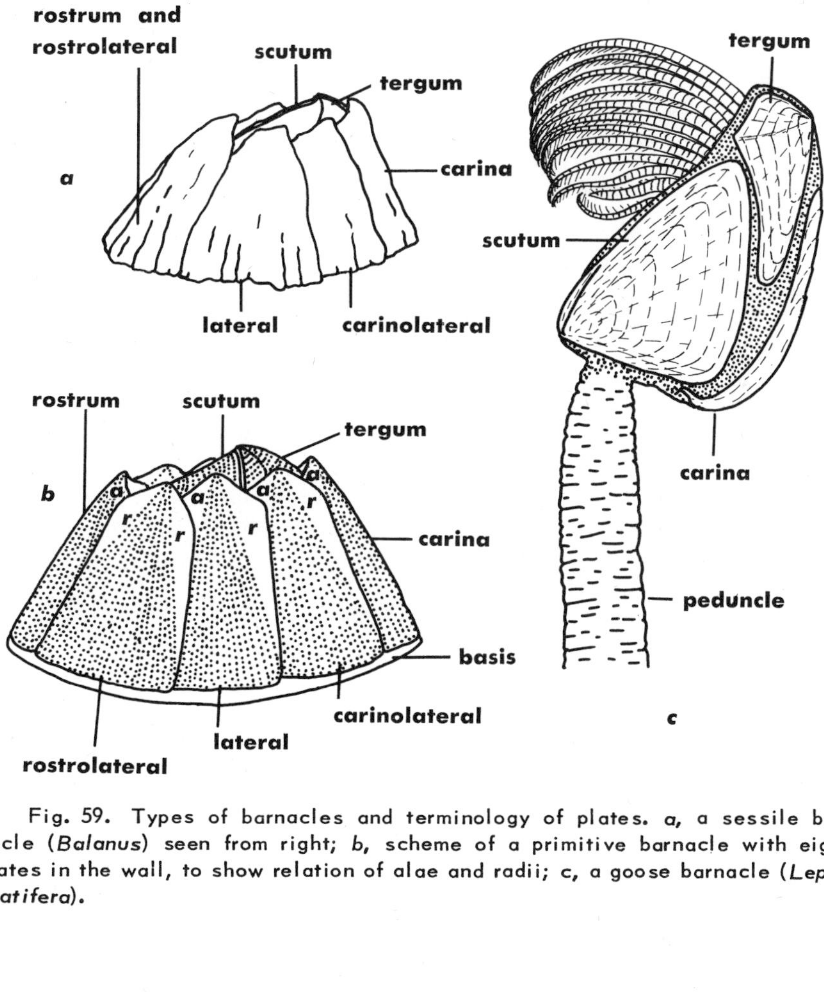

Fig. 59. Types of barnacles and terminology of plates. *a*, a sessile barnacle (*Balanus*) seen from right; *b*, scheme of a primitive barnacle with eight plates in the wall, to show relation of alae and radii; *c*, a goose barnacle (*Lepas anatifera*).

are known as **laterals**. Various combinations and eliminations of these primitive eight plates have occurred in different genera. In the genus *Balanus* the rostrolaterals are fused with the rostrum (fig. 60, *b*). In *Chthamalus* (fig. 60, *a*) the carinolaterals are lacking, and in *Tetraclita* there are only four plates, the rostrolaterals being fused with the rostrum and the carinolaterals absent. The exposed median triangle of each compartment is known as a paries (plural parietes). The edges of a compartment that overlap another compartment are known as radii if the lines of growth are different from the paries; otherwise they receive no special designation. The edges of a compartment that are overlapped are known as alae. The lateral compartments are symmetrical, each with a radius on the posterior (carinal) side and an ala on its anterior (rostral) side. In *Chthamalus* the rostrum is normal with two alae (fig. 60, *a*). In *Balanus* (fig. 60, *b*) and *Tetraclita*, however, it seemingly has two radii, but these are really the radii of the rostrolaterals which are

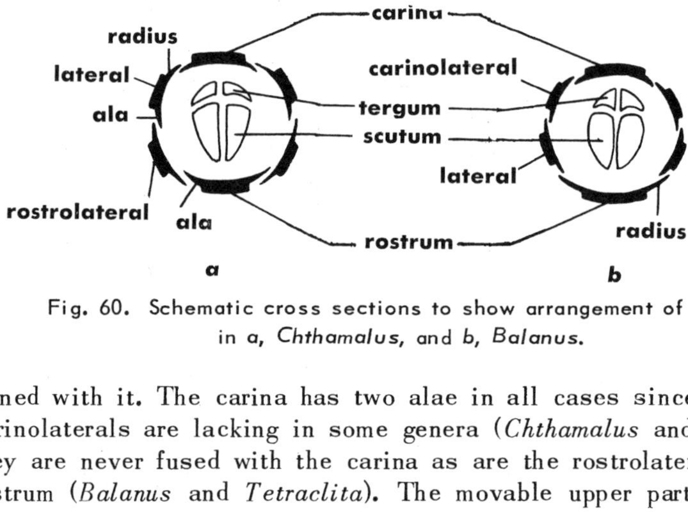

Fig. 60. Schematic cross sections to show arrangement of plates in *a*, *Chthamalus*, and *b*, *Balanus*.

joined with it. The carina has two alae in all cases since, though the carinolaterals are lacking in some genera (*Chthamalus* and *Tetraclita*), they are never fused with the carina as are the rostrolaterals with the rostrum (*Balanus* and *Tetraclita*). The movable upper part consists of two pairs of plates, two longer anterior scuta, and two shorter posterior terga. Characters important in classification are found on the inner surfaces of these plates (fig. 61).

Key to Barnacles (Thoracica)

1. Stalked forms 2
1. Sessile forms 4

2. At most, 5 valves; floating or on floating objects 3
2. With many small plates surrounding the 5 primary valves; on rocks . . . . *Mitella polymerus*

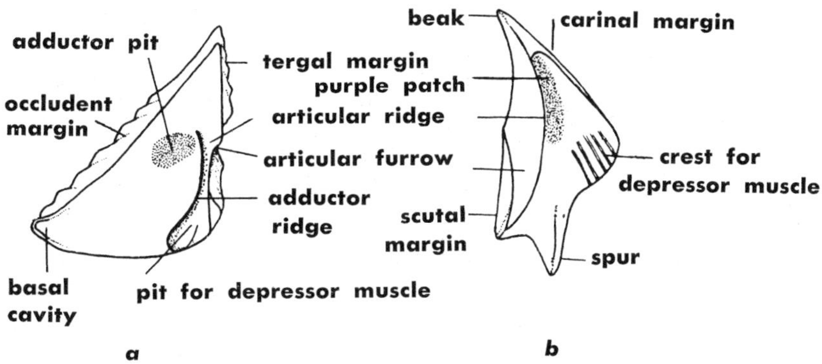

Fig. 61. *Balanus nubilis*, interior of opercular plates (after Henry): *a*, left scutum; *b*, left tergum.

3. Valves reduced to remnants of the scuta; attached to tops of large medusae . . . . *Alepas pacifica*
3. Valves thin and papery; carina ending below in flat disc; animal forms its own float, often in clusters . . . . *Lepas fascicularis*
3. Valves thicker; carina forked below; on floating timbers, and so on . . . . (usually) *Lepas anatifera*

4. Both carina and rostrum overlapped at sides by adjacent plates (fig. 60, *a*); small, often brownish; opercular plates darker than wall . . . . *Chthamalus* 5
4. Carina overlapped by adjacent plates, but rostrum (fused with rostrolaterals) overlaps adjacent plates (fig. 60, *b*); often larger; opercular plates same color as wall; divisions between compartments often obscured externally 6

5. Adductor ridge of scutum prominent and reflexed toward basal margin; depressor muscle pit crossed by several shallow furrows or crests . . . . *Chthamalus dalli*
5. Adductor ridge not prominent, may be lacking; depressor muscle pit simple . . . . *Chthamalus fissus*
5. Adductor ridge may be prominent but is not reflexed; depressor muscle pit extends up under lamina of the articular furrow; this species differs from the preceding two in having a smaller aperture and a very low, broad, and more irregular outline . . . . *Chthamalus microtretus*

6. Wall finely ridged externally, red, consisting of but 4 compartments, the divisions between which are hardly discernible externally . . . . *Tetraclita squamosa rubescens*
6. Wall variously ridged, consisting of 6 compartments, the margins of which may not be discernible externally . . . . *Balanus* 7

7. Walls rose-colored, regular; radii broad, more or less grooved transversely; parietes relatively narrow, smooth or vertically ridged, and vertically striped with white . . . . . *Balanus tintinnabulum*
7. Walls gray or white; radii narrow or lacking; parietes broad, smooth, or rough, white or with longitudinal purplish lines 8

8. Outside of wall with numerous fine ridges, mostly ending below as projecting points in unworn specimens; suture between scuta strongly sinuate; spur of lower end of terga longer than basal width . . . . *Balanus cariosus*
8. Outside of wall without, or with few (less than 30), ridges, which are relatively heavy and rarely project as points below; suture between scuta feebly or minutely sinuate 9

9. Scutum with distinct pit internally immediately below adductor ridge, so that with the 2 muscle impressions there is a series of 3 more or less distinct pits in a diagonal line across scutum; spur of lower end of tergum broader than long, broadly rounded; common species of higher littoral, commonly with distinct ridges on wall, which is often punctate . . . . *Balanus glandula*
9. Scutum without pit immediately below adductor ridge; wall smooth or ridged, not punctate 10

10. Scutum without adductor ridge; spur of tergum broader than long, broadly rounded; small, low form, occurring usually on under sides of boulders . . . . *Balanus crenatus*
10. Scutum with distinct adductor ridge; spur of tergum at least as long as broad; small to very large. 11

11. Small; spur of tergum not tapering, placed close to basiscutal angle; ridges for lateral depressor muscle prominent, extending beyond margin of tergum; compartments externally smooth, regular; cone usually rather low, with relatively small, regular opening; in bays or in brackish water at river mouths . . . *Balanus improvisus*
11. Small or medium-sized; spur of tergum not tapering, situated near basiscutal angle; ridges for lateral depressor muscle not extending beyond margin of tergum; compartments externally smooth, regular; parietes with longitudinal purplish lines (sometimes absent, indistinct, or present only near basis in older animals); in bays, brackish waters, or saline lakes . . . . *Balanus amphitrite*
11. Large; spur of tergum tapering, situated near middle of basal margin; compartments with few heavy ridges, more or less irregular; cone elevated with large opening, irregular in larger individuals; in mussel beds, and under rocks or ledges in low intertidal . *Balanus nubilis*

List of Cirripedia

Alepas pacifica Pilsbry, 1907 (Found attached to large medusae, usually *Phacellophora*)
**Balanus aquila* Pilsbry, 1907
Balanus amphitrite Darwin, 1854 (Several varieties; in brackish waters)
Balanus cariosus (Pallas, 1788)
Balanus crenatus Bruguière, 1789
**Balanus flos* Pilsbry, 1907 (Usually overgrown by sponges)
Balanus glandula Darwin, 1854
Balanus improvisus Darwin, 1854 (In brackish waters)
Balanus nubilis Darwin, 1854
Balanus tintinnabulum (Linnaeus, 1758) (Low intertidal, below mussel beds and on floats)
Chthamalus dalli Pilsbry, 1916
Chthamalus fissus Darwin, 1854
Chthamalus microtretus Cornwall, 1937 (Apparently rare)
Lepas anatifera (Linnaeus, 1758) (Common on floating timbers)
**Lepas hilli* (Leach, 1818) (Rare; probably many reports of this species refer to young *anatifera*)
Lepas fascicularis Ellis and Solander, 1786
Mitella polymerus (Sowerby, 1833) (Occurs in dense masses associated *Mytilus californianus*)
**Peltogaster* sp. (On hermit crabs)
**Sacculina* sp. (On brachyurans)
Tetraclita squamosa rubescens Darwin, 1854 (In summer commonly found with masses of larvae in mantle cavity)
**Tetraclita squamosa elegans* Darwin, 1854
**Trypetesa* sp. Tomlinson (in MS)

CLASS MALACOSTRACA

The great class Malacostraca includes most of the large, common, and conspicuous crustaceans. Although its members are well-known, they show a diversity of features which makes the group rather hard to define, and which makes it necessary to recognize a number of rather distinct orders. Among the characteristics that unite the various orders of the Malacostraca, the most striking is perhaps the division of the body into well-marked thoracic and abdominal regions; the thorax (often to some extent fused with the head) of eight somites and bearing appendages variously adapted for swimming, feeding, or walking; and an abdomen of six somites, all of which usually bear appendages except when the abdomen is reduced, as in crabs, the last pair of these appendages forming together with the telson a caudal fan. Other features usually present include the possession of stalked eyes, a carapace covering part or all of the thorax, and biramous first antennae. When we consider the *more primitive* members of the various major groups, we see a sort of unity of form, an approach to a common type that is shrimplike in form and is said to exhibit the "caridoid facies." The mysids perhaps most closely approach this generalized concept of a malacostracan. Such types as amphipods and isopods seem to diverge widely in general form and appearance, although a detailed study will soon reveal the malacostracan ground plan in them also.

Subclass Leptostraca

This group includes small forms that in some ways resemble certain branchiopods, since they possess flattened limbs and have a bivalved carapace not fused to the thorax. But they have the characteristic stalked eyes, the eight-segmented thorax, and other features which give them a place as primitive malacostracans. They are somewhat aberrant in having seven rather than the usual six abdominal segments. The Leptostraca are represented locally by the small *Epinebalia pugettensis*, very abundant in certain years among green algae in Elkhorn Slough.

Subclass Stomatopoda (Hoplocarida)

Here belong the weird mantis shrimps. of which we have no local intertidal representatives, though *Pseudosquilla lessonii* (Guérin) may be turned up occasionally in the low intertidal south of Point Conception, and *Squilla polita* Bigelow has been dredged in Monterey Bay.

Subclass Syncarida

This is another group not found locally. These creatures are known only from fresh-water lakes in Tasmania and Australia and from subterranean waters in Europe. They resemble small shrimps without a carapace and with little distinction between thorax and abdomen. Apparently the existing types are relicts of an ancient primitive stock.

Subclass Peracarida

An extremely abundant, successful, and varied group, the peracaridans include two orders, the Mysidacea and Cumacea, which are shrimplike and possess most of the features of the Malacostraca in more or less typical form. In contrast, the orders Chelifera, Isopoda, and Amphipoda, with reduced carapaces and sessile eyes, have departed widely from the typical shrimplike form. The feature which unites these diverse orders is the possession of a brood pouch or "marsupium" beneath the thorax of the female, formed by oöstegites borne upon the thoracic limbs.

Key to Orders of Peracarida

1. Body possessing the caridoid (shrimplike) form, with a distinct carapace over the thorax and an elongated abdomen 2
1. Body having thorax and abdomen not sharply distinguishable; carapace lacking or very small 3
2. Eyes stalked when present; carapace covering all or most of thorax (fig. 51, *g*). . . . . Mysidacea (p. 135)
2. Eyes sessile when present; carapace covering only 3 or 4 thoracic segments and inflated into a branchial chamber on each side . . . . Cumacea (p. 136)
3. A small carapace present, covering 2 thoracic segments; resemble small isopods but have 1st pair of legs chelate . . . . Chelifera (p. 136)
3. Carapace lacking 4
4. Body usually dorsoventrally flattened; thoracic legs (except for maxilliped) essentially alike; abdominal limbs modified for respiration or swimming . . . . Isopoda (p. 137)
4. Body usually laterally compressed; thoracic limbs of more than one form, with the 2d and 3d usually prehensile . Amphipoda (p. 155)

Order Mysidacea

The small shrimplike mysids are commonly encountered, often in very large numbers, although they are individually inconspicuous. Species likely to be collected include:

Archaeomysis maculata (Holmes), a tiny form in the sand of open beaches.

Neomysis mercedis Holmes, often occurring in swarms in brackish-water bays and estuaries, as well as in fresh water.

Neomysis costata (Holmes), found in tide pools and kelp beds of the Monterey area.

Detailed information will be found in the works of Banner and of the Tattersalls.

Order Cumacea

These tiny shrimplike forms (fig. 62, *a*) with sessile eyes and a curiously inflated carapace are found subtidally in mud or silt, but are not encountered in general collecting.

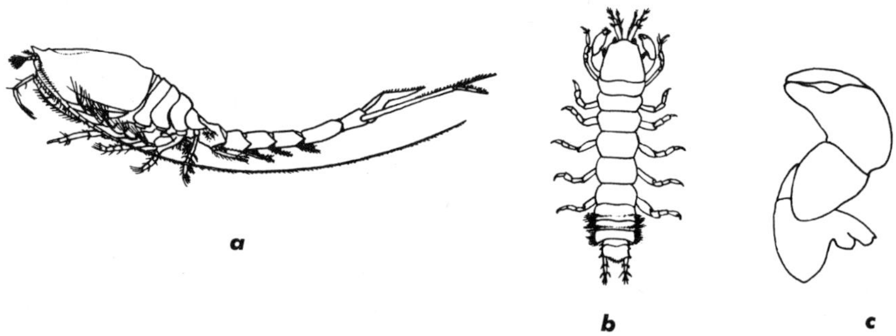

Fig. 62. *a*, *Diastylis*, a cumacean; *b*, *Tanais*, a cheliferan, dorsal view; *c*, chela of same enlarged (*a* after Sars; *b*, *c* after Richardson).

Order Chelifera (Tanaidacea)

This group is sometimes classed as a suborder of the Isopoda, but its members differ from true isopods in possessing a small carapace which covers and is fused with the first two thoracic somites. The most obvious point of difference from isopods is that the first pair of legs is provided with heavy pincers (fig. 62 *b*, *c*). Cheliferans are tiny creatures, commonly found crawling, in a slow and awkward fashion, in clusters of hydroids and bryozoans. One, *Pagurapseudes* sp., has taken up life in tiny empty snail shells and resembles a hermit crab, even to the asymmetrical abdomen—a remarkable example of convergence in evolution.

Order Isopoda

Isopoda are dorsoventrally flattened crustaceans, which, in their lack of a carapace, their sessile eyes, and departure from the caridoid form, are atypical but very successful malacostracans, on land and in fresh water as well as in the sea. Special attention should be given to the features of this order, and comparisons made with the equally important order Amphipoda. The body has a characteristically flattened form, sessile compound eyes, and a long thorax with (usually) seven free segments, the first of the eight being fused with the head. Correspondingly, there are seven pairs of walking legs (peraeopods) which are rather alike (iso-pod). The anterior thoracic limbs are modified to form a pair of maxillipeds. The abdomen is shortened, with at least the last segment united with the telson. Ventrally it bears five pairs of pleopods which function for swimming and/or respiration. The sixth pair of abdominal appendages are the uropods, which may take the form of a pair of caudal furcae (fig. 63, *a*), or may form with the telson a caudal fan (fig. 63, *e*, *f*), or may form a protective case over the pleopods (fig. 63, *b*, *c*). Owing to fusions with head and telson, the free areas of thorax and telson may not quite represent the full extent of these body divisions. The seven free segments of the thorax are sometimes called the peraeon (or perion), and the first five abdominal segments constitute the pleon. The entire abdomen plus the telson is referred to as the pleotelson.

The edges of the body somites may in some forms be expanded laterally into pleural plates. In the thoracic region, these pleural plates may be supplemented or replaced by separate epimeral plates (fig. 63, *b*) or epimera (singular, epimeron), which represent the coxal articles of the peraeopods and hence are often called coxal plates. Female isopods bear oöstegites (fig. 63, *a*, *d*), which arise from the inner bases of the thoracic limbs and form the egg pouch or marsupium in which the eggs are carried during their development.

It would be well for the student contemplating any serious study of isopods to examine carefully the head and its appendages, either on prepared whole mounts of the separated head appendages, or by means of careful dissection with fine forceps or needles. These are, in order, first antennae (antennules), second antennae (antennae), mandibles, first maxillae (maxillules), and second maxillae (maxillae). The maxillipeds, although clearly mouthparts in function, are the first appendages of the thorax. A labrum (anterior lip) is borne beneath a broader and more sclerotized plate, the clypeus. Just dorsal to the clypeus, and lying between the bases of the antennules, there may be a small frontal process. The clypeus may have one or more lamelliform extensions which are called frontal laminae.

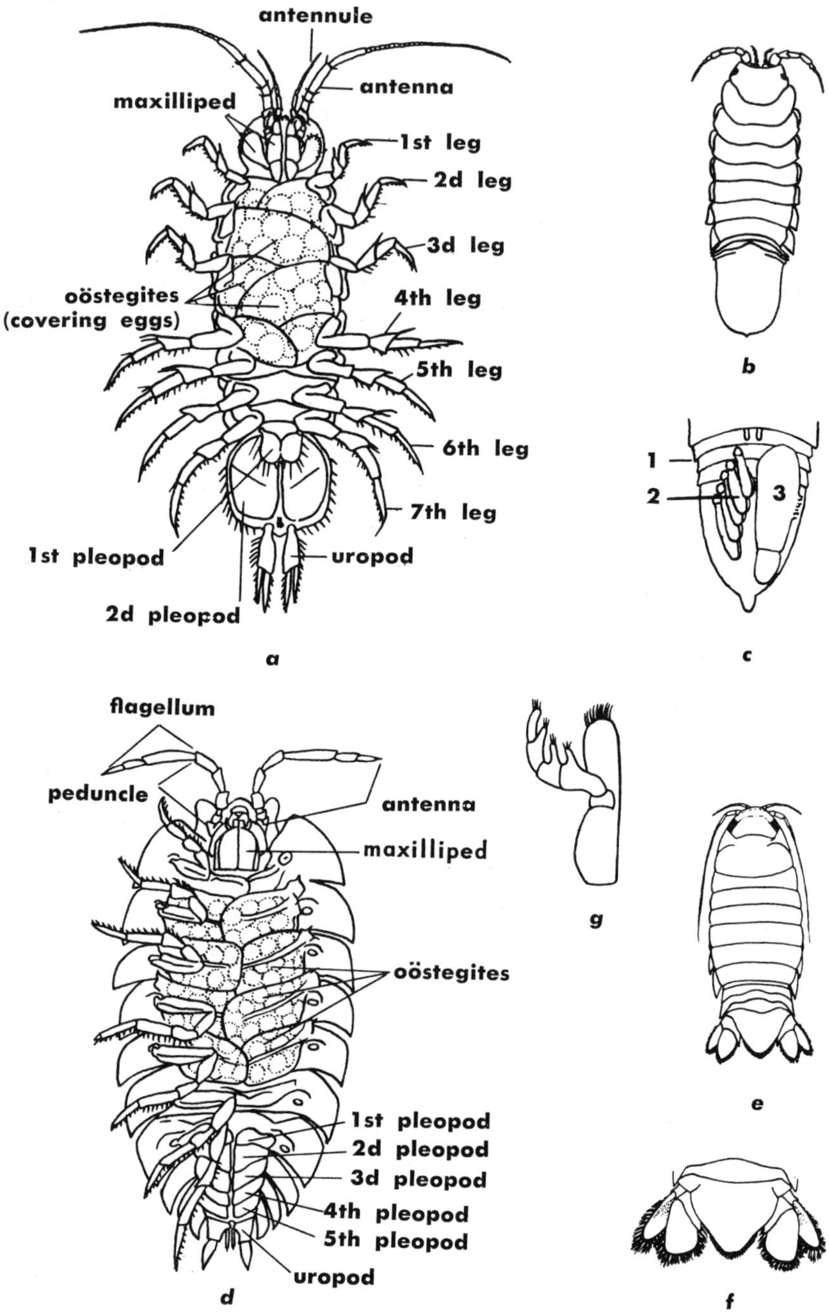

Fig. 63. Isopod anatomy. (Legend on facing page)

Key to Chelifera and the Suborders of the Isopoda

by Robert J. Menzies and Milton A. Miller
Scripps Institution of Oceanography, La Jolla, and University of California College of Agriculture, Davis

1. First antenna larger than 2d antenna; 1st pair of legs (peraeopods) chelate (called gnathopods); 1st 2 thoracic somites united with cephalon to form a carapace (leaving 6 free thoracic segments); eyes when present usually separated from cephalon by fine incision; uropods terminal and composed of small cylindrical articles (fig. 62, *b*, *c*); respiration branchial . . . . Order Chelifera (p. 140)
1. First antenna smaller than 2d antenna; gnathopods chelate or not; in nonparasitic forms 7 free thoracic somites are not usually united with cephalon; eyes usually present, never separated from cephalon; uropods lateral, ventral, or terminal, articles usually dorsoventrally flattened; respiration abdominal . . . . . . Order Isopoda 2

2. Parasitic on crustacea . . I. Bopyroidea (Epicaridea) (p. 141)
2. Not parasitic on crustacea (some Asellota are ectocommensal on crustacea) 3

3. Uropods lateral (fig. 63, *e*, *f*) or ventral (fig. 63, *c*) 4
3. Uropods terminal (figs. 63, *a*; 71) 5

4. Uropods lateral, visible dorsally, with flattened branches forming with telson a caudal fan (fig. 63, *e*, *f*). . II. Flabellifera (p. 142)
4. Uropods ventral, not visible dorsally, modified to form opercular plates hinged laterally and covering the pleopods (fig. 63, *c*). (The Tylidae of the Oniscoidea have similar opercular plates. They differ markedly from the Valvifera, however, in lacking an elongate, multi-articulate flagellum on the 2d antenna, and of course are terrestrial in habit.) . . . . III. Valvifera (p. 144)

5. Aquatic forms; pleon consisting of 2 somites (figs. 69, 70) . . . . IV. Asellota (p. 147)
5. Terrestrial forms; pleon consisting of 6 somites (fig. 71). . . . . V. Oniscoidea (p. 150)

Fig. 63. Isopod anatomy. *a*, *Asellus communis* Say, suborder Asellota, ventral view of female to show structures referred to in key; *b*, *Idothea (Pentidotea) wosnesenskii*, suborder Valvifera, dorsal view showing fusion of telson and posterior pleon segments; *c*, ventral view of *Idothea baltica* showing valvelike uropods (3), covering pleopods (2), note three free pleon segments (1); *d*, *Porcellio scaber* Latreille, suborder Oniscoidea; *e*, *f*, *Cirolana harfordii*, suborder Flabellifera, and caudal fan, enlarged; *g*, Maxillary palp of *Exosphaeroma*, showing lobes (*a*, *d*, after van Name; *b*, *e*-*g*, after Richardson; *c* after Eales).

Order Chelifera (Tanaidacea)

1. First antenna with a biramous flagellum 2
1. First antenna with a uniramous flagellum 3

2. Pleon with 6 somites . . . . *Pagurapseudes laevis*
2. Pleon with 3 somites (fig. 64) . . . . . *Synapseudes intumescens*

3. Pleon with 3 somites plus lateral incisions of a 4th . . . . *Pancolus californiensis*
3. Pleon with 5–6 somites 4

4. Pleon with 5 somites . . . . *Tanais* sp.
4. Pleon with 6 somites 5

5. Three pairs of pleopods . . . . *Anatanais normani*
5. Five pair of pleopods *Leptochelia* sp.

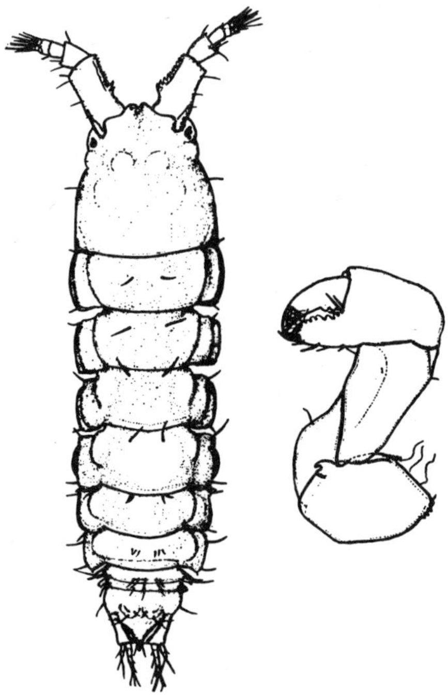

Fig. 64. **Chelifera.** *Synapseudes intumescens* and gnathopod (from Menzies, 1949).

Order Isopoda

I. Bopyroidea (based on female only)

1. Female lacks segmentation; internal parasites 7
1. Female segmented; external parasites of shrimps and anomurans . . . . BOPYRIDAE 2
2. Only 1 leg present on the swollen side of the body; other side with 7 legs . . . . *Phryxus* sp.
2. Seven legs present on each side of the body 3
3. Abdomen without enlarged pleural lamellae (fig. 65, *c*, *d*). . . . 4
3. Abdomen with enlarged pleural lamellae (fig. 65, *a*, *b*) . *Ione* sp.
4. Pleopods triramous . . . . *Stegophryxus* sp.
4. Pleopods biramous 5
5. Branches of the pleopods of adult female similar in size and shape, narrow, elongated . . . . *Phyllodurus abdominalis*
5. Branches of pleopods of adult female unlike 6
6. Exopod of pleopods narrow and elongated; endopod oval, small . . . . *Argeia pugettensis*
6. Exopod of pleopods oval, small; endopod narrow, elongated . . . . *Pseudione* sp.
7. Parasitic on crabs (*Hemigrapsus oregonensis* in the Bay Region). . . . . a genus closely related to *Portunion*
7. Parasitic on barnacles (*Balanus* and others) . . . *Cryptothir* sp.

Fig. 65. Bopyroidea. *a*, *Ione brevicauda*, adult female, dorsal view; *b*, same, male, relatively much more enlarged; *c*, *Argeia pugettensis*, immature female, dorsal view; *d*, same, ventral view showing oöstegites (after Richardson 1905).

II. Flabellifera

1. Uropodal exopod arching over telson; length of body exceeds 7 times the width . . . . ANTHURIDAE 2
1. Uropodal exopod not arching over telson; length of body seldom, if ever, exceeding 5 times the width 4
2. Free part of maxilliped consists of 1 article; mandibular palp lacking (fig. 66, *a*). . . . . *Colanthura squamosissima*
2. Free part of maxilliped consisting of 2–3 articles; mandibular palp present 3
3. Free part of maxilliped with 2 articles; mouth parts adapted for piercing and sucking (fig. 66, *b*) . . . . . *Paranthura elegans*
3. Free part of maxilliped with 3 articles; mouth parts adapted for chewing (fig. 66, *c*) . . . . *Cyathura munda*
4. Pleotelson consists of 6 somites 5
4. Pleotelson consists of less than 6 somites . SPHAEROMIDAE 14
5. Exopod of uropod clawlike and much smaller than endopod, which is styliform but with a blunt apex . . . . . . LIMNORIIDAE 6
5. Exopod of uropod not clawlike, usually slightly smaller than endopod; both rami leaflike and similar in appearance 8
6. Dorsal surface of telson lacks symetrically arranged tubercles . . . . *Limnoria* sp.
6. Dorsal surface of telson with symmetrically arranged tubercles . 7
7. Four tubercles on telson, e.g. (::). . . . *Limnoria quadripunctata*
7. Three tubercles on telson, e.g. (∴) . . . . *Limnoria tripunctata*
8. Plumose setae present on margins of at least the 1st and 2d pairs of pleopods 9
8. Plumose setae not present on margins of pleopods . . . . CYMOTHOIDAE 12
9. Maxillipedal palp with 5 articles . . . . . CIROLANIDAE 10
9. Maxillipedal palp with 2 articles (AEGIDAE) . . *Rocinela belliceps*
10. Setae on posterior border of telson stout, not plumose; frontal margin of head with inconspicuous median tubercle (fig. 63, *e, f*). . . . . *Cirolana harfordi*
10. Setae on posterior border of telson long, narrow, and plumose; head with long narrow apically knobbed rostrum 11
11. Posterior border of telson triangulate; 3d article of peduncle of first antenna much longer than second . . . . . *Excirolana chiltoni*
11. Posterior border of telson rounded; articles of peduncle of first antenna subequal . . . . *Excirolana linguifrons*

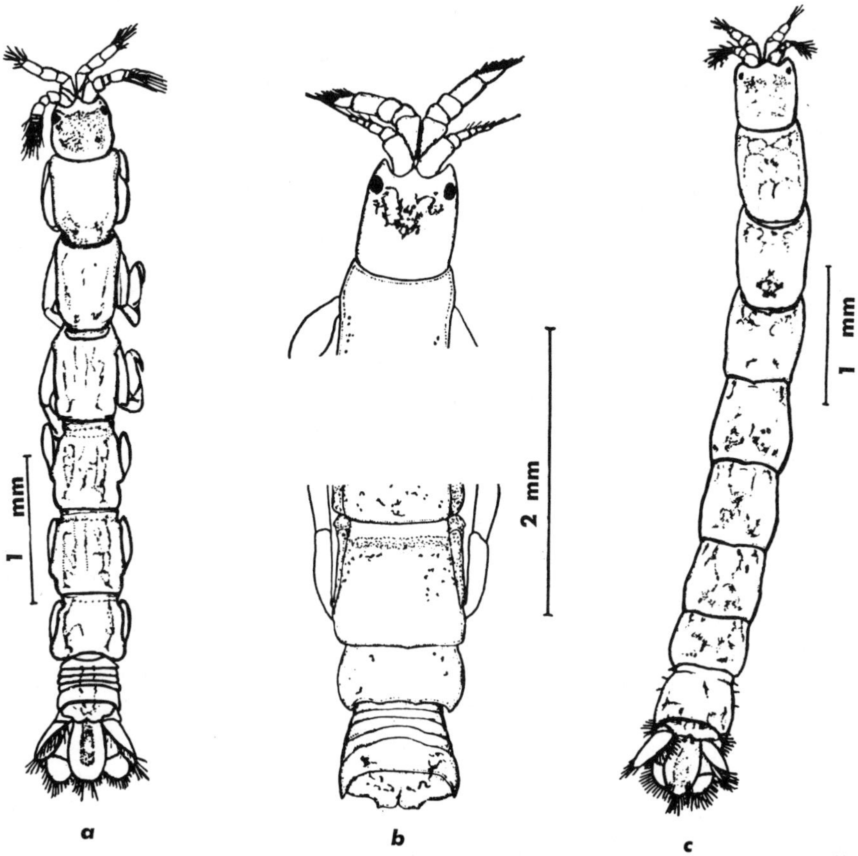

Fig. 66. **Flabellifera.** *a*, *Colanthura squamosissima*; *b*, *Paranthura elegans*, head and pleotelson only; *c*, *Cyathura munda* (after Menzies; scales as indicated).

12. Seventh peraeopod lacks setae (spines) on inferior margin of carpal and propodal articles 13
12. Seventh peraeopod with a few setae on inferior margin of carpal and propodal articles . . . . *Nerocila californica*

13. Telson nearly twice as broad as long . . . . . *Livoneca vulgaris*
13. Telson about as long as broad . . . . . *Livoneca californica*

14. First pair of peraeopods subchelate . . . *Tecticeps convexus*
14. First pair of peraeopods not subchelate 15

15. Endopods and exopods of all pleopods membranous, thin, lacking deep folds . . . . *Neosphaeroma oregonensis*
15. Endopods of pleopods 4 and 5 fleshy, with deep folds; exopods with or without folds 16

16. Exopods fleshy, with deep folds (much like accordion pleats) . 17
16. Exopods fleshy, but lack deep folds 18

17. Last somite of pleon (exclusive of telson) with a transverse row of 3 large tubercles; exopod of 3d pleopod with 2 articles . . . . *Paracerceis cordata*
17. Last somite of pleon (exclusive of telson) without a row of tubercles, but with 3 longitudinal ridges on telson; exopod of 3d pleopod with 1 article . . . . *"Dynamene" dilatata*

18. Maxillipeds with 2d, 3d, and 4th articles of palp not produced into lobes . . . . *Sphaeroma pentodon*
18. Maxillipeds with 2d, 3d, and 4th articles of palp produced into lobes (fig. 63, *g*) 19

19. Exopod of 3d pleopod biarticulate 20
19. Exopod of 3d pleopod uniarticulate . . *"Dynamenella" benedicti*

20. Extremity of telson produced in a rhomboid process . . . . *"Exosphaeroma" rhomburum*
20. Extremity of telson not produced in a rhomboid process, being pointed or notched 21

21. Dorsum of body tuberculate; apex of telson pointed . . . . *Exosphaeroma amplicauda*
21. Dorsum of body smooth; apex of telson with an evident notch . . . . *"Dynamenella" glabra*

III. Valvifera

1. Head fused with 1st somite of peraeon (ARCTURIDAE) . . . . *Idarcturus hedgpethi*
1. Head not fused with 1st somite of peraeon . . IDOTHEIDAE 2

2. All somites of peraeon lack dorsally visible epimera; pleotelson composed of single somite with lateral sutures of another partly coalesced somite . . . . *Synidotea* 3
2. Second to 7th peraeonal somites with very evident, dorsally visible epimera; pleotelson composed of 3 distinct somites plus lateral sutures of another partly coalesced somite . . . . *Idothea* 5

3. Apex of superior preorbital processes blunt, directed toward the frontal margin and upwards; lateral areas of peraeonal somites with heavy rugosities . . . . *Synidotea ritteri*
3. Apex of superior preorbital processes acute, directed laterally . 4

4. Three rows of large sharply pointed tubercles run the length of peraeon . . . . *Synidotea consolidata*

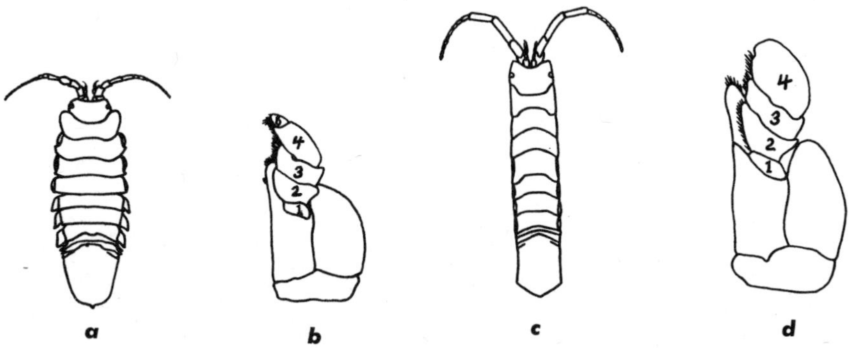

Fig. 67. **Valvifera.** *a*, *Idothea (Pentidotea) wosnesenskii*; *b*, maxillipedal palp of same showing five articles; c, *Idothea (Idothea) urotoma*; *d*, maxillipedal palp of same showing four articles (*a*, *c*, after Richardson).

4. Peraeon with minute tubercles that are not arranged in 3 rows . . . . *Synidotea* sp.

5. Maxillipedal palp with 4 articles (fig. 67, *d*) . *Idothea (Idothea)* 6
5. Maxillipedal palp with 5 articles (fig. 67, *b*) . *Idothea (Pentidotea)* 8

6. Superior margin of clypeus (1st frontal lamina) with a pronounced median concavity (fig. 68, *a*) . . . . . *Idothea (Idothea) urotoma*
6. Superior margin of clypeus evenly rounded, lacking a concavity . 7

7. Frontal process apically blunt; posterior margin of telson concave in outline . . . . *Idothea (Idothea) rufescens*
7. Frontal process apically pointed; posterior margin of telson with a pronounced median tooth or projection (fig. 68, *b*) . . . . *Idothea (Idothea) fewkesi*

8. Each maxilliped with 1 coupling hook 9
8. Each maxilliped with 2 coupling hooks (fig. 68, *c*) . . . . *Idothea (Pentidotea) stenops*

9. Apex of frontal process entire, not notched 10
9. Apex of frontal process with a median notch . . . . *Idothea (Pentidotea) aculeata* (a few specimens; see also 12)

10. Frontal process blunt or widely angulate, not extending forward beyond the frontal extent of the clypeus 11
10. Frontal process narrow and pointed and considerably exceeds the forward extent of the first (dorsal) frontal protuberance of the clypeus 13

11. Posterolateral margin of epimeral plate of 7th peraeonal somite evenly convex, not acute; eyes somewhat pyriform . . . . *Idothea (Pentidotea) schmitti*

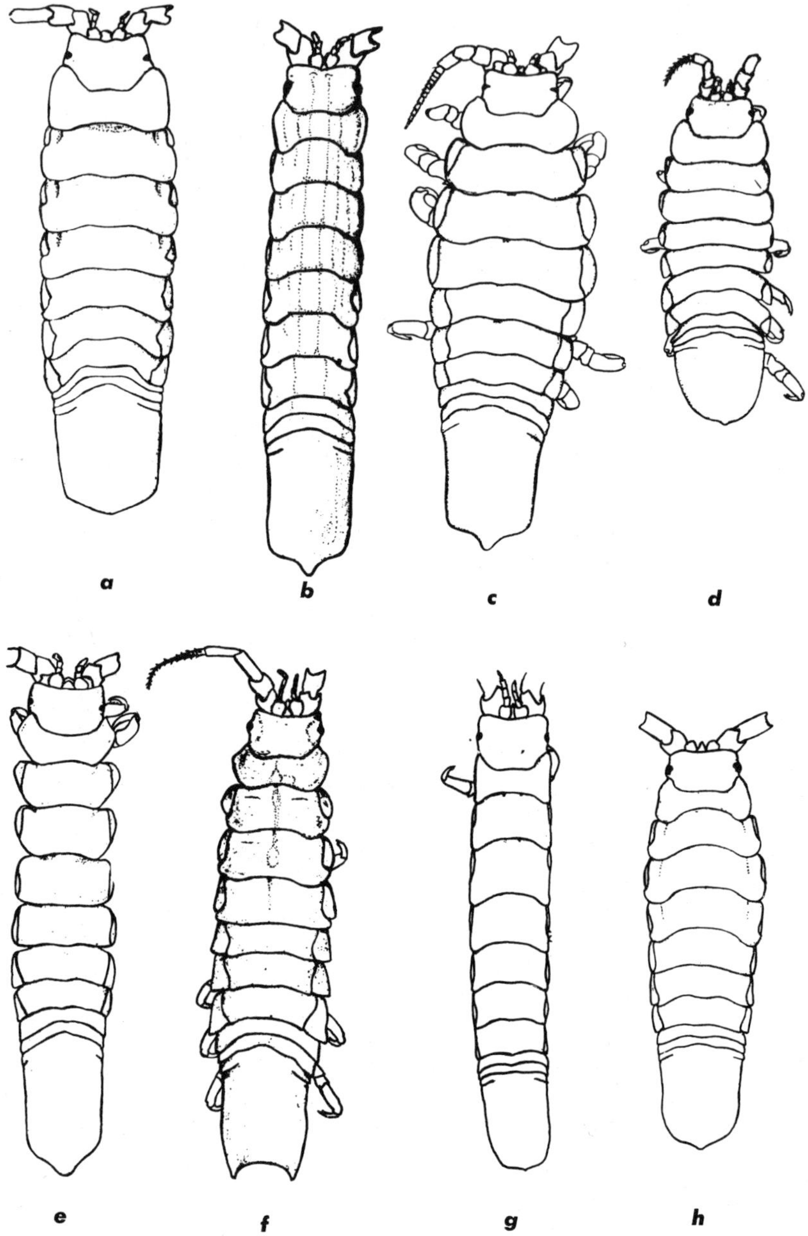

Fig. 68. Valvifera. *Idothea. a, I. (Idothea) urotoma; b, I. (Idothea) fewkesi; c, I. (Pentidotea) stenops: d, I. (Pentidotea) wosnesenskii; e, I. (Pentidotea) aculeata; f, I. (Pentidotea) resecata; g, I. (Pentidotea) montereyensis*, male; *h*, same, female (from Menzies, 1950; scales various).

11. Posterolateral margin of epimeral plate of 7th peraeonal somite acute; eyes (in lateral view) reniform (kidney or bean-shaped) or circular 12

12. First somite of pleon with acute lateral borders; eyes reniform (fig. 68, *d*). . . . . *Idothea* (*Pentidotea*) *wosnesenskii*
12. First somite of pleon with wide lateral borders; eyes circular (fig. 68, *e*) . . . . . . *Idothea* (*Pentidotea*) *aculeata* (see also 9)

13. Posterior margin of telson deeply concave, posterolateral angles acute, each angle with a small but noticeable dorsal carina (fig. 68, *f*) . . . . *Idothea* (*Pentidotea*) *resecata*
13. Posterior margin of telson usually convex, with a small but distinct median tooth; when concave only slightly so, and lacking acute posterolateral angles (fig. 68 *g*, *h*) . . . . *Idothea* (*Pentidotea*) *montereyensis*

IV. Asellota

1. Eyes situated on lateral borders of head, not dorsally located; dactyls of legs 2–7 with 1 or 2 claws 2
1. Eyes dorsal; dactyls of legs 2–7 with 3 claws 6

2. Uropods very small (less than 1/8 length of telson) and with no apparent peduncle . . . . *Munna* 3
2. Uropods greater than 1/2 length of telson; peduncle very evident, being about as long as the exopod and endopod 18

3. Uropods small, leaflike, lacking spinelike protuberances; male first pleopods (sympods) apically pointed (fig. 69, *a*, *b*, *c*) . . . . *Munna ubiquita*
3. Uropods rounded in cross section; male 1st pleopods with apices laterally expanded (fig. 69, *e*, *h*, *k*) 4

4. Uropods with at least 1 large spinelike protuberance (fig. 69, *f*, *i*); dentate shelf below uropods lacking 5
4. Uropods lack large spinelike protuberances; dentate shelf present below uropods (fig. 69, *j*, *k*, *l*). . . . . *Munna halei*

5. Lateral borders of telson lack large 2-pointed setae (spines) (fig. 69, *g*, *h*, *i*) . . . . *Munna chromatocephala*
5. Lateral borders of telson with 2–3 large 2-pointed spines (fig. 69, *d*, *e*, *f*). . . . . *Munna stephenseni*

6. Second antenna with a very evident squama or scale attached at 3d article 7
6. Second antenna with no squama 9

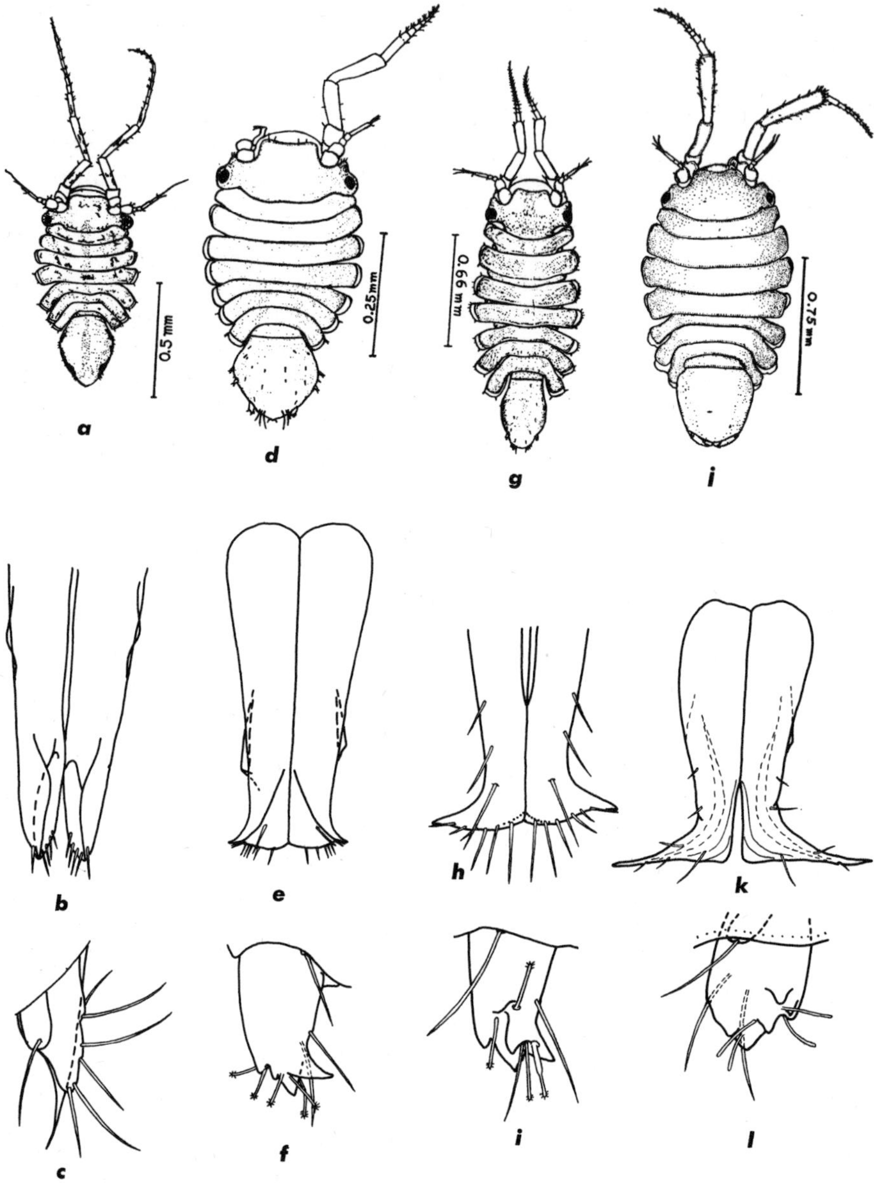

Fig. 69. **Asellota.** *Munna.* Top row, dorsal view of bodies; middle row, male first pleopods; bottom row, uropods; *a-c, Munna ubiquita; d-f, M. stephenseni; g-i, M. chromatocephala; j-l; M. halei* (after Menzies, 1952).

7. Inferior border of the proximal 1/3 of propodus of 1st leg with a conspicuously serrated margin. . . . . *Janiralata* 8
7. Inferior border of proximal 1/3 of propodus of 1st leg smooth, lacking serrations . . . . *Ianiropsis* 12

8. Pleotelson with distinct medially recurved, spinelike posterolateral angles (fig. 70, *a*) . . . . *Janiralata occidentalis*
8. Posterolateral areas of pleotelson evenly curved, lacking distinct angles or spinelike processes (fig. 70, *b*). . . . *Janiralata davisi*

9. Peraeon without dorsally visible epimeral (coxal) plates (fig. 70, *c*) . . . . *Jaeropsis* 10
9. Peraeon with evident dorsally visible epimeral plates (fig. 70, *d*) . . . . *Iais californica*

10. Pleotelson with 5–7 spines on each lateral border *Jaeropsis dubia*
10. Pleotelson with 3 or no spines on each lateral border . . . 11

11. Pleotelson with 3 spines on each lateral border . . . . *Iaeroposis dubia* var. *paucispinis*
11. Pleotelson lacking spines on each lateral border . *Jaeropsis lobata*

12. Lateral borders of pleotelson with spinelike serrations 13
12. Lateral borders of pleotelson smooth, without spinelike serrations (some fine setae may be present) 15

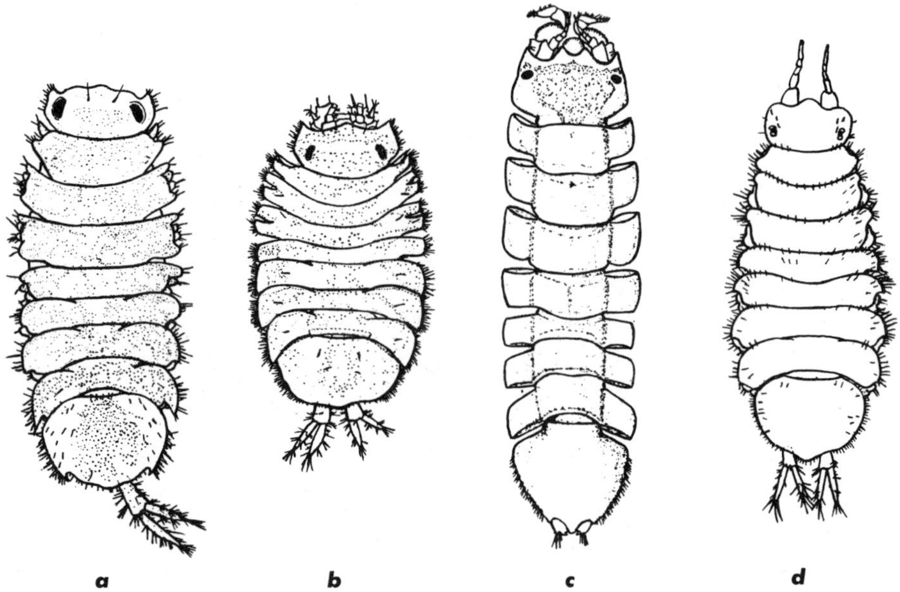

Fig. 70. Asellota. *a, Janiralata occidentalis; b, Janiralata davisi; c, Jaeropsis dubia; d, Iais californica* (from Menzies).

13. Pleotelson with 4 or more spinelike serrations on each side . . . . *Ianiropsis analoga*
13. Pleotelson with 2 or 3 spinelike serrations on each side . . . 14
14. Pleotelson with 2 spinelike serrations on each side . . . . *Ianiropsis epilittoralis*
14. Pleotelson with 3 spinelike serrations on each side . . . . *Ianiropsis tridens*
15. Uropods half or less than half the length of pleotelson . . . 16
15. Uropods considerably exceeding half the pleotelson length . . 17
16. Pleotelson with distinct posterolateral angles lateral to uropod insertions . . . . *Ianiropsis kincaidi derjugini*
16. Pleotelson lacking posterolateral angles lateral to uropod insertions . . . . *Ianiropsis minuta*
17. Apex (lateral) of each 1st male pleopod bifurcating; uropod exceeds telson length . . . . *Ianiropsis montereyensis*
17. Apex (lateral) of each 1st male pleopod not bifurcating; uropods not exceeding pleotelson length . . *Ianiropsis kincaidi kincaidi*
18. Dactyls of all legs with one claw; fresh water *Asellus tomalensis*
18. Dactyls of legs 2–7 with two claws; marine . . . *Antias hirsutus*

V. *Oniscoidea*

1. Uropods underneath abdomen (not visible in dorsal view), forming paired opercular plates covering pleopods (TYLIDAE) . *Tylos* sp.
1. Uropods terminal, visible in dorsal view (not forming operculum) 2
2. Flagellum of second antennae (first is rudimentary in all Oniscoidea) multiarticulate (more than 4 articles); head without anterolateral lobes . . . . LIGIIDAE 3
2. Flagellum of second antennae not multiarticulate, with not more than 4 articles; head with anterolateral lobes 6
3. Uropods with process on basal article at inner distal angle for articulation with endopod; telson without posterolateral projections (fig. 71, *e*). 4
3. Uropods without process at inner distal margin; telson with more or less pointed posterolateral projections (fig. 71, *d*, *f*) 5
4. Surface of body smooth and shiny . . . . . . *Ligidium gracilis*
4. Surface of body rough and scaly . . . . *Ligidium latum*
5. Eyes separated in front by distance equal to twice length of 1 eye; basal article (peduncle) of uropods about as broad as long (fig. 71, *f*) . . . . *Ligia pallasii*

5. Eyes separated in front by distance equal to length of 1 eye; peduncle of uropods several times longer than broad (fig. 71, *d*) . . . . *Ligia occidentalis*

6. Uropods not reaching beyond posterior border of telson; body markedly convex, capable of being rolled into a ball . . . . ARMADILLIDIIDAE 11
6. Uropods reaching beyond posterior border of telson; body slightly, or in some markedly, convex; may not be capable of being rolled up into a ball 7

7. Flagellum of second antenna composed of 4 articles 8
7. Flagellum of second antenna composed of 2 or 3 articles. . . . . ONISCIDAE 12

8. Abdomen abruptly narrower than thorax; eyes lacking (TRICHONISCIDAE) . . . . *Protrichoniscus heroldi*
8. Abdomen not abruptly narrower than thorax; eyes present (SCYPHACIDAE) . . . . *Armadilloniscus* 9

9. Median projection of head truncate when viewed from above (fig. 71, *c*); 4 ocelli; capable of being rolled up into a ball . . . . *Armadilloniscus lindahli*
9. Median projection of head pointed when viewed from above (fig. 71, *a*, *b*); 4 (5?) ocelli; not capable of being rolled up into a ball . . 10

10. Posterior border of body segments closely set with minute tubercles, giving borders beaded appearance (fig. 71, *a*) . . . . *Armadilloniscus coronacapitalis*
10. Posterior border of body segments smooth *Armadilloniscus holmesi*

11. Exopod of uropods minute; peduncle large, flattened, extending to tip of telson . . . . *Cubaris microphthalma*
11. Exopod (in ventral view) of uropods large, flattened, extending to tip of telson and inserted at apex of similarly flattened peduncle . . . . *Armadillidium vulgare*

12. Flagellum of second antennae triarticulate; exopods of pleopods lack respiratory trees ("white bodies") 13
12. Flagellum of second antennae biarticulate; 1st and 2d pairs of pleopods (exopod) have respiratory organs 14

13. Abdomen abruptly narrower than thorax . *Philoscia richardsonae*
13. Abdomen not abruptly narrower than thorax *Alloniscus perconvexus*

14. Abdomen abruptly narrower than thorax; frontal lobe of head absent . . . . *Porcellionides pruinosus*
14. Abdomen not abruptly narrower than thorax; frontal lobe of head present. . . . . *Porcellio* 15

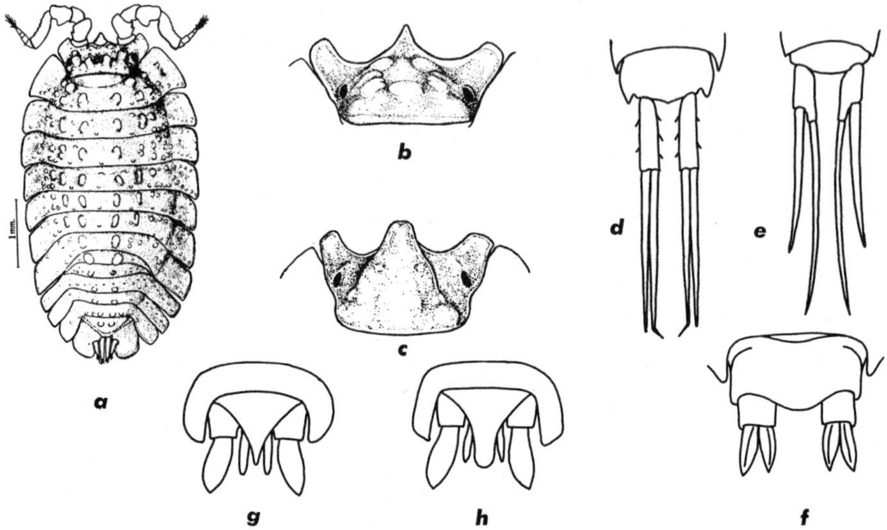

Fig. 71. Oniscoidea. *a*, *Armadilloniscus coronacapitalis*; *b*, head of *A. holmesi*; *c*, head of *A. lindahli*; *d*, telson and uropods of *Lygia occidentalis*; *e*, of *Lygidium* sp.; *f*, of *Lygia pallasii*; *g*, of *Porcellio scaber occidentalis*; *h*, of *Porcellio scaber scaber* (*a-c* from Menzies, 1950; *d-f* after Richardson, 1905; *g*, *h* after Miller, 1936; not to scale).

15. Hinder angles of 1st thoracic segment not produced posteriorly . . . . *Porcellio* (*Mesoporcellio*) 16
15. Hinder angles of 1st thoracic segment produced posteriorly . . . . *Porcellio* (*Porcellio*) *scaber* 17

16. First article of flagellum of second antenna less than 1/2 length of 2d . . . . *Porcellio* (*Mesoporcellio*) *littorina*
16. First article of flagellum of second antenna generally longer than 2d . . . . *Porcellio* (*Mesoporcellio*) *laevis*

17. Apex of telson spatulate (fig. 71, *h*) . . . . *Porcellio* (*Porcellio*) *scaber occidentalis*
17. Apex of telson triangulate (fig. 71, *g*). 18

18. Inferior margin of ischium of 7th leg of male straight or slightly concave; pleopods of neither sex pigmented . . . . *Porcellio* (*Porcellio*) *scaber scaber*
18. Inferior margin of ischium of 7th leg of male markedly concave; posterior 3 pairs of pleopods pigmented in both sexes . . . . *Porcellio* (*Porcellio*) *scaber americanus*

List of Species

Order Chelifera (Tanaidacea)
- *Anatanais normani* (Richardson, 1905)
- *Leptochelia* sp.
- *Pagurapseudes laevis* Menzies, 1953
- *Pancolus californiensis* Richardson, 1905
- *Synapseudes intumescens* Menzies, 1949
- *Tanais* sp.

Order Isopoda

Suborder Bopyroidea
- *Argeia pugettensis* Dana, 1853 (Branchial chamber of *Crago*)
- *Cryptothir* sp. (Internal parasite of barnacles)
- *Ione* sp. (Branchial chamber of *Callianassa*)
- *Phryxus* sp. (Abdomen of *Spirontocaris*)
- *Phyllodurus abdominalis* Stimpson, 1857 (Abdomen of *Upogebia*)
- *Portunion* (?) sp. (Hepatic tissue of *Hemigrapsus oregonensis*)
- *Pseudione* sp. (Branchial chamber of *Pagurus*)
- *Stegophryxus* sp. (Abdomen of *Pagurus*)

Suborder Flabellifera
- *Cirolana harfordi* (Lockington, 1877)
- *Colanthura squamosissima* Menzies, 1951
- *Cyathura munda* Menzies, 1951
- "*Dynamenella*"[3] *benedicti* (Richardson, 1899)
- "*Dynamene*" *dilatata* Richardson, 1899
- "*Dynamenella*" *glabra* (Richardson, 1899)
- *Excirolana linguifrons* (Richardson, 1899)
- *Excirolana chiltoni* (Richardson, 1905)
- *Exosphaeroma amplicauda* (Stimpson, 1857)
- *"*Exosphaeroma*"[3] *octoncum* (Richardson, 1897)
- "*Exosphaeroma*" *rhomburum* (Richardson, 1899)
- *Limnoria quadripunctata* Holthuis, 1949
- *Limnoria* sp.
- *Limnoria tripunctata* Menzies, 1951
- *Livoneca californica* Schioedte and Meinert, 1883-1884
- *Livoneca vulgaris* Stimpson, 1857
- *Neosphaeroma oregonensis* (Dana, 1854-1855)
- *Nerocila californica* Schioedte and Meinert, 1881-1883
- *Paracerceis cordata* (Richardson, 1899)
- *Paranthura elegans* Menzies, 1951
- *Rocinela belliceps* (Stimpson, 1864)
- *Sphaeroma pentodon* Richardson, 1904
- *Tecticeps convexus* Richardson, 1899

[3] Quotation marks refer to generic designations which are in doubt.

Suborder Valvifera

Idarcturus hedgpethi Menzies, 1951
Idothea (Idothea) fewkesi Richardson, 1905
Idothea (Idothea) rufescens Fee, 1926
Idothea (Idothea) urotoma Stimpson, 1864 (The female was formerly called *I. rectilinea*, also *I. rectilineata*)
Idothea (Pentidotea) aculeata (Stafford, 1913)
Idothea (Pentidotea) montereyensis Maloney, 1933 (=*Idothea gracillima*, of Richardson, 1905)
Idothea (Pentidotea) resecata (Stimpson, 1857)
Idothea (Pentidotea) schmitti Menzies, 1951 (=*Pentidotea whitei* of Richardson, not Stimpson)
Idothea (Pentidotea) stenops (Benedict, 1898)
Idothea (Pentidotea) wosnesenskii (Brandt, 1851)
Synidotea consolidata (Stimpson, 1856) (=*S. pettibonae* Hatch, 1947)
**Synidotea macginitiei* Maloney, 1933
Synidotea ritteri Richardson, 1904
Synidotea sp.

Suborder Asellota

Antias hirsutus Menzies, 1951
Asellus tomalensis Harford, 1877
Iais[4] *californica* (Richardson, 1904)
Ianiropsis analoga Menzies, 1952
Ianiropsis epilittoralis Menzies, 1952
Ianiropsis kincaidi kincaidi Richardson, 1904
Ianiropsis kincaidi derjugini Gurjanova, 1933
Ianiropsis minuta Menzies, 1952
Ianiropsis montereyensis Menzies, 1952
Ianiropsis tridens Menzies, 1952
Jaeropsis[4] *dubia* Menzies, 1951
Jaeropsis dubia var. *paucispinis* Menzies, 1951
Jaeropsis lobata Richardson, 1899
Janiralata davisi Menzies, 1951
Janiralata occidentalis (Walker, 1898)
Munna chromatocephala Menzies, 1952
Munna halei Menzies, 1952
Munna stephenseni Gurjanova, 1933
Munna ubiquita Menzies, 1952

Suborder Oniscoidea

**Alloniscus cornutus* Budde-Lund, 1885

---

[4]In the use of the initial I or J in certain names in this group of isopods, this list follows the original (taxonomically correct) spelling. In Roman usage the initial consonant J was written with the same sign as the vowel I, leading to some confusion among modern writers. Some of these have named genera using the initial I, others with the initial J, although basically the same consonant (J) is involved. These names should not be pronounced with the "I" sound, but rather with the "J" (or "Y") sound. (Ed.)

**Alloniscus mirabilis* (Stuxberg, 1875)
Alloniscus perconvexus Dana, 1854
Armadillidium vulgare (Latreille, 1804)
Armadilloniscus caronacapitalis Menzies, 1950
Armadilloniscus holmesi Arcangeli, 1933
Armadilloniscus lindahli (Richardson, 1905)
**Cubaris affinis* (Dana, 1854-1855)
**Cubaris californica* (Budde-Lund, 1885)
Cubaris microphthalma Arcangeli, 1932
**Ligia exotica* Roux, 1828
Ligia occidentalis Dana, 1853
Ligia pallasii Brandt, 1833
Ligidium gracilis (Dana, 1854-1855)
**Ligidium hypnorum* (Cuvier, 1792)
Ligidium latum Jackson, 1932
Philoscia richardsonae Holmes and Gay, 1909
**Porcellio formosus* Stuxberg, 1875
Porcellio (*Mesoporcellio*) *laevis* Latreille, 1804
Porcellio (*Mesoporcellio*) *littorina* Miller, 1936
Porcellio (*Porcellio*) *scaber americanus* Arcangeli, 1932
Porcellio (*Porcellio*) *scaber occidentalis* Miller, 1936
[=*P.* (*P*). *dilatatus* Brandt and Ratzeburg, 1833]
Porcellio (*Porcellio*) *scaber scaber* Latreille, 1804
Porcellionides pruinosus (Brandt, 1833)
Protrichoniscus heroldi Arcangeli, 1932
**Tylos punctatus* Holmes and Gay, 1909
Tylos sp.

Order Amphipoda

The Amphipoda are divided into three suborders: Gammaridea (fig. 72, *a-c*), Hyperiidea (fig. 72, *d*), and Caprellidea (fig. 72, *e*), the first two with seven free thoracic segments, the third with six. The Caprellidea are further distinguished by a very narrow body (except for the Cyamidae, or whale lice) and a rudimentary abdomen. The Hyperiidea, commonly pelagic or commensal with medusae and not usually encountered, are further characterized by a very large head and large eyes. The Gammaridea are represented by the familiar "sand fleas" of the seashore and a large number of species living on seaweed, under rocks, and in a variety of other niches.

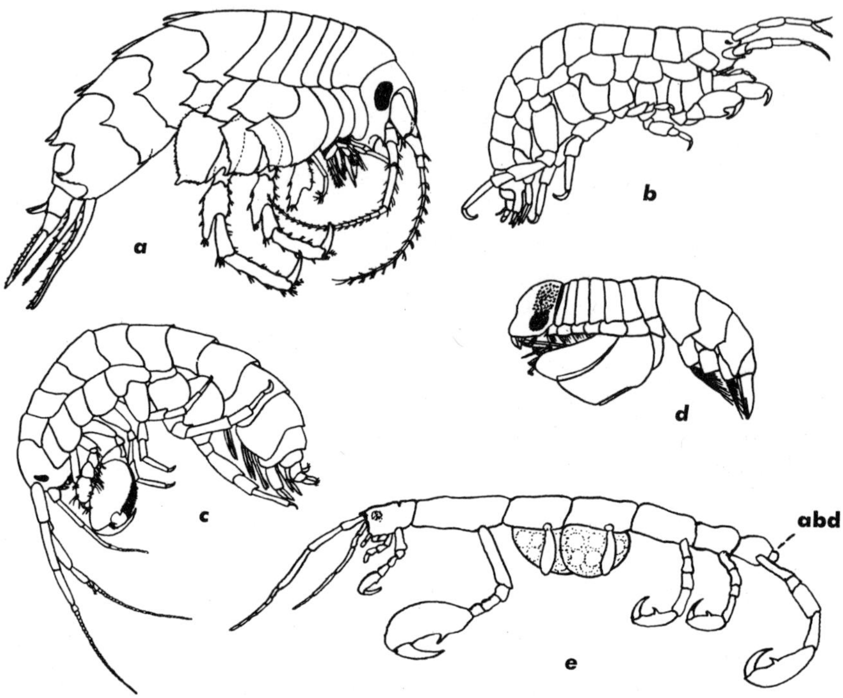

Fig. 72. Amphipod types. *a-c*, Gammaridea. *a*, *Panoplea* (family Acanthonotozomatidae); *b*, *Elasmopus*; *c*, *Microprotopus*; *d*, Hyperiidea, *Platycelus*; *e*, Caprellidea, schematic female showing oöstegites.

Suborder Gammaridea

The anatomy of a typical gammarid (fig. 73) may profitably be compared with that of an isopod. Close study of the smaller parts requires the use of a dissecting microscope, beneath which the specimen may be dissected as much as necessary with needles. Separated parts may be mounted in glycerine under a cover slip for closer observation with the compound microscope.

The limbs of a gammarid include first and second legs (second and third thoracic appendages) which are chelate or subchelate and known as the first and second gnathopods. In these the sixth segment (propodus) forms an expanded palm. The seventh or terminal segment (dactylus) commonly closes across the end of the palm, or back against its side, the condition being spoken of as subchelate (fig. 74, *b*). In some forms the propodus is extended into a rigid finger against which the movable dactylus closes; this is the chelate condition (fig. 74, *a*). Behind the

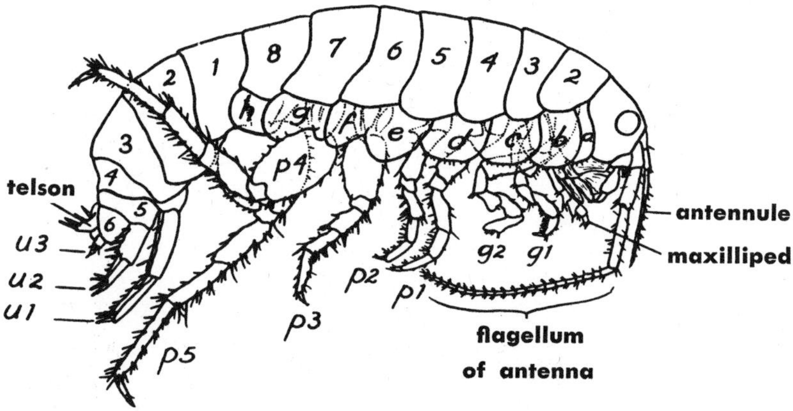

Fig. 73. *Talitrus sylvaticus* Haswell, a gammarid amphipod labeled to show characteristic structures: Pleopods 1-3, are omitted. Segments of thorax and abdomen are numbered; G1, G2 = first and second gnathopods; p1-p5 = walking legs or peraeopods; U1-U3 = the three posterior abdominal appendages termed uropods; a-h = the side plates or coxal plates of the thoracic appendages. (After Shoemaker).

gnathopods are five pairs of simple walking legs (peraeopods). The abdomen or pleon is composed of six segments (the last three of which are called the urosome). It bears two types of legs (hence amphi-pod), namely, three pairs of slender biramous pleopods with slender multi-articulate rami, and three pairs of biramous uropods ("springing feet"). The flat extensions of the chitinous bases of the thoracic appendages are known as side plates or coxal plates (fig. 73).

The appendages of the head include antennules (first antennae) and antennae (second antennae). The elongate basal joints form the peduncle; the small distal joints the flagellum. In some genera the antennule bears a small accessory flagellum (fig. 76, *e*). The most posterior of the mouthparts are the maxillipeds (actually the first thoracic appendages); these are united by their two basal segments. Each extends distally as a palp and bears internally two endites or lobes, known respectively as the inner and outer plates (fig. 74, *c*). More anteriorly lie two very small appendages, the paired second and first maxillae (fig. 74, *e*, *f*), which may or may not bear a small palp. The mandibles are heavily chitinized structures, sometimes having a palp (fig. 74, *g*) as in *Gammarus*, or lacking one as in *Allorchestes*. In front of and behind the mandibles are fleshy upper and lower lips.

Sexual dimorphism is frequently marked, and it is often necessary to recognize the sex in order to identify. This may sometimes be done on the basis of the size of the gnathopods, which are usually larger in the male. In sexually mature females the egg mass on the ventral surface, enclosed in the marsupial plates or oöstegites, is a distinguishing character.

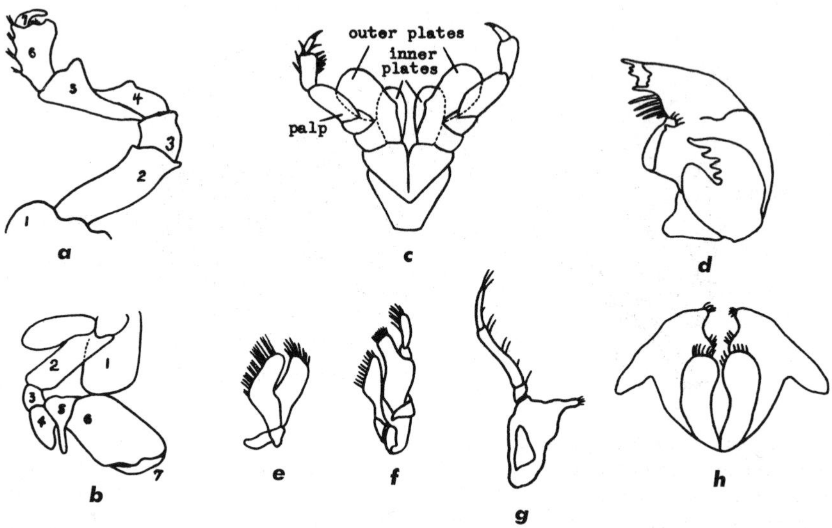

Fig. 74. Mouth parts of various amphipods. *a*, first gnathopod of *Allorchestes* with segments numbered, showing chelate condition; *b*, second gnathopod of same, subchelate; c, maxilliped ot a *gammarid; d*, mandible of *Allorchestes* (lacking palp); *e-g*, second maxilla, 1st maxilla, and mandible of a gammarid (mandible with palp); *h*, lower lip of *Ampithoe*, showing notched front lobes.

Key to the Gammaridea

by J. Laurens Barnard
California State Polytechnic College, San Dimas

1. Pleon depressed, platelike 2
1. Pleon compressed, cylindrical 7

2. Uropods greatly enlarged; found boring in wood (CHELURIDAE) . . . . *Chelura terebrans*
2. Uropods small; occurring in tubes . . . . . . COROPHIIDAE 3

3. Telson small, without lateral lobes, surface not spinulose (fig. 75, *i, j*); second antenna larger than first, often conspicuously so (fig. 75, *h*); mandibular palp 2-jointed, each article with a long, curved, distal seta . . . . *Corophium* 5
3. Telson, short, broad, with spinulose lateral lobes (fig. 75, *k*); second antenna not conspicuously larger than first; mandibular palp 3-jointed, 3d article heavily setose . . . . *Ericthonius* 4

4. Fourth article of male 2d gnathopod with 1 tooth . . . . *Ericthonius hunteri*
4. Fourth article of male 2d gnathopod with 2 teeth . . . . *Ericthonius brasiliensis*

5. Segments of urosome distinct (fig. 75, *g*) . . *Corophium spinicorne*
5. Segments of urosome fused 6

6. Rostrum of male long, slender (fig. 75, *c*) . . *Corophium insidiosum*
6. Rostrum of male absent (fig. 75, *a*, *b*, *h*) . *Corophium acherusicum*

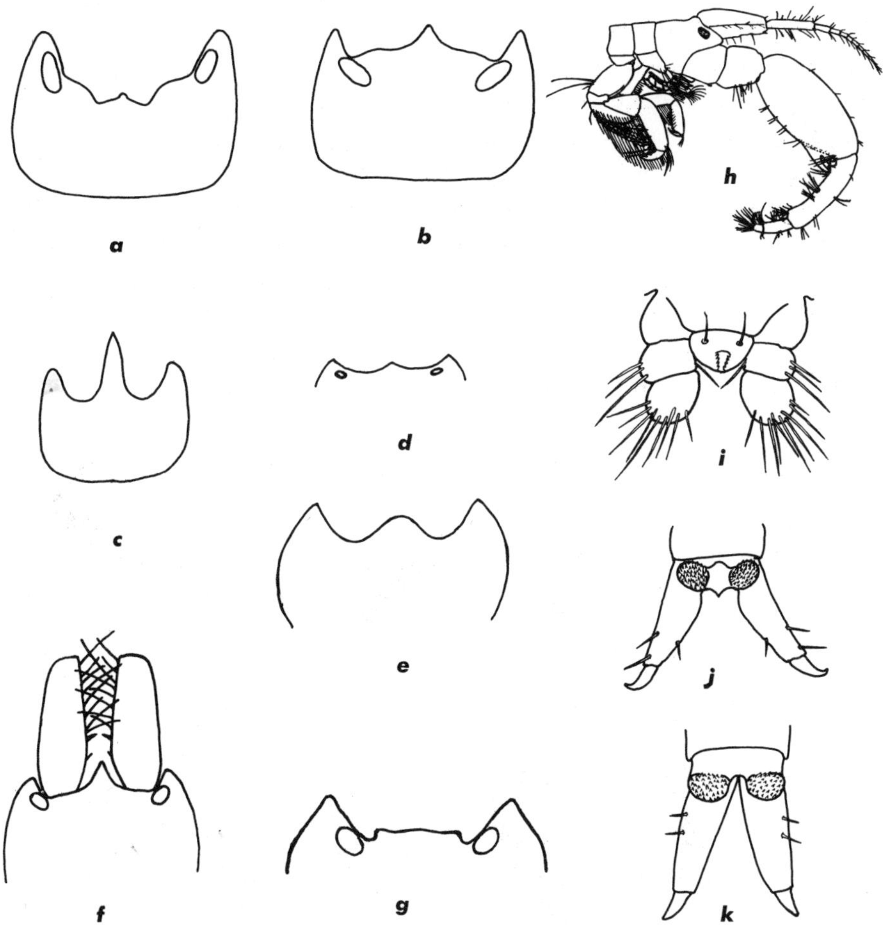

Fig. 75. COROPHIIDAE. *a-g*, *Corophium*, dorsal aspect of rostral region: *a*, *acherusicum* (male); *b*, *acherusicum* (female); *c*, *insidiosum* (male); *d*, *brevis* (male); *e*, *stimpsoni* (male); *f*, *oaklandense* (hermaphroditic); *g*, *spinicorne* (male); *h*, *Corophium acherusicum*, anterior body of male; *i*, *Corophium insidiosum*, telson and uropods; *j*, *Ericthonius hunteri*, telson and uropods; *k*, *Ericthonius brasiliensis*, telson and uropods (*a-f*, *h*, *i* after Shoemaker; *j* after Sars).

7. Coxae 2–4 greatly enlarged, shieldlike (STENOTHOIDAE) . . . . *Metopa* sp.
7. Coxae small 8

8. First gnathopod of male much longer than 2d, complex (AORIDAE) (fig. 76, *g, h*) . . . . *Aoroides columbiae*
8. First gnathopod smaller than or equal to 2d 9
9. Integument with large processes (fig. 72, *a*); mouth parts prolonged as if for piercing . . . . ACANTHONOTOZOMATIDAE
9. Integument without large processes 10
10. Third uropod vestigial, seemingly absent . . PODOCERIDAE 11
10. Third uropod functional, easily seen 12
11. Fourth joint of male 2d gnathopod produced into a large lobe (fig. 76, *d*) . . . . *Podocerus spongicolus*
11. Fourth joint of male 2d gnathopod not produced into a lobe . . . . *Podocerus brasiliensis*
12. First antenna short, stout, with long accessory flagellum (fig. 76, *e, f*) 4th peraeopod much longer than 5th (PHOXOCEPHALIDAE) . . . . *Pontharpinia obtusidens*
12. First antenna slender; 5th peraeopod as long as 4th 13

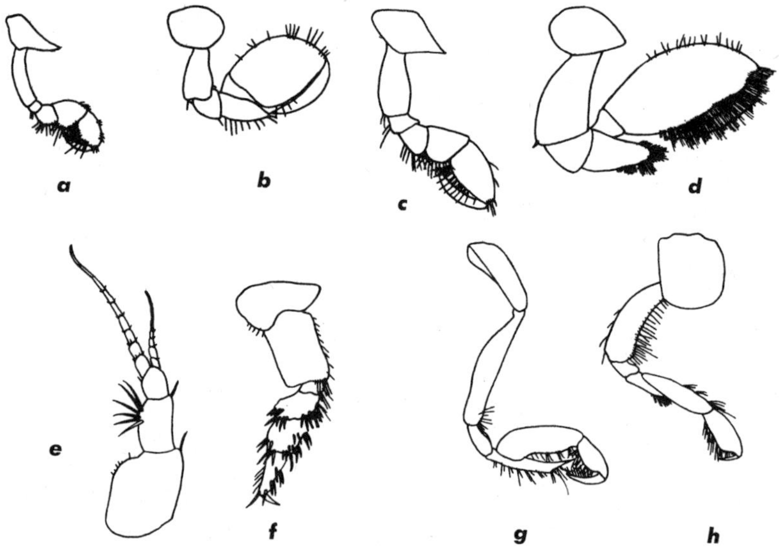

Fig. 76. *a-d, Podocerus spongicolus; a-b,* first and second gnathopods of female; *c-d,* same of male; *e-f, Pontharpinia obtusidens,* antennule and third peraeopod; *g-h, Aoroides columbiae,* first and second gnathopods of male.

13. Pleon segments 5–6 coalesced 14
13. Pleon segments not coalesced 15
14. Mandible with palp (ATYLIDAE) (fig. 77, *a, b, c*) . *Nototropis tridens*
14. Mandible without palp (DEXAMINIDAE) . . *Polycheria antarctica*

15. Fourth coxa strongly excavated behind or produced into a tooth . 16
15. Fourth coxa straight behind 36

16. First antenna with distinct accessory flagellum . GAMMARIDAE 17
16. First antenna with rudimentary or no accessory flagellum. . . 23

17. Second gnathopod of male always larger than 1st. 18
17. Second gnathopod of male subequal or slightly smaller than 1st . . . . *Anisogammarus* 22

18. Rami of 3d uropod very unequal . . . . *Melita* 19
18. Rami of 3d uropod equal in size 20

19. Fourth pleon segment with 1 tooth; palm of male 2d gnathopod large . . . . *Melita sulca*
19. Fourth pleon segment with 3 teeth; palm of male 2d gnathopod of medium size (fig. 77, *d, e, f*) . . . . *Melita californica*
19. Pleon segments without teeth or cusps; 5th segment armed with 6–8 minute spinules . . . . *Melita nitida*

20. Second gnathopod of male with palm transverse . *Maera inaequipes*
20. Second gnathopod of male with palm oblique (fig. 72, *b*) . . . . *Elasmopus* 21

21. Second gnathopod of male with 2–3 teeth inside palm *Elasmopus rapax*
21. Second gnathopod of male with no teeth in palm, fringed with long setae . . . . *Elasmopus antennatus*

22. Fifth pleon segment with a large tooth *Anisogammarus pugettensis*
22. Fifth pleon segment with spines only . *Anisogammarus confervicolus*

23. Mandible without palp . . . . TALITRIDAE 24
23. Mandible with palp 33

24. First antenna shorter than peduncle of second antenna . . . 25
24. First antenna longer than peduncle of second antenna . . . 27

25. First gnathopod subchelate in male (fig. 77, *g*) *Orchestia traskiana*
25. First gnathopod simple in male . . . . . *Orchestoidea* 26

26. Second antenna of male with enlarged 2d article and flagellum shorter than peduncle (fig. 78, *d*); whole antenna about 1/2 as long as body; 2d gnathopod with prominent tooth in palm (fig. 78, *c*). Female with small spine at distal angle of 4th article of 2d gnathopod (fig. 78, *a*). Uropods of both sexes colored (red, brown, purple, or blue); general body color pinkish or brownish. . . . . *Orchestoidea corniculata*
26. Second antenna of male with 2d article not conspicuously enlarged (fig. 78, *h*); whole antenna longer than body; 2d gnathopod lacking prominent tooth in palm (fig. 78, *g*). Female with prominent process at distal angle of 4th article of 2d gnathopod (fig. 78, *e*). Uropods

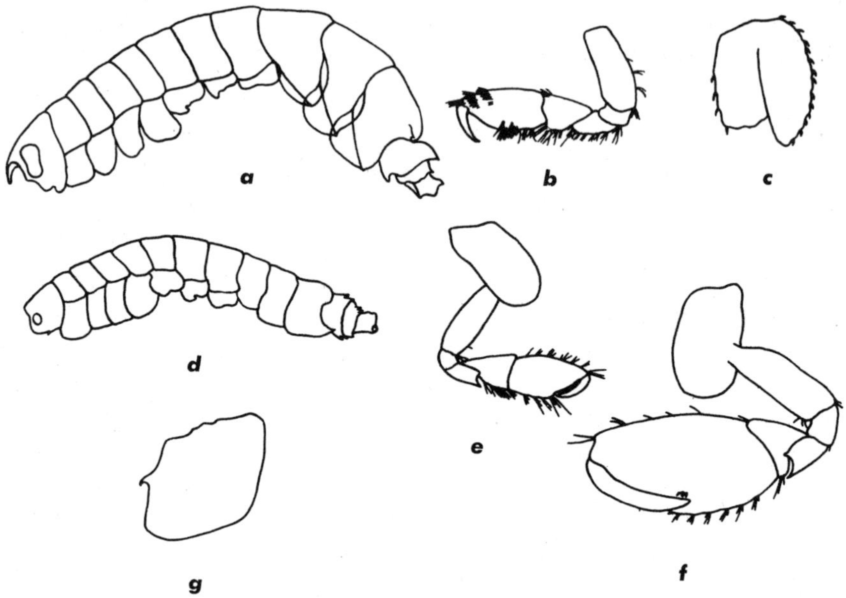

Fig. 77. *a-c, Nototropis tridens; a,* side view; *b,* first gnathopod; c, second joint of fifth peraeopod. *d-f, Melita californica; d,* side view; e, first gnathopod; *f,* second gnathopod. *g,* side plate four of *Orchestia traskiana* (*a-f,* after Alderman).

of both sexes semitranslucent whitish to faint bluish; general body color grayish or bluish. . . . . . . *Orchestoidea californiana*

27. Inhabiting fresh water . . . . *Hyalella azteca*
27. Marine 28

28. Third uropod with a minute inner ramus *Parallorchestes ochotensis*
28. Third uropod with a single ramus 29

29. Fifth article of male second gnathopod produced between 4th and 6th articles . . . . *Allorchestes angustus*
29. Fourth and 6th articles of male 2d gnathopod in contact . *Hyale* 30

30. Second antenna with thick tufts of setae . . . *Hyale plumulosus*
30. Second antenna with relatively few setae 31

31. Second antenna definitely less than half as long as body, flagellum with less than 20 joints; body color gray or black; size relatively large . . . . *Hyale pugettensis*
31. Second antennal flagellum with 20 or more joints; body size relatively small 32

32. Second antenna at least 1/2 as long as body; flagellum of second antenna in male with 40–60 joints, in female with about 36 joints; body color pink . . . . *Hyale rubrc*

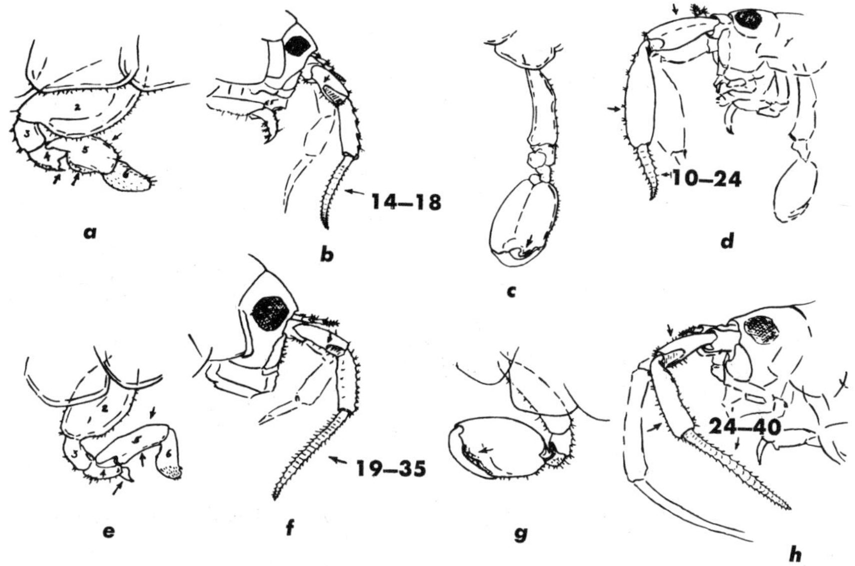

Fig. 78. *Orchestoidea.* Top row, *O. corniculata*: *a*, female, right second gnathopod; *b*, female, second antenna; *c*, male, left second gnathopod; *d*, male, second antenna. Bottom row, *O. californiana*: *e*, female right second gnathopod; *f*, female second antenna; *g*, male, left second gnathopod; *h*, male, second antenna (original sketches by Darl E. Bowers).

32. Second antenna 1/2 as long as body or less; flagellum of second antenna with 20–36 joints; body variously colored but not pink . . . . *Hyale frequens*

33. Telson lobed (PONTOGENEIIDAE) . . . *Pontogeneia inermis*
33. Telson entire, not lobed 34

34. Rostrum large; pleon segments with 3 low ridges running longitudinally (fig. 79, *d*, *e*). . . . . *Pleustes depressus*
34. Rostrum small; pleon segments smooth 35

35. First 4 coxal plates with spines on posterior edges . . . . *Parapleustes bairdi*
35. First 4 coxal plates lacking spines . . *Parapleustes pugettensis*

36. Third uropod with hooks 37
36. Third uropod without hooks . . . . PHOTIDAE 44

37. Lower lip with principal lobes notched . . AMPITHOIDAE 38
37. Lower lip with lobes rounded 43

38. Fifth article of male 2d gnathopod longer than 6th . . . . *Ampithoe humeralis*
38. Fifth article of male 2d gnathopod shorter than 6th 39

39. Second antenna with dense tufts of setae (fig. 80, *a*) . . . . *Ampithoe plumulosa*

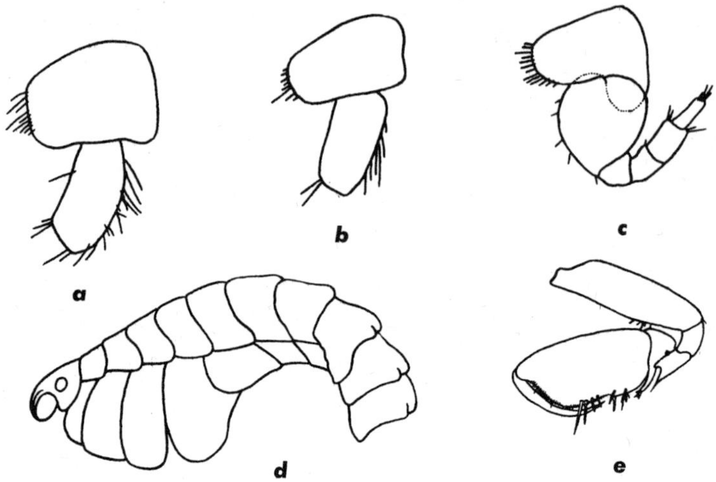

Fig. 79. *a-c*, *Photis conchicola*: *a*, side plate three and second segment of first peraeopod; *b*, side plate four; *c*, side plate five and third peraeopod; *d-e*, *Pleustes depressus*: *d*, side view; *e*, second gnathopod (after Alderman).

39. Second antenna with relatively few setae (fig. 80, *b*) 40

40. Second gnathopod of male having palm transverse, with small medial tooth . . . . *Ampithoe valida*

40. Second gnathopod of male having palm oblique, with distal tooth. 41

41. Second gnathopod of male with 6th article produced and setose at base of 7th article. . . . . *Ampithoe ramondi*

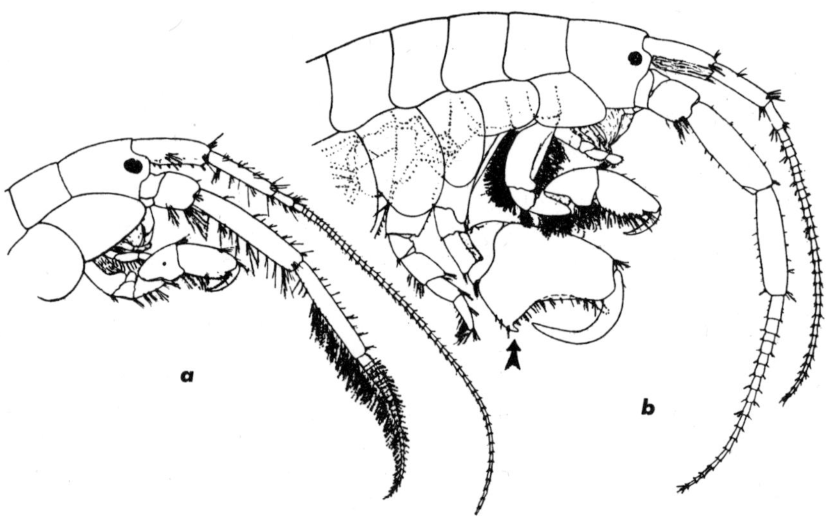

Fig. 80. *a*, *Ampithoe plumulosa*, male; *b*, *A. simulans*, male (from Shoemaker, 1938).

41. Second gnathopod of male with 6th article not produced or greatly setose 42

42. Second gnathopod of male with tooth of palm short (fig. 80, *b*) . . . . *Ampithoe simulans*

42. Second gnathopod of male with tooth of palm reaching to end of 6th article . . . . *Ampithoe pollex*

43. Male with a large tooth on palm of 2d gnathopod; female with 2d gnathopod much larger than 1st . . . . *Jassa falcata*

43. Male lacking gnathopod tooth; female with 2d gnathopod not much larger than 1st. . . . . *Ischyrocerus parvus*

44. Third uropod with a single ramus (fig. 72, *c*) . . *Microprotopus* sp.

44. Third uropod with 2 rami 45

45. Third uropod with subequal rami . . . *Eurystheus tenuicornis*

45. Third uropod with unequal rami 46

46. Male with 1st and 2d coxae much shallower than others; female with 4th coxa broader than front lobe of 5th coxa (fig. 79, *a*, *b*, *c*) . . . . *Photis conchicola*

46. Male with 1st and 2d coxae not much shallower than others; female with 4th coxa narrower than front lobe of 5th coxa . . . . *Photis californica*

List of Species

Pacific Coast gammarids are little known, and the following list probably includes only part of the species occurring in the central California intertidal. Specific determinations usually cannot be made from a key alone without recourse to full descriptions. This is so not only because undescribed species occur but also because several genera are represented by a series of closely similar species.

ACANTHONOTOZOMATIDAE (See Shoemaker, 1931)
Allorchestes angustus Dana, 1854
**Ampelisca macrocephala* Lilljeborg, 1852
**Ampelisca typica* (Bate, 1856)
**Amphilochus neopolitanus* Della Valle, 1893
Ampithoe humeralis (Stimpson, 1864)
Ampithoe plumulosa Shoemaker, 1938
**Ampithoe lacertosa* (Bate, 1858)
Ampithoe pollex Kunkel, 1910 [= *A. indentata* (Stout, 1913)]
Ampithoe ramondi (Audouin, 1826)

Ampithoe simulans Alderman, 1936 (= *A. dalli* Shoemaker, 1938)
Ampithoe valida (S. I. Smith, 1873) (See Alderman, 1936)
Anisogammarus (*Anisogammarus*) *pugettensis* (Dana, 1853)
Anisogammarus (*Eogammarus*) *confervicolus* (Stimpson, 1857)
Aoroides columbiae Walker, 1898 (= *A. californica* Alderman, 1936)
**Aruga macromerus* Shoemaker, 1916
**Calliopius laeviusculus* (Krøyer, 1838)
Chelura terebrans Philippi, 1839
Corophium acherusicum (Costa, 1857)
**Corophium brevis* Shoemaker, 1949
Corophium insidiosum Crawford, 1937
**Corophium oaklandense* Shoemaker, 1949 (This species is said to be hermaphroditic)
Corophium spinicorne Stimpson, 1857 (In brackish water)
**Corophium stimpsoni* Shoemaker, 1941 (May be very abundant in nearly fresh water of estuaries)
Elasmopus antennatus (Stout, 1913)
Elasmopus brasiliensis (Dana, 1853) (See Alderman, 1936)
Elasmopus rapax Costa, 1853
Ericthonius brasiliensis (Dana, 1853)
Ericthonius hunteri (Bate, 1862)
Eurystheus tenuicornis (Holmes, 1904) (See Shoemaker, 1931)
Hyale frequens (Stout, 1913) (Possibly a color variant of *H. rubra*)
**Hyale hawaiensis* (Dana, 1853)
Hyale plumulosus (Stimpson, 1857)
Hyale pugettensis (Dana, 1853)
Hyale rubra (Thompson, 1879)
Hyalella azteca (Saussure, 1858)
Ischyrocerus parvus Stout, 1913
Jassa falcata (Montagu, 1808)
**Leucothoe spinicarpa* (Abildgard, 1789) (?)
**Maera* sp.
Maera inaequipes (Costa, 1851) (See Alderman, 1936)
Melita californica Alderman, 1936
Melita sulca (Stout, 1913) (Has been called *M. palmata* Montagu, 1804)
Melita nitida S. I. Smith, 1873
Metopa sp.
Microprotopus sp.
Nototropis tridens Alderman, 1936
Orchestia traskiana Stimpson, 1857
Orchestoidea californiana (Brandt, 1851) (Characteristic of clean exposed beaches with fine sand)
Orchestoidea corniculata Stout, 1913 (Found on smaller, sheltered beaches with coarse sand and algal debris)

Parallorchestes ochotensis (Brandt, 1851) (See Shoemaker, 1941)
Parapleustes bairdi (Boeck, 1872)
Parapleustes pugettensis (Dana, 1853)
Photis californica Stout, 1913
Photis conchicola Alderman, 1936
Pleustes depressus Alderman, 1936
Podocerus spongicolus Alderman, 1936
Podocerus brasiliensis (Dana, 1853-1855)
Polycheria antarctica (Stebbing, 1875) (See Alderman, 1936; = *P. osborni* Calman, 1898?)
Pontharpinia obtusidens Alderman, 1936
Pontogeneia inermis (Krφyer, 1838)

Suborder Caprellidea

Caprellidea are bizarre creatures characterized by such general amphipod features as two pairs of subchelate gnathopods (representing the appendages of the second and third thoracic segments; the first bears the maxillipeds), by gills or branchiae on certain segments, and by marsupial plates or oöstegites. Caprellidea differ, however, in having a vestigial abdomen or pleon, usually with rudiments of uropods, and in one genus (*Cercops*), rudiments of pleopods; the first two thoracic segments are fused with the head to form a cephalon, while the six free thoracic segments (peraeods) making up the peraeon are well developed. Often the sides of the peraeods project ventrally as pleura (sing. pleuron; fig. 82, *c*).

The family CAPRELLIDAE (figs. 72, *e*; 82) is composed of free-living, elongated, slender forms with cylindrical bodies, found on algae in mud flats (e.g., *Caprella californica*), on algae and hydroids in the low rocky intertidal ("*acutifrons*" group, etc.), and on various sessile organisms in deeper waters. They are sometimes referred to as "skeleton shrimps." In California, species of the genus *Caprella* are by far the most common in the habitats accessible to the average collector. The second family, CYAMIDAE or "whale lice," including several genera ectoparasitic on cetaceans, have very broad and flattened bodies (fig. 81) but otherwise resemble the caprellids proper.

A number of genera, including *Caprella*, lack appendages on the third and fourth peraeods, although fully formed legs occur there in a few genera. Legs and rudiments of legs variously developed (fig. 82, *f*) occur on third to fifth peraeods in still other genera. Branchiae are found on second to fourth peraeods in certain genera, but are restricted to third and fourth peraeods in most, including *Caprella*. Oöstegites (fig.

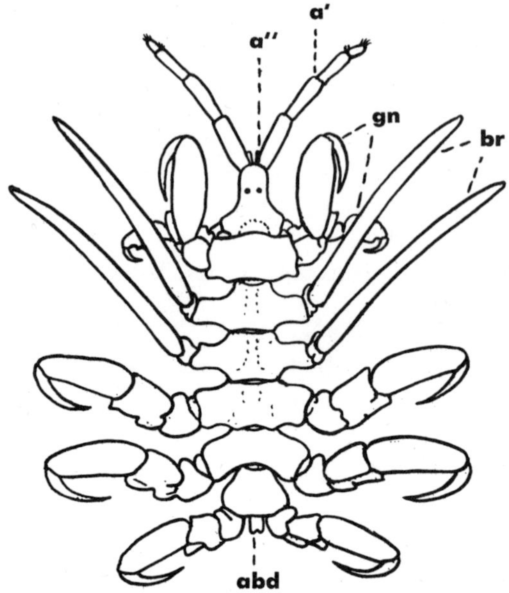

Fig. 81. **Caprellidea:** CYAMIDAE. *Paracyamus*, a whale-louse. a', first antenna; a', second antenna; gn, gnathopods; br, branchiae; abd, abdomen (after Sars).

72, *e*) are uniformly found on peraeods three and four in the female. Rudimentary appendages on the vestigial abdomen range from four pairs in *Cercops* to none in *Pseudaeginella*. Great sexual dimorphism exists (fig. 82, *b*), as well as marked difference in the proportion of parts between young and old males. This morphological variation makes the group very difficult to characterize by means of a key.

The known California fauna now consists of eighteen named species, of which all but one (*Perotripus brevis*) have been found in central California. There are without doubt a number of undescribed forms, possibly many.

Key to the Caprellidea of California[5]

by Ellsworth C. Dougherty and Joan E. Steinberg,
University of California, Berkeley

1. Body much depressed; parasitic on whales . . Family CYAMIDAE
1. Body slender and cylindrical; free-living on algae and hydroids . . . . Family CAPRELLIDAE 2

2. Appendages or rudiments of appendages on all 7 peraeods, including the 1st peraeod, which is fused with the head 3

[5] Most reliable for adult males.

2. Appendages completely lacking on peraeods 3 and 4 7

3. Rudimentary peraeopods on peraeods 3 to 5 . . *Perotripus brevis*
3. Rudimentary peraeopods on peraeods 3 and 4, normal on peraeod five 4

4. Peraeopods 3 and 4 each with 2 free segments *Deutella californica*
4. Peraeopods 3 and 4 each with only 1 free segment . . *Tritella* 5

5. Body very elongate, with long slender appendages; a deep-water form . . . . *Tritella tenuissima*
5. Body and appendages not elongate; found in intertidal zone . . 6

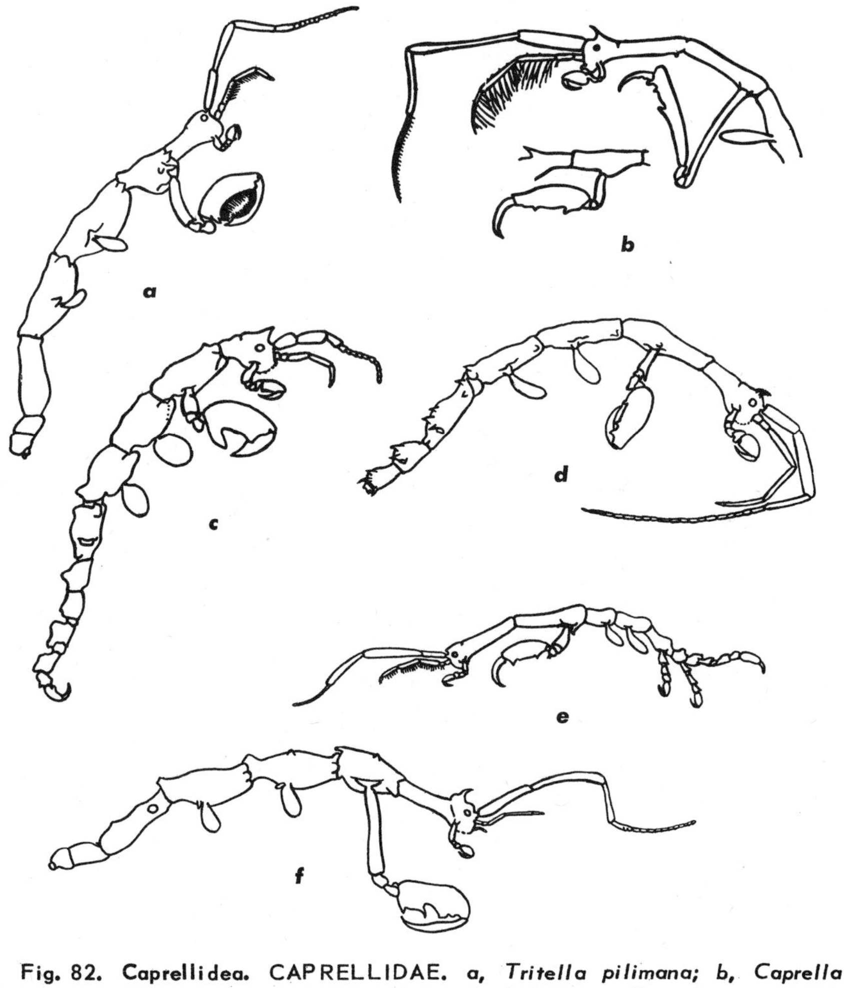

Fig. 82. Caprellidea. CAPRELLIDAE. a, *Tritella pilimana*; b, *Caprella californica*, with second gnathopod of female below; c, *Caprella incisa*; d, *Metacaprella anomala*; e, *Caprella equilibra*; f, *Deutella californica* (a, c-f after Mayer; B after Johnson and Snook).

6. Anterior dorsal margins of peraeods 2 to 4 projecting laterally as spines; propodus and dactylus of 2d gnathopod very hairy . . . . *Tritella pilimana*
6. Anterior dorsal margins of peraeods 2 to 4 not projecting laterally; sides of peraeods smooth; propodus and dactylus moderately hairy . . . . *Tritella laevis*

7. Frons of cephalon with 2 anteriorly directed spines . *Metacaprella* 8
7. Frons with 1 or no anteriorly directed spine 10

8. Spines on cephalon very small; peduncle of first antenna extremely long in relation to flagellum . . . . *Metacaprella ferrea*
8. Spines on cephalon prominent; peduncle of first antenna not extremely long in relation to flagellum 9

9. Peraeods 2 to 4 smooth; peraeod 5 with a single spine posterodorsally . . . . *Metacaprella anomala*
9. All peraeods bearing spines which are mostly paired . . . . *Metacaprella kennerlyi*

10. Frons with no spine 11
10. Frons with a single medially situated spine 13

11. Subchela of 2d gnathopod shorter than basal segment (coxa) . . . . *Caprella gracilior*
11. Subchela of 2d gnathopod as long as, or longer than, coxa . . 12

12. Poison tooth on subchela of 2d gnathopod enormous . . . . *Caprella laeviuscula*
12. Poison tooth not enormous. . . . . *Caprella equilibra*

13. Basal joint of 2d gnathopod long and equal to at least 1/2 of entire leg 14
13. Basal joint of 2d gnathopod distinctly less than 1/2 as long as entire leg 15

14. Spine on frons of cephalon long and directed anteriorly; all peraeods much elongate . . . . *Caprella californica*
14. Spine on frons of cephalon very small but distinct; cephalon and peraeod 2 much elongate; peraeods 3 to 7 quite short. . . . . *Caprella brevirostris*

15. Coxa of 2d gnathopod thin and more than twice as long as 2d and 3d leg segments combined. . . . . *Caprella uniforma*
15. Coxa of 2d gnathopod thick and only slightly longer than 2d and 3d leg segments combined . . . *"Caprella acutifrons"* group, 16

16. Palm of subchela densely covered with hairs . . *Caprella pilipalma*
16. Not so. 17

17. Peraeon of both sexes and of immature forms with very obvious tubercles . . . . *Caprella verrucosa*

17. Peraeon without tubercles or with slight protuberences only, usually restricted to posterior peraeods 18
18. Second gnathopod with large poison tooth at base of palm equal or almost equal in size to dactylus, giving subchela a double chelate appearance . . . . *Caprella incisa*
18. Second gnathopod without great development of poison tooth on palm . . . . *Caprella angusta*

List of California Caprellidae

Caprella angusta Mayer, 1903
Caprella brevirostris Mayer, 1903
Caprella californica Stimpson, 1857
Caprella equilibra Say, 1818
Caprella gracilior Mayer, 1903
Caprella incisa Mayer, 1903
Caprella laeviuscula Mayer, 1903
Caprella pilipalma Dougherty and Steinberg, 1953
Caprella uniforma La Follette, 1915
Caprella verrucosa Boeck, 1872
Deutella californica Mayer, 1890
Metacaprella anomala (Meyer, 1903)
Metacaprella ferrea (Mayer, 1903)
Metacaprella kennerlyi (Stimpson, 1864)
Perotripus brevis (La Follette, 1915) (See Dougherty and Steinberg, 1953)
Tritella laevis Mayer, 1903
Tritella pilimana Mayer, 1890
Tritella tenuissima Dougherty and Steinberg, 1953

SUBCLASS EUCARIDA

This subclass includes two orders, one a small group of small pelagic shrimps, order **Euphausiacea**, the other the large order **Decapoda** including the familiar shrimps, prawns, crabs, lobsters, hermit crabs, and other higher crustaceans. Various texts employ different divisions or suborders, the relationships of which are shown in the table below. It is not so important which system is followed, but it is well to understand the groupings as used by different authors.

Parker and Haswell system	Borradaile and Potts system	Pratt system	Schmitt system	
Suborders	Suborders	Tribes or sections	Tribes	Suborders
	Penaeidea (Primitive prawns)		Peneides	Natantia (Swimming types)
	Caridea (Shrimps and prawns)		Carides	
Macrura (Shrimp and crayfishlike forms)	Palinura (Spiny lobsters)		Palinura	Reptantia (Creeping types)
	Astacura (True lobsters and fresh-water crayfish)		Astacura	
Anomura	Anomura (Hermit crabs and allies)		Anomura	
Brachyura	Brachyura (True crabs)		Brachyura	

For our purposes it is important to be able to recognize on sight the Brachyura, the Caridea, the fresh-water Astacura, and the Anomura. Only the last group presents much difficulty, owing to the fact that some of its members (brachyurous anomurans) closely resemble the true crabs, whereas others have well-developed abdomens (macrurous anomurans) and resemble the true Macrura. Still others have the familiar twisted abdomen characteristic of hermit crabs (anomurous anomurans). If this seems confusing or even absurd, we should remember that the Anomura are defined not only by the character of the abdomen which gives the group its name, but by an array of characters, some quite obvious and others obscure.

The least specialized (or most generalized) local decapods are the macruran shrimps such as *Spirontocaris* or *Crago*, whose bodies are said to exhibit the "caridoid facies" (or generalized malacostracan aspect, seen also in mysids). Figure 83 illustrates the main macruran features, familiar to most students from earlier studies on crayfish or lobster. The

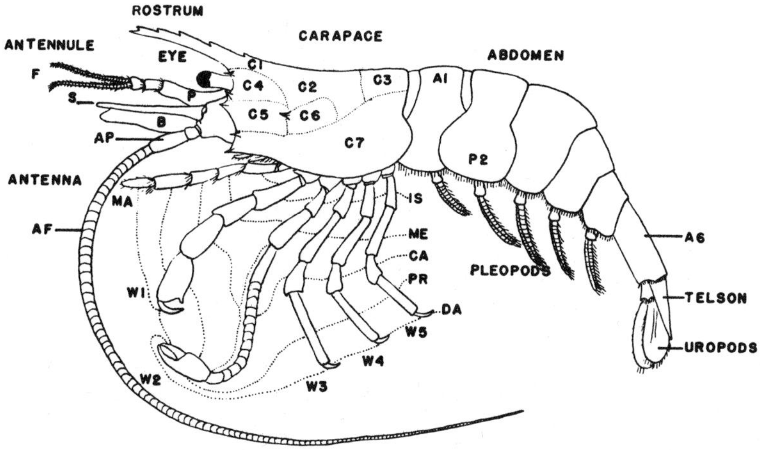

Fig. 83. Lateral view of a macruran decapod (redrawn from Schmitt). The antennule includes a basal peduncle (P), two smaller segments and two flagella (F). There is a large antennal scale made up of a spine (S) and a blade (B). The antenna has a basal peduncle (AP) and a flagellum (AF). The third maxilliped (MA) is followed by five legs (W1-W5), the first of which (W1) is subchelate and is known as the cheliped. The second leg is chelate. The segments indicated on the appendages are referred to as the dactylus or dactylopodite (DA), the propodus or propodite (PR), carpus or carpopodite (CA), merus or meropodite (ME) and ischium or ischiopodite (IS). The dactylus forms the movable finger and the propodus, the "palm" of the first and second legs. The carpus is jointed in the second leg to form a long "wrist." There are five pairs of pleopods or swimmerets and a pair of uropods. The eye includes a cornea borne on a stalk. (The rostrum is measured to the posterior margin of the eye orbit.) The carapace may be divided into seven regions: frontal (C1); gastric (C2); cardiac (C3); orbital (C4); antennal (C5); hepatic (C6); and branchial (C7). There are six distinctive spines on the carapace: the supra- and sub-orbital at the upper and lower anterior margin of the orbital region; the antennal spine at the upper margin of the antennal region; the branchiostegal and pterygostomian spines at the upper and lower anterior margins of the branchial region; and the hepatic spine at the anterior margin of the hepatic region. The abdomen has six segments (A1-A6), which bear on the sides the pleural plates (P2, pleural plate of the second abdominal segment). The body terminates in a telson which, with the uropods, makes up the tail fan.

features of a generalized brachyuran are shown in figures 84 and 85. There is an extreme (but symmetrical) reduction of the abdomen and complete absence of uropods. The abdomen of males is narrow and has reduced pleopods, except for the specialized anterior gonopods or copulatory legs. The female crab has a broadly rounded abdomen, with well-developed fringed pleopods for the attachment of eggs. Anomurans present much more diversity of form. The abdomen may vary from a fully developed, symmetrical condition (*Callinassa*), to the twisted form of the hermit crabs (*Pagurus*), down to the reduced and very crablike condition found in *Petrolisthes*, *Cryptolithodes*, and the like, or to the

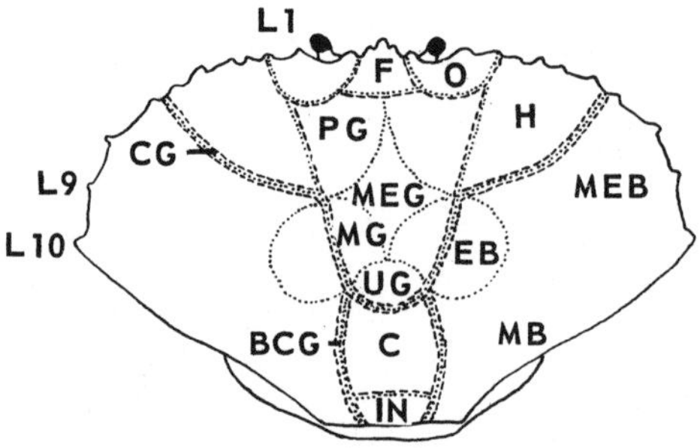

Fig. 84. Dorsal view of a brachyuran carapace showing the areas referred to in descriptions (schematic, modified after Schmitt). The regions of the carapace are the: frontal (F); orbital (O); hepatic (H); branchial, including the meso- (MEB), meta- (MB), and epi-branchial (EB) areas; gastric, including the meso- (MEG), peri- (PG) and uro-gastric (UG) areas; cardiac (C); and intestinal (IN). There are two major grooves, the branchial cardinal (BCG) and the cervical (CG). The latter separates the branchial region from the hepatic and gastric regions and the gastric from the cardinal region. The spination along the margin of the carapace is that of *Cancer magister*. L 1 is the first anterior-lateral tooth (post-obital spine); L9 and L 10 are the ninth and tenth anterior-lateral teeth.

intermediate condition in *Emerita*. All these anomurans have in common the characteristic that the last (fifth) pair of legs (pereiopods) differs from the ones in front of them. In most anomurans these legs are very small and are folded up behind the bases of the pair ahead. In *Callianassa*, where these legs are not reduced in size, they are held differently, and the sternite to which they are attached is not joined to the other sternites of the thorax, a feature shared by the other anomurans.

Fortunately, Schmitt, *Marine Decapod Crustacea of California* is a monograph that covers local decapods very well in almost all cases. Students should run known specimens of the several groups down to species in Schmitt, going through all the steps in classification, in order

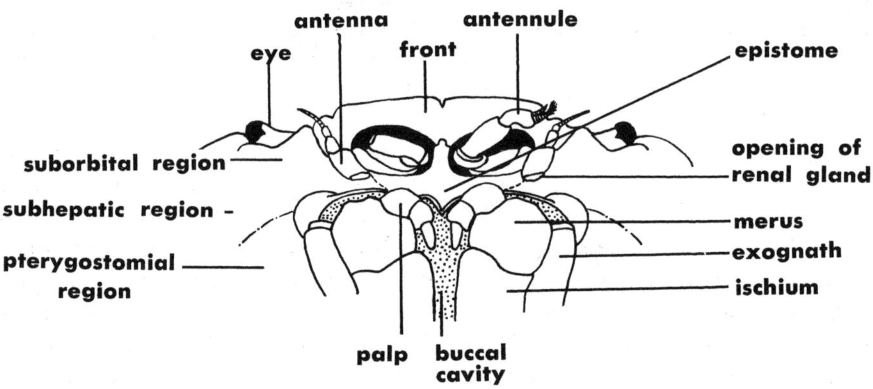

Fig. 85. Ventral view of anterior part of body of a brachyuran decapod (after Schmitt.)

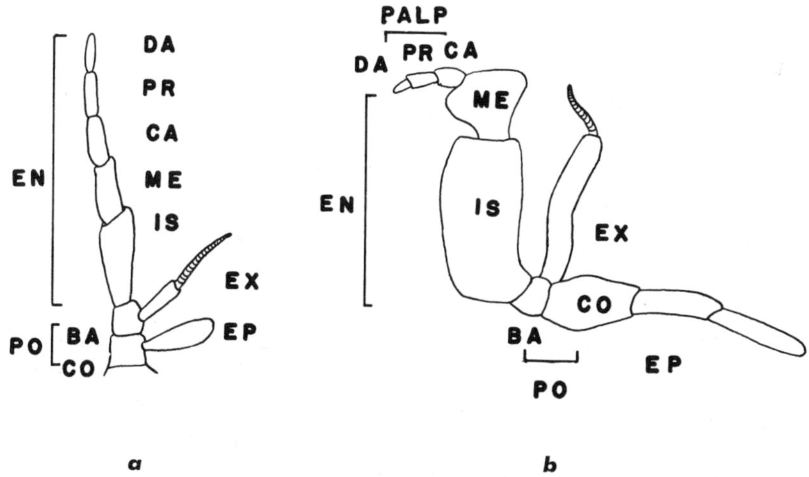

Fig. 86. Typical third maxilliped of: *a*, a macruran and *b*, a brachyuran decapod. BA, basipodite; CA, carpopodite; CO, coxopodite; DA, dactylopodite; EN, endopodite (endognath); EP, epipodite; EX, exopodite (exognath); IS, ischiopodite; ME, meropodite; PO, protopodite; PR, propodite.

to become thoroughly familiar with this work. With a little practice, many students will find the monograph, when available, much more satisfactory than the following necessarily abbreviated keys.

Key to Decapod Tribes

I. Abdomen macrurous, with well-developed tail fan:
   A. Almost entirely marine; body usually laterally compressed, shrimp-like in form; pleura (side plates) of 2d abdominal segment overlap those of 1st; abdomen usually with a sharp bend . . . . Tribe **Carides** (p. 176)
   B. Fresh-water (in this area); body heavy, with abdomen dorsoventrally flattened; chelipeds very large and strong Tribe **Astacura** (p. 177)

II. Abdomen small, folded under thorax, fully symmetrical; uropods always absent; last pair of legs not markedly reduced; antennae internal to eyes; the true crabs . . . Tribe **Brachyura** (p. 178)

III. Abdomen usually asymmetrical and/or reduced; uropods usually present; last pair of legs almost always of reduced size and carried folded up above bases of preceding pair; if abdomen macrurous, posterior sternite of thorax not fused to others; antennae external to eyes (fig. 90, *b, c*). . . . . Tribe **Anomura** (p. 182)

IA. Key to Tribe Carides

1. Rostrum very short, dorsally flattened; eyes free; hands subchelate (fig. 87, *a*, *b*) 2
1. Rostrum very small or wanting; eyes covered by carapace; hands chelate, one or both powerfully developed 4
1. Rostrum distinct, usually well-developed and spinose (fig. 88, *b—e*); eyes free; hands weakly chelate 5

2. Hand of cheliped slender and elongated, with finger turned back almost longitudinally when closed . . . . *Crago franciscorum*
2. Hand of cheliped not slender, less than three and one half times as long as wide; finger when closed more or less transverse . . 3

3. Fifth segment of abdomen with a median dorsal keel; tail usually black; antennal scale hardly twice as long as wide (fig. 87, *b*) . . . . *Crago nigricauda*
3. Fifth segment of abdomen without a median dorsal keel; a prominent dark spot on each side of 6th abdominal segment; antennal scale nearly 3 times as long as wide (fig. 87, *a*) . *Crago nigromaculata*

4. One chela greatly enlarged and complex, with dactyl above (fig. 87, *c*; the Pistol Shrimp) . . . . *Crangon* sp.
4. Chelae about equal, simple; dactyl below . . . *Betaeus harfordi*

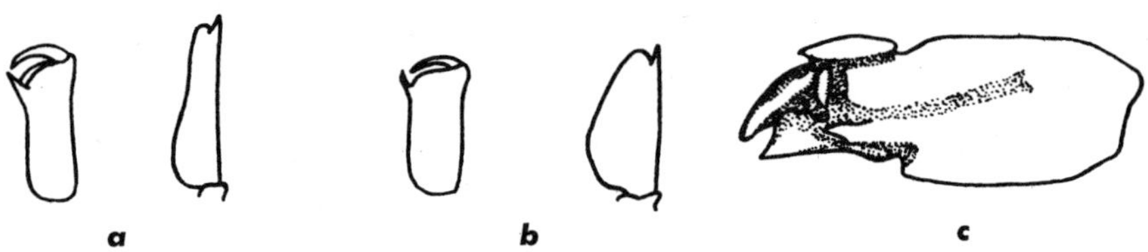

Fig. 87. *a-b, Crago*: *a*, hand and scale of *C. nigromaculata*; *b*, hand and scale of *C. nigricauda*; *c*, chela of *Crangon dentipes* (after Schmitt.)

5. Medium-sized; both legs of 1st pair simple; 2d pair of legs very unequal, both with multiarticulate carpus . . . *Pandalus danae*
5. Smaller; both legs of 1st pair chelate (chelae small); 2d pair equal or nearly so 6

6. Carpus of 2d legs with 7 annulations; marine . . *Spirontocaris* 7
6. Carpus of 2d legs with 3 annulations; marine . . . . *Hippolyte*
6. Carpus of 2d legs not annulated; in fresh water . *Syncaris pacifica*

7. Supraorbital spines present; rostrum high, like vertical leaf; body opaque; in lower-tide pools . . . . . . *Spirontocaris prionota*
7. Lacking supraorbital spines 8

8. Rostrum elongate (fig. 88, *d, e*), reaching beyond middle of antennal scale 9
8. Rostrum short, not reaching middle of antennal scale 10
9. Rostrum reaching beyond middle of antennal scale but not into distal quarter; greenish, semitransparent, with oblique reddish bands on carapace and crimson bars on legs; common in tide pools (fig. 88, *d*) . . . . *Spirontocaris picta*
9. Rostrum longer, reaching to or beyond end of antennal scale; uniform green; common in eelgrass and on mud flats (fig. 88, *e*) . . . . *Spirontocaris paludicola*
10. Rostrum not reaching as far as cornea of eye . *Spirontocaris taylori*
10. Rostrum reaching cornea or beyond . . . *Spirontocaris palpator*

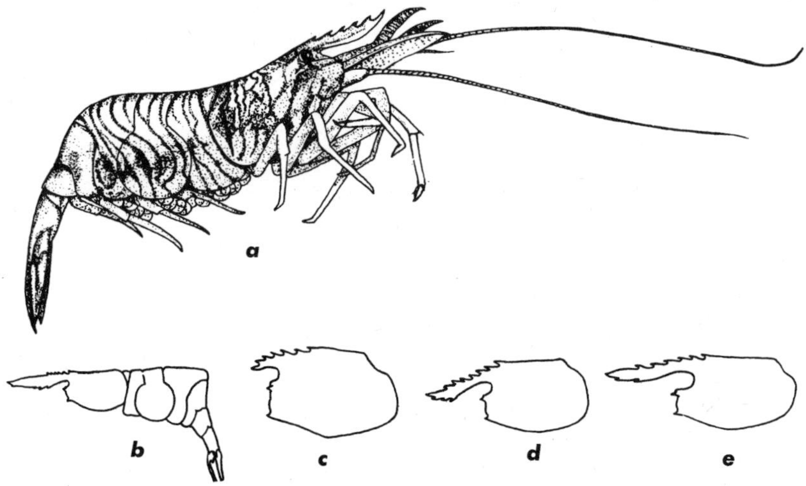

Fig. 88. *a, Spirontocaris paludicola* (after Ricketts and Calvin). *b-e,* rostra of common species of *Spirontocaris* (after Schmitt): *b, S. carinata; c, S. brevirostris; d, S. picta; e, S. paludicola.*

IB. Key to Tribe Astacura

The following key is adapted from that in Hay's synopsis (1899), which should be consulted if species of *Cambarus* are encountered. These are introduced and should be referred to a competent authority for verification. Crayfish are, of course, not intertidal animals, but may be encountered at the heads of estuaries.

1. Last thoracic segment with a gill on each side, making a total of 18 gills in each branchial chamber; copulatory legs of males neither

toothed nor bifid at tips; female lacking an *annulus ventralis* (see below); 5 species native to West Coast states . . . *Astacus* 2

1. Last thoracic segment lacking gills, making a total of 17 in each branchial chamber; copulatory legs of male bifid and hooked or toothed; female with conspicuous ringlike *annulus ventralis* for reception of sperm between bases of last 2 pairs of walking legs; many species east of Rockies, but introduced into western states . . . . *Cambarus*

2. Margins of rostrum smooth (except for a large tooth on each side) 4
2. Margins of rostrum denticulate 3

3. Chelae with a patch of soft setae on outer face . *Astacus gambelii*
3. Chelae bare on outer face . . . . *Astacus nigrescens*

4. Postorbital ridges with posterior spine or tubercle 5
4. Postorbital ridges without posterior spine or tubercle . . . . *Astacus klamathensis*

5. Space between lateral teeth of rostrum less than the length of rostrum anterior to these teeth . . . . *Astacus leniusculus*
5. Space between lateral teeth of rostrum not less than length of rostrum anterior to these teeth. . . . . *Astacus trowbridgii*

II. Key to Tribe Brachyura

1. Mouth field triangular, narrowed in front; carapace either rounded and with posterior spines or ovate with large, straight, lateral spine and more than 12 anterolateral teeth 2
1. Mouth field square; carapace square, triangular, ovate, or rounded, but without spines on posterior margin if round and with not more than 10 anterolateral teeth if having a long lateral spine . . . 3

2. Carapace orbicular with 2 short prominent spines posteriorly; color white with purple patches . . . . *Randallia ornata*
2. Carapace ovate with a pronounced lateral spine and about 15 small teeth on anterolateral margin; color reddish . *Mursia gaudichaudii*

3. Carapace nearly square; sides nearly parallel; anterior edge nearly transverse; eyes at anterolateral corners 11
3. Carapace not square but may be triangular, oval, or nearly round; sides not parallel 4

4. Forepart of body narrow, forming a single or bifid rostrum . . 5
4. Forepart of body broad; rostrum usually reduced or absent . . 13

5. Chelipeds much longer and heavier than other legs; rostrum single; carapace broadly triangular; on sandy substrate (fig. 89, *g*) . . . . *Heterocrypta occidentalis*
5. Chelipeds not greatly longer and heavier than other legs; rostrum bifid; carapace triangular or squarish 6

6. Carapace about as broad as long with lateral margins flattened and produced; upper surface smooth (fig. 89, *d*) . . *Mimulus foliatus*
6. Carapace longer than broad; lateral margins not markedly flattened or produced; upper surface smooth or rough 7

7. Rostrum decurved; carapace and legs hairy; carapace with 9–12 tubercles (fig. 89, *c*) . . . . *Loxorhynchus crispatus*
7. Rostrum flat, not decurved; carapace may be smooth or hairy and tuberculate 8

8. No spines projecting from posterolateral margin of carapace; color gray or tan, never reddish . . . . *Scyra acutifrons*
8. Prominent posterolateral projections; color greenish, reddish, or reddish brown 9

9. Surface of carapace smooth; distance between eyes less than 1/3 the width of carapace; the kelp crab . *Pugettia* (= *Epialtus*) *producta*
9. Surface of carapace tuberculated or spiny; distance between eyes about 1/2 greatest width of carapace 10

10. Smaller; carapace not expanded posteriorly; anterolateral tooth broad, anteriorly directed (fig. 89, *a*) . . . . . *Pugettia gracilis*
10. Larger; carapace distinctly broader posteriorly; anterolateral tooth narrow, laterally directed (fig. 89, *d*) . . . . . *Pugettia richii*

11. Surface of carapace with numerous transverse lines . . . . *Pachygrapsus crassipes*
11. Surface without transverse lines 12

12. Color whitish, red, or purple; no hair on legs: under rocks (fig. 89, *f*) . . . . *Hemigrapsus nudus*
12. Color dull greenish; legs hairy; muddy areas . . . *H. oregonensis*

13. Free-living; front (i.e., area between eyes) either 5-toothed or bilobed; carapace sclerotized; anterolateral border toothed . . 14
13. Small commensal crabs; front between eyes entire; carapace often membranaceous, frequently rounded or may be much wider than long; margin of carapace entire, lacking sharp teeth or spines . . . 20

14. Front 5-toothed; carapace broadly oval; antennules fold back longitudinally 15
14. Front divided by a median notch; carapace broader anteriorly; antennules fold back transversely or obliquely 27

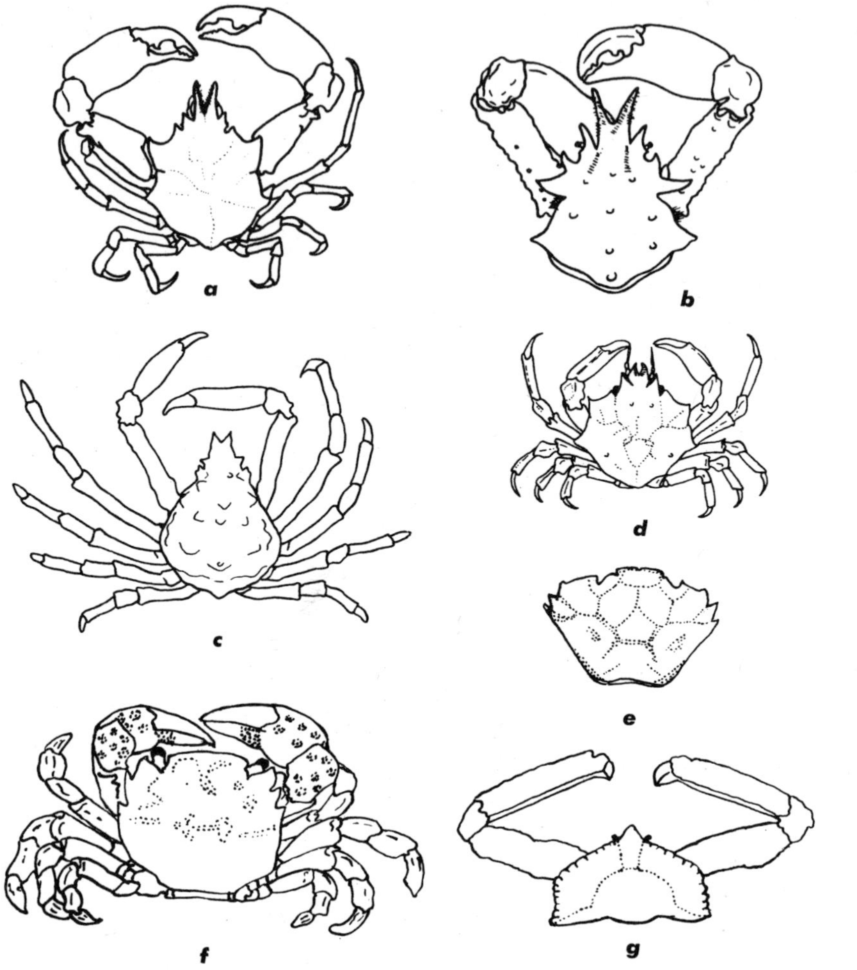

Fig. 89. *a, Pugettia gracilis; b, Pugettia richii; c, Loxorhynchus crispatus; d, Mimulus foliatus;* e, Carapace of *Lophopanopeus bellus; f, Hemigrapsus nudus; g, Heterocrypta occidentalis.* (c, after Rathbun; others after Schmitt).

15. Front markedly produced beyond outer orbital angles forming 5 subequal teeth; fingers of chelipeds dark tipped . . *Cancer productus*
15. Front not markedly produced, formed of 5 unequal teeth . . . 16
16. Carapace widest at 10th anterolateral tooth, posterior margin of 10th tooth forming a short lateral projection; fingers of chelipeds without dark color . . . . *Cancer magister*
16. Carapace widest at 8th and 9th tooth 17
17. Carapace widest at 8th tooth; 10th and 11th teeth distinct; teeth curving forward with entire edges; red spotting beneath; black on fingers of chelipeds . . . . *Cancer antennarius*

17. Carapace widest at 9th tooth; not red spotted beneath . . . 18

18. Upper surface of carapace pubescent; teeth sharp, curving, with edges entire . . . . *Cancer jordani*
18. Upper surface of carapace glabrous; teeth short, blunt, with serrate posterior margins 19

19. Chelipeds without dark color on fingers; merus of external maxillipeds rounded anteriorly . . . . *Cancer gracilis*
19. Chelipeds with dark color on fingers; merus of external maxillipeds truncate anteriorly. . . . . *Cancer anthonyi*

20. Carapace distinctly wider than long, 3d walking legs the longest 21
20. Carapace suborbicular or subquadrate, never markedly wider than long 25

21. Carapace hard; legs slender and round; 3d walking legs only slightly longer than others; anterolateral margin of carapace curving gradually into posterolateral margin; commensal in burrows of *Urechis* and *Callianassa* . . . . *Scleroplax granulata*
21. Carapace membranaceous; legs flattened; 3d walking legs much larger and longer than others; anterolateral margin of carapace forming an angle with posterolateral margin; a difficult genus, commensal with molluscs or tube-dwelling annelids . . . . . *Pinnixa* 22

22. Dactyl (terminal segment) of 3d walking leg strongly hooked . . 23
22. Dactyl of 3d walking leg straight or moderately curved . . . 24

23. Carapace oblong, not pointed at sides, about 1½ times as wide as long; commensal in bivalve molluscs . . . . *Pinnixa faba*
23. Carapace pointed at sides, about twice as wide as long; commensal in bivalve molluscs . . . . *Pinnixa littoralis*

24. Carapace about twice as wide as long; 3d walking legs longer than others but not excessively so; reported only from San Francisco bay. . . . . *Pinnixa franciscana*
24. Carapace nearly 3 times as wide as long; 3d walking legs relatively enormous; one of the smallest of all crabs, and approached by none in its great relative width and overdevelopment of 3d walking legs; commensal in tubes of *Axiothella* (fig. 90, *a*) . . *Pinnixa longipes*

25. Carapace of female with 2 grooves leading back from orbits; carapace smooth and glossy, but membranaceous and yielding; male with rounded carapace with anterior and anterolateral margins defined by a ridge of setae; female commensal in *Mytilus*, male often free-living in mussel beds. . . . . *Fabia subquadrata*
25. Carapace without longitudinal grooves 26

26. Carapace thin, but hard and stiff, very smooth and rounded; walking legs subequal in length; both sexes with carapace spotted with red;

abdomen of female may be rather narrow; commensal in *Megathura, Cryptochiton, Schizothaerus* . . . . *Opisthopus transversus*

26. Carapace thin but membranaceous and yielding, smooth and rounded; 4th pair of walking legs noticeably shorter than the others; abdomen of female broad; commensal in bivalve molluscs . . . . *Pinnotheres concharum*

27. Chelipeds with numerous prominent rounded tubercles . . . . *Paraxanthias taylori*

27. Chelipeds without such tubercles 28

28. Fingers same color as rest of chela; introduced from the Atlantic, now in Coos Bay and San Francisco Bay areas, in brackish water . . . . *Rhithropanopeus harrisii*

28. Fingers black; found on ocean coast 29

29. Hand of cheliped with a tooth on inner side of upper margin . . . . *Lophopanopeus heathii*

29. Hand without tooth (fig. 89, *e*). . . . . . *Lophopanopeus bellus*

III. Key to Tribe Anomura

1. Abdomen short, reflexed beneath thorax 2

1. Abdomen elongate, not reflexed, may be twisted 9

2. Second to 4th legs with last joint curved and flattened; sandy-beach dwellers 3

2. Second to 4th legs with last joint ending in sharp pointed dactyl; rocky areas 4

3. First pair of legs simple; carapace without sharp spines along anterolateral margin . . . . *Emerita analoga*

3. First pair of legs chelate; carapace with sharp spines along anterolateral margin . . . . *Blepharipoda occidentalis*

4. Uropods present; abdomen folded against body; carapace nearly round in outline 5

4. Uropods absent; abdomen soft or (in *Cryptolithodes*) firm; carapace wider posteriorly or completely covering appendages 15

5. Body and chelae thick; chelae unequal, tuberculate or granulate; carpus of chelipeds as broad as long . . . . . *Pachycheles* 6

5. Body and chelae flattened; chelae equal or subequal, smooth; carpus of chelipeds longer than broad . . . . *Petrolisthes* 7

6. Telson with 5 plates, no small plates at anterior margin of lateral telson plates (fig. 90, *d*) . . . . *Pachycheles rudis*

6. Telson with 7 plates, a small plate at anterior margin of each lateral telson plate (fig. 90, *e*) . . . . . . *Pachycheles pubescens*

7. Carpus of chelipeds about 1½ times as long as wide, sides converging toward distal end; palp of external maxilliped orange or reddish (fig. 90, *f*). . . . . *Petrolisthes cinctipes*
7. Carpus twice as long as wide, sides parallel; outer edge of palp of external maxilliped blue 8

8. Carpus rough; cheliped a solid color on upper surface; proximal inner base of dactyl of cheliped blue . . . *Petrolisthes eriomerus*
8. Carpus smooth; cheliped with a median row of blue dots on hand; proximal inner base of dactyl of cheliped orange . . . . *Petrolisthes manimaculis*

9. Abdomen symmetrical, extended, externally segmented; burrowing in mud or sand. 10
9. Abdomen asymmetrical, soft and twisted, externally nonsegmented; living in snail shells 11

10. First pair of legs subequal, subchelate; others simple; burrowing in soft mud . . . . *Upogebia pugettensis*
10. First pair of legs very unequal, chelate; 2d pair small and chelate; burrowing in sandy mud . . . . . . *Callianassa californiensis*

11. Chelae equal or subequal in size; sandy substrate . . . . *Holopagurus pilosus*
11. Chelae unequal, right larger than left; usually in rocky intertidal 12

12. Antennae same color as the body . . . . *Pagurus hirsutiusculus*
12. Antennae red 13

13. Walking legs banded with blue or white . . . *Pagurus samuelis*
13. Walking legs not banded with blue or white 14

14. Rostrum rounded and blunt; eyes without yellow circles . . . . *Pagurus granosimanus*
14. Rostrum with short median spine (fig. 90, *b*); eyes with yellow circles. . . . . *Pagurus hemphillii*

15. Carapace wider than long, completely covering legs and body; abdomen small and flattened . . . . *Cryptolithodes sitchensis*
15. Carapace as long as, or longer than, broad; abdomen thick and fleshy 16

16. Legs and carapace hairy, flattened . . *Hapalogaster cavicauda*
16. Legs and carapace roughly tuberculate; legs subcylindrical . . . . *Oedignathus inermis*

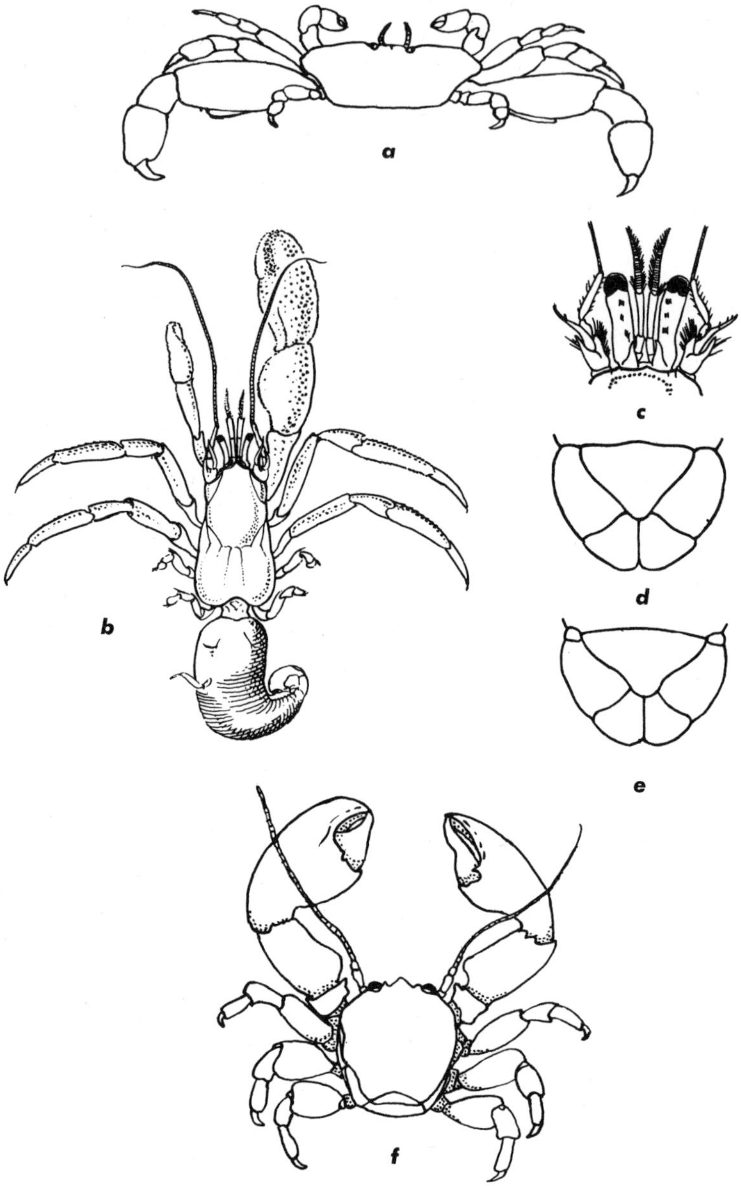

Fig. 90. *a, Pinnixa longipes; b, Pagurus hemphillii;* c, head of *Pagurus californiensis;* d, telson of *Pachycheles rudis;* e, telson of *Pachycheles pubescens;* f, *Petrolisthes cinctipes* (after Schmitt)

List of Species of Decapod Crustaceans

Carides

Betaeus harfordi (Kingsley, 1878)
Crago alba (Holmes, 1900)
Crago franciscorum (Stimpson, 1859)
Crago nigricauda (Stimpson, 1856)
Crago nigromaculata (Lockington, 1877)
**Crago stylirostris* (Holmes, 1900)
**Crangon bellimanus* (Lockington, 1877)
**Crangon dentipes* (Guérin, 1832)
**Hippolyte californiensis* Holmes, 1895
Pandalus danae Stimpson, 1857 (The "coon-stripe" shrimp)
Spirontocaris palpator (Owen, 1839)
Spirontocaris paludicola (Holmes, 1900)
Spirontocaris picta (Stimpson, 1871)
Spirontocaris prionota (Stimpson, 1864)
Spirontocaris taylori (Stimpson, 1857)
Syncaris pacifica (Holmes, 1895) (In streams of Marin and Sonoma counties)

Brachyura

Cancer antennarius Stimpson, 1856
Cancer anthonyi Rathbun, 1897
Cancer gracilis Dana, 1852
Cancer jordani Rathbun, 1900
Cancer magister Dana, 1852
Cancer productus Randall, 1839
Fabia subquadrata Dana, 1851
Hemigrapsus nudus (Dana, 1851)
Hemigrapsus oregonensis (Dana, 1851)
Heterocrypta occidentalis (Dana, 1854)
Lophopanopeus bellus (Stimpson, 1860) (1862)
Lophopanopeus heathii Rathbun, 1900
Loxorhynchus crispatus Stimpson, 1875
Mimulus foliatus Stimpson, 1860 (1862)
Mursia gaudichaudii (Milne-Edwards, 1837)
Opisthopus transversus Rathbun, 1893
Pachygrapsus crassipes Randall, 1839
Paraxanthias taylori (Stimpson, 1860) (1862)
Pinnixa faba (Dana, 1851)
Pinnixa franciscana Rathbun, 1918
Pinnixa littoralis Holmes, 1895
Pinnixa longipes (Lockington, 1877)

Pinnixa occidentalis Rathbun, 1893
Pinnixa tubicola Holmes, 1895
Pinnixa weymouthi Rathbun, 1918
Pinnotheres concharum (Rathbun, 1893)
Pugettia gracilis Dana, 1851
Pugettia (= *Epialtus*) *producta* (Randall, 1839)
Pugettia richii Dana, 1851
Randallia ornata (Randall, 1839)
Rhithropanopeus harrisii (Gould, 1841)
Scleroplax granulata Rathbun, 1893
Scyra acutifrons Dana, 1852

Anomura

Blepharipoda occidentalis Randall, 1839
Callianassa californiensis Dana, 1854
Cryptolithodes sitchensis Brandt, 1853
Emerita analoga (Stimpson, 1857)
Hapalogaster cavicauda Stimpson, 1859 (1862)
Holopagurus pilosus Holmes, 1900
Oedignathus inermis (Stimpson, 1860)
Pachycheles pubescens Holmes, 1900
Pachycheles rudis Stimpson, 1859 (1860)
Pagurus granosimanus (Stimpson, 1859)
Pagurus hemphillii (Benedict, 1892)
Pagurus hirsutiusculus (Dana, 1851)
Pagurus samuelis (Stimpson, 1857)
Petrolisthes cinctipes (Randall, 1839)
Petrolisthes eriomerus Stimpson, 1871
Petrolisthes manimaculis Glassell, 1945
Upogebia pugettensis (Dana, 1852)

Astacura

Astacus gambelii (Girard, 1852) (Utah, Idaho, Montana, Wyoming; not yet reported from California)
Astacus klamathensis Stimpson, 1857 (Klamath Lake and River region)
Astacus leniusculus Dana, 1852 (Native to Oregon; introduced into central coastal California)
Astacus nigrescens Stimpson, 1857 (Coastal, San Francisco County to Alaska)
Astacus trowbridgii Stimpson, 1857 (Columbia River)
**Cambarus clarkii* Girard, 1852 (Has been reported from Pasadena, and the editor has taken what may be a young female at Castroville, Monterey County)
Cambarus sp.

TERRESTRIAL ARTHROPODS AND INTERTIDAL INSECTS

Although the emphasis in this manual is chiefly marine, we should remember that, aside from the Crustacea, the arthropods are predominately a terrestrial group, of which a number have returned to aquatic habitats. We may consider three main assemblages: I, the "myriapods" (a heterogeneous group), II, the Insecta, and III, the Arachnida (see page 195).

This enormous assemblage of animals can have but brief and fragmentary consideration in a manual devoted mainly to marine forms, but in any course that attempts to be at all general, at least one excursion should be made to some area such as a canyon where varied environments, including pond, stream bed, stream banks (riparian), tangled thickets, open forested hillsides, and grassy hilltops support a varied fauna, largely of molluscs and arthropods.

Groups that may be encountered include the garden centipedes (class Symphyla), true centipedes (class Chilopoda), and millipedes (class Diplopoda), insects (class Insecta), and, of the class Arachnida, scorpions (order Scorpionida), spiders (order Araneida), mites and ticks (order Acarina), harvestmen (order Phalangida), and occasionally the false scorpions (order Pseudoscorpionida).

The food habits of terrestrial arthropods are extremely varied. Certain groups, however, have relatively uniform habits. Thus the Symphyla and Diplopoda are plant feeders, whereas the Chilopoda and practically all Arachnida are predators. The Insecta run the gamut of food habits. Many are predatory carnivores, others are parasites, and a vast number are plant feeders and serve as key industry animals in important food chains. Such, for example, are the aphids.

I. "Myriapods" (part of superclass Labiata)

Like the Crustacea, the Labiata are mandibulate arthropods (having mandibles as the third appendages), but the labiates differ in having only one pair of antennae (the first) and in having (in most) the bases of the second maxillae fused into a characteristic median labium or "lower lip." All the labiates, except insects, are many-legged and were at one time united as the class Myriapoda. Recent views, however, separate the myriapods into two groups. One has the genital pores near the front of the body (progoneate) and includes the Diplopoda, Symphyla, and Pauropoda. The other, the Chilopoda, has genital openings at the rear (opisthogoneate) and in this and other features seems closer to insects than to other "myriapods."

Class Diplopoda (Millipedes)

The familiar millipedes are vegetarian forms inhabiting dampish situations. The elongate body is cylindrical in most, flattened in some, and is protected by very stiff ringlike tergites. The first few anterior segments bear one pair of legs apiece, but over most of the body the segments are fused in pairs, each apparently bearing two pairs of legs and respiratory apertures. The genital pores open behind the second pair of legs, and in the male the seventh segment bears strangely modified copulatory appendages. The types commonly encountered are the relatively large, dark-brown, cylindrical millipedes of the genus *Tylobolus* and the flattened members of the family Polydesmidae.

Class Symphyla (Garden Centipedes)

These tiny white wormlike creatures may be found in loam and among decaying leaves. They feed on vegetation and cause damage of great importance to some crops. *Scutigerella* is a common genus. The Symphyla are progoneate, as are the Diplopoda, but differ from them in many important features. They possess many structural indications of relationship to the primitive stock from which the insects have arisen. There are only twelve leg-bearing somites, with peculiar coxal sacs at the bases of the legs below. There is but one pair of tracheae, opening at the bases of the antennae.

Class Chilopoda (True Centipedes)

These are familiar to all, with depressed bodies of many segments, each bearing a pair of appendages. The limbs of the first postcephalic segment are modified as poison fangs (maxillipeds). Those of the last pair are tactile. Reproductive apertures (not easily made out) open on the last true segment. Most commonly encountered in the San Francisco Bay area is a stout, flattened, reddish species, *Otocryptops sexspinosus* Say, an inhabitant of decaying wood, characterized by having twenty-three leg-bearing segments and two prominent spines on the first segment of the anal legs.

II. Class Insecta (part of superclass Labiata)

by Robert L. Usinger
University of California, Berkeley

Despite their abundance in fresh waters and on land, insects are scarce in marine habitats, so much so that they are usually overlooked entirely. The only insects that inhabit the open ocean are the marine water striders (*Halobates*) of which one species, *H. sericeus* Esch., occurs in the warm offshore waters fifty miles or more from the coast as far north as San Francisco.

A richer and more varied but nevertheless inconspicuous insect fauna is found along the coast in brackish pools, salt marshes, and intertidal rocks. Sandy beaches are the last stage in the transition to typical terrestrial habitats where insects become dominant. Beaches are inhabited by so many and such varied groups of insects that it would be impractical to treat them in detail in this manual. The Brues and Melander key to families of insects, general textbooks, and Essig's *Insects of Western North America* are recommended as aids in the study of the insect fauna of beaches. In general, canaceid flies (*Canaceoides*), staphylinid beetles (including the pictured rove beetle, *Thinopinus pictus* Le Conte), tiger beetles (*Cicindela*), and shore bugs (*Saldula*) are common on the lower hard-packed sand, whereas bembicid wasps nest in the sand-dune areas where hairy ground beetles (*Coelus ciliatus* Esch.) and burrowing bugs [*Aethus testudinatus* (Uhler)], feed on the roots of plants.

Of the truly aquatic insects, perhaps the most obnoxious is the salt-marsh mosquito [*Aedes squamiger* (Coq.)] which breeds in the brackish marshes surrounding San Francisco Bay. This species may fly or be blown for miles, and is therefore one of the main problems of mosquito-abatement agencies.

Brackish ponds are the preferred habitat of at least two groups of insects, water boatmen (*Trichocorixa reticulata* and *verticalis*) and brine flies (*Ephydra gracilis*). These insects are so perfectly adjusted to saline conditions that they live with the brine shrimp, *Artemia salina*, in the brine pools of southern San Francisco Bay.

Intertidal rocks present the most hazardous conditions to insect life and yet at least nine families of insects occur on the rocks between high and low tides along the Pacific Coast. Least conspicuous of these are the beetles, which retreat into cracks in the rocks at low tide and can best be collected by splitting the rocks apart with a crowbar.

The Diptera are more conspicuous and are the most highly modified for intertidal existence. The marine midges (Chironomidae-Clunioninae) are the most remarkable. Living on marine algae (*Ulva*, *Enteromorpha*, and others) in the larval stage, these midges emerge at low tide, mate, lay their eggs, and die before they are swamped by the next high tide. Adaptations for this hazardous existence include reduction or loss of wings, enormous development of claws for holding to rocks, a curious scampering gait quite unknown to other nematocerous Diptera, and loss of the plumose antennae in the males. This last modification is correlated with loss of the swarming habit which is so characteristic of other midges during mating and which would be a disadvantage on a windswept seacoast.

Detailed references to insects of intertidal areas are given in University of California Syllabus SS, *Biology of Aquatic and Littoral Insects*, by Usinger, La Rivers, Chandler and Wirth, 1948. The treatment of marine midges is based on Wirth (1949). The only general work on intertidal insects is Saunders' paper (1928).

Key to the Adult Stages of Intertidal Insects of the Pacific Coast

1. Wings absent; abdomen with only 6 segments. Intertidal rocks. Order Collembola, fam. PODURIDAE . *Anurida maritima* (Guerin, 1836)
1. Wings present, though sometimes concealed by leathery front wings (elytra), or reduced to small but articulated lobes. Abdomen with 8-10 visible segments 2

2. Front wings overlapping apically when at rest, the basal part divided by converging sutures which form a triangle on the dorsum. Brackish pools. Water boatmen, order Hemiptera, fam. CORIXIDAE. . . 3
2. Front wings not as above, either entirely free and membranous, reduced to vestigial stumps, or covering the dorsum without overlapping 4

3. Width of interocular space at narrowest point distinctly exceeding width of eye along hind margin as seen from above. Brackish pools . . . . *Trichocorixa reticulata* (Guerin, 1857)
3. Width of interocular space subequal to or less than width of eye along hind margin. Brackish pools . *Trichocorixa verticalis* (Fieber, 1851)

4. Front wings free and membranous or reduced to vestigial stumps; hind wings absent, the metathorax with a pair of balancers (halteres). Flies and midges, order Diptera 5
4. Front wings leathery, completely or sometimes only partly covering dorsum, concealing hind wings when at rest. Beetles, order Coleoptera

5. Antennae longer than thorax, consisting of a flagellum of 6 or more free segments 6
5. Antennae short, 3-segmented, the 3d segment with a style or arista (bristle) 15
6. Mesonotum with a V-shaped transverse suture beginning on each side in front of root of wings, the pointed middle part close to the scutellum. Marine algae. Crane flies, fam. TIPULIDAE . . . . *Limonia signipennis* (Coquillett, 1905)
6. Mesonotum without such a transverse V-shaped suture, or suture interrupted 7
7. Costa (thickened leading edge of wing) continuing around wing margin, although often weaker along hind margin; proboscis long and slender. Salt marshes, San Francisco Bay. Mosquitoes, fam. CULICIDAE . . . . . *Aedes squamiger* (Coquillett, 1902)
7. Costa disappearing beyond tip of wing. Intertidal rocks on marine algae. Marine midges, fam. CHIRONOMIDAE 8
8. Pronotum not, or only slightly, notched anteriorly on median line; male antennae usually plumose. Subfam. HYDROBAENINAE . . 9
8. Pronotum widely divided into lateral lobes; male antennae almost bare, never plumose. Subfam. CLUNIONINAE 11
9. Male antenna with only 8 segments, not at all plumose; nonswarming. British Columbia . . . *Camptocladius clavicornis* Saunders, 1928
9. Male antenna with 12 or 13 segments, plumose, the males swarming 10
10. Male antenna 12-segmented. British Columbia . . . . *Camptocladius marinus* Saunders, 1928
10. Male antenna 13-segmented. British Columbia . . . . *Camptocladius pacificus* Saunders, 1928
11. Wings complete; hind tarsi with 2d segment longer than third; all tarsi with 4th segment heart-shaped, the 5th simple or deeply trilobed at tip 12
11. Wings vestigial or straplike; hind tarsi with 2d segment not longer than 3d; all tarsi with 4th segment cylindrical and simple, the 5th simple and never trilobed 14
12. Front legs of male modified, the femora swollen, with an angular projection near apex which interlocks with a basal projection of the tibia; hairs of legs strong, sometimes flattened as appressed scales . . . . *Paraclunio* 13
12. Legs unmodified in both sexes; hairs of legs weak. Point Cabrillo, Mendocino County . . . . *Telmatogeton macswaini* Wirth, 1949
13. Mesonotum, femora, and tibiae with rows of strong bristly hairs;

tarsal claws slender, those of male deeply cleft. Pacific Coast from Alaska to San Diego . *Paraclunio alaskensis* (Coquillett, 1900)

13. Mesonotum, femora, and tibiae with rows of hairs mostly replaced by scales; tarsal claws flattened and broadened, those of male shallowly cleft. Central California coast *Paraclunio trilobatus* Kieffer, 1911

14. Wings straplike, reaching to 4th segment of abdomen; halteres present. Point Lobos . . . . . *Eretmoptera browni* Kellogg, 1900

14. Wings vestigial, not reaching to abdomen; halteres absent (fig. 91A, no. 6). Central California coast . . *Tethymyia aptena* Wirth, 1949

15. Frontal suture or lunule entirely absent; front uniformly sclerotized. Fam. DOLICHOPODIDAE . . . . *Aphrosylus* 16

15. Frontal suture well developed as a horseshoe-shaped groove over the antennae, continuing down to separate the center of the face from the sides; frontal lunule a crescentic sclerite between antennae and frontal suture 18

16. Arista pubescent; posterior cross vein forming an obtuse angle with 4th vein. Pacific Grove . . *Aphrosylus grassator* Wheeler, 1897

16. Arista of antenna naked; posterior cross vein at right angles to 4th vein 17

17. Wing with a black blotch covering distal end of discal cell; fore and hind femora ciliated above. Pacific Grove . . . . *Aphrosylus direptor* Wheeler, 1897

17. Wing without a black blotch; fore and hind femora without cilia. Pacific Grove, Point Loma, La Jolla, British Columbia . . . . *Aphrosylus praedator* Wheeler, 1897

18. No costal fracture near humeral cross vein; 2d basal and anal cells complete. Intertidal rocks on green algae. Fam. CANACEIDAE. Pacific Coast, Washington to Los Angeles . . . . *Canaceoides nudata* (Cresson, 1926)

18. Costal fracture present near humeral cross vein; 2d basal and anal cells not formed. Fam. EPHYDRIDAE. Brine pools, Pacific Coast . . . . *Ephydra gracilis* Packard, 1871

19. First ventral abdominal segment divided at middle by hind coxal cavities. Suborder Adephaga, fam. CARABIDAE. Central California coast . . . . . . *Thallasotrechus nigripennis* Van Dyke, 1918

19. First ventral abdominal segment extending for its entire breadth behind coxal cavities. Suborder Polyphaga 20

20. Tarsal formula 5-5-4 (tarsi of front and middle pairs of legs 5-segmented, tarsi of hind pair 4-segmented). Fam. EURYSTETHIDAE 21

20. Tarsal formula not as above 22

21. Elytra striate. Mendocino coast and Farallon Islands . . . . *Eurystethes fuchsi* (Horn, 1892)
21. Elytra not striate. Marin and San Mateo coasts . . . . *Eurystethes subopacus* Van Dyke, 1918

22. Tarsi 4-4-4 or 4-4-5; thorax without extensile membranous vesicles. Fam. STAPHYLINIDAE 23
22. Tarsi 5-5-5; extensile membranous vesicles on prothorax and between metathorax and abdomen. Fam. MELYRIDAE 26

23. Eyes hairy; mandibles asymmetrically developed and without serrations between apex and median tooth; anterior margin of labrum strongly rounded . . . . *Diaulota* 24
23. Eyes hairless; mandibles symmetrically developed and with a series of distinct serrations between apex and median tooth; anterior margin of labrum truncate . . . . *Liparocephalus* 25

24. Body coloration uniformly dark throughout (fig. 91A, no. 8). Moss Beach . . . . *Diaulota densissima* Casey, 1894
24. At least head and abdominal tip reddish. Moss Beach . . . . *Diaulota brevipes* (Casey, 1894)

25. Head conspicuously broader than prothorax and distinctly broader than long, lighter in color than rest of body. Moss Beach. . . . . *Liparocephalus cordicollis* Leconte, 1880
25. Head and thorax practically equal in width; length of head equal to width or but very slightly greater; of a uniformly dark color (fig. 91A, no. 7). Moss Beach. . *Liparocephalus brevipennis* Mäklin, 1853

26. Elytra nearly 1/2 as long as abdomen, black with pale bases . 27
26. Elytra much less than 1/2 as long as abdomen, of one color throughout 28

27. Abdomen yellow or fuscous. Southern California coast . . . . *Endeodes basalis* (Leconte, 1852)
27. Abdomen black. Southern California coast . . . . *Endeodes abdominalis* (Leconte, 1852)

28. Head black. Pacific Coast . . *Endeodes collaris* (Leconte, 1852)
28. Head reddish. California coast *Endeodes rugiceps* Blackwelder, 1932

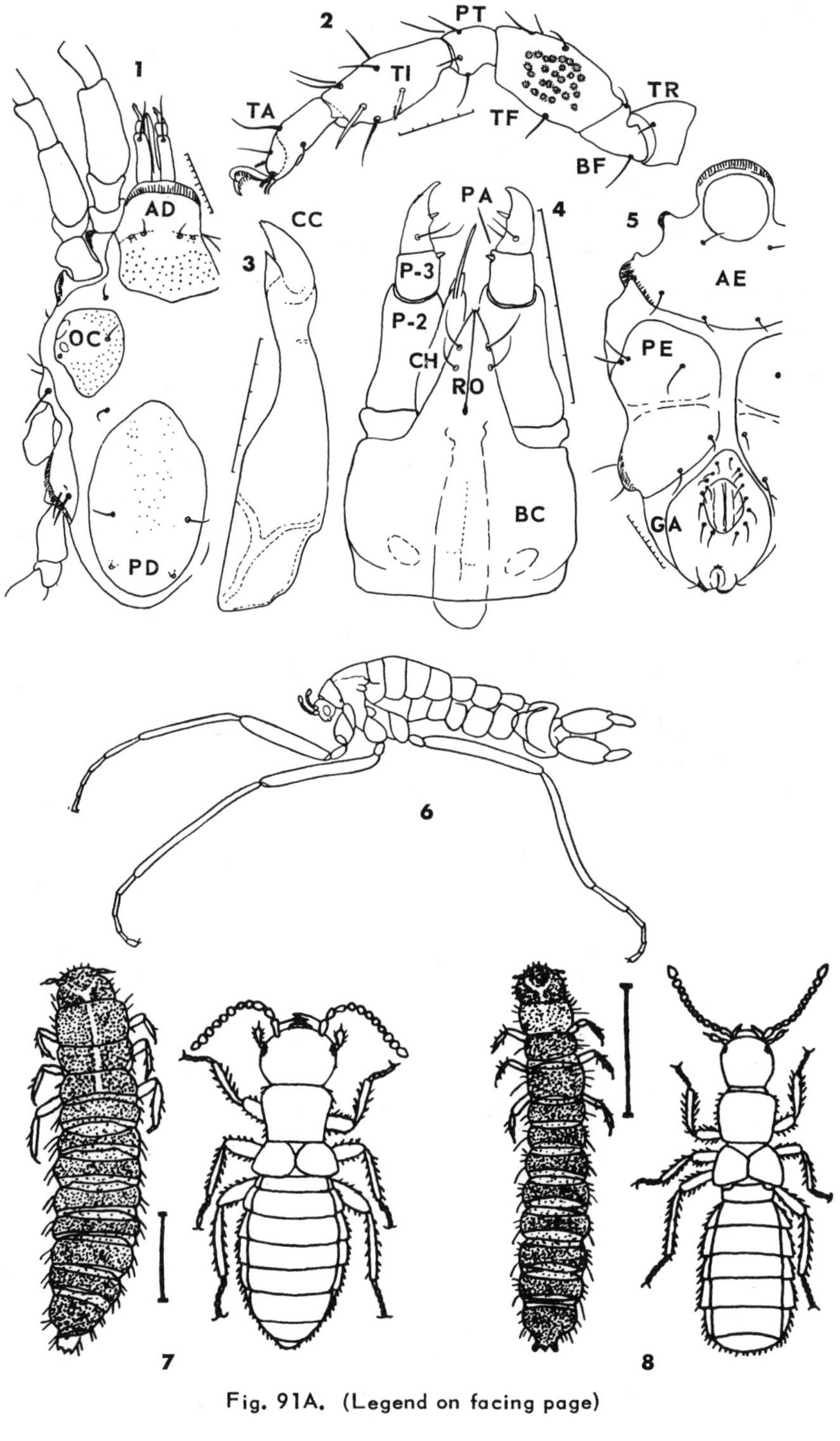

Fig. 91A. (Legend on facing page)

CLASS ARACHNIDA

The class Arachnida falls within the great subphylum Chelicerata, defined on page 112, and includes the orders mentioned below.

Order Scorpionida (Scorpions)

Scorpions are perhaps the most primitive land arachnids, now exclusively terrestrial, although the earliest known scorpions (of the Silurian) were marine. The segmented body is made up of (1) the prosoma, covered by a carapace and bearing the six pairs of appendages characteristic of of arachnids, the chelicerae, pedipalpi modified as powerful chelae, and four pairs of legs; (2) the mesosoma, consisting of seven broad somites, the first of which bears the reproductive openings covered by the genital operculum (seen between bases of last two pairs of legs); the second, the pectines; and the next four, stigmata opening into book lungs; (3) the metasoma, consisting of five segments, and the telson modified as a caudal sting. Mesosoma and metasoma together constitute the opisthosoma.

Order Araneida (Spiders proper)

Although abundant, varied, and widespread, the spiders are not nearly so familiar as they should be. However, their availability, remarkable adaptations, and complex behavior make them especially favorable objects for study. The large garden spider *Argiope* is an excellent type for dissection in class work (described in Brown *et al*, *Selected Invertebrate Types*). Spiders have a cephalothorax and an unsegmented abdomen. There are usually eight simple eyes. Of the six pairs of appendages, the first (chelicerae) is modified as poison fangs. The second (pedipalps) is leglike and tactile in females, but in males is curiously modified for use in transfer of sperm. These are followed by four pairs of walking legs. Well forward on the underside of the abdomen are the openings of the

---

Fig. 91A. 1-5, **Halacaridae.** 1, *Agaue bradypus* Newell, female, dorsum; 2, *Copidognathus curtus* (Hall), male, leg I; 3, *Copidognathus unalaskensis* Newell, chelicera; 4, *Actacarus illustrans*, Newell, capitulum, ventral view; 5, *Agaue bradypus*, female, venter. AD, anterodorsal plate; AE, anterior epimeral plate; BC, base of capitulum; BF, basifemur; CC, cheliceral claw; CH, chelicera; GA, genito-anal plate; OC, ocular plate; P-3, segment 3 of palp; PA, palp; PD, posterior dorsal plate; PE, posterior epimeral plate; PT, patella; RO, rostrum, TA, tarsus; TF, telofemur; TI, tibia; TR, trochanter (Newell, original).

6-8, **Insecta.** 6, *Tethymyia aptena* Wirth, wingless marine midge; 7, *Liparocephalus brevipennis* Mäklin, larva and adult; 8, *Diaulota densissima* Casey, larva and adult; (6 after Wirth, 1949; 7 and 8 after Saunders, 1928).

book lungs, and just posterior to them is the genital aperture. Near the tip of the abdomen is a cluster of three pairs of spinnerets. There are several excellent books on spiders listed in the bibliography. Of these, *The Spider Book*, by Comstock, may be the most generally useful.

Order Phalangida (Harvestmen or Daddy Longlegs)

These are spiderlike, but differ from spiders in that cephalothorax and abdomen are not separated, the chelicerae are chelate (pincerlike), and the abdomen is segmented. Phalangids are easily recognized by their extremely long legs and are commonly encountered under leaves and trash. There seems to be a general lack of information about local species.

Order Solpugida

These animals are characteristic of the warm dry southern deserts, with one species commonly occuring northward along the California coast. They are nocturnal hunters, moving rapidly above ground in search of active prey. During the day they lie beneath logs, rocks, cow chips, or other cover, generally in a shallow burrow. They are tracheate and lack spinning organs. The body appears to be divided into three regions: a large, elongate, segmented abdomen, usually pale buff in color; a mid-region formed from the posterior segments of the cephalothorax; and what at first glance appears to be a large pointed head. On closer observation this "head" will be found to include, anteriorly, the large, closely appressed chelicerae held rigidly before the fused anterior segments of the cephalothorax. The pedipalps are long and nonchelate, terminating in tactile or adhesive structures. The first pair of walking legs is relatively slender and is carried forward, serving as tactile organs. The last three pairs of walking legs are ambulatory in function; this plus the apparent division of the body into three regions gives the solpugids a rather insectlike aspect.

Order Pseudoscorpionida (Chelonethida)

This group includes minute (1–3 mm.) tracheate arachnids that resemble scorpions in general structure but lack the posterior extension of the abdomen, which is segmented, depressed, and somewhat disc-shaped. The pedipalps are relatively enormous pincers. Egg and molting webs are spun by pseudoscorpions, the spinning glands opening at the bases of the chelicerae. Like the solpugids, they are active hunters. Woodland members of the group may be found in leaf mold and under bark. According to J. W. Hedgpeth, the two species most likely to be encountered along the shore are *Garypus californicus* (Banks), which is found in stony beaches above high-tide marks from southern California to Tomales Bay, and *Halobisium occidentale* Beier, under *Salicornia* and logs in marshes from San Francisco to Alaska. Chamberlin's monograph (1931) includes keys to genera.

Order Acarina (Ticks and Mites)

In this widespread group the body is fused into a single ovoid mass, without segmentation or subdivision. The obnoxious bloodsucking ticks are familiar to most people, and the group as a whole contains many species of great economic and medical importance. There are numerous aquatic mites (Hydracarina), especially in fresh water, where they are often conspicuous because of their general activity and brilliant coloration. Baker and Wharton (1952) give a good general account of the order. For identification of fresh-water mites see Ward and Whipple. Marine mites, although very common, are usually passed over in despair by the average invertebrate zoölogist; hence the following discussion and key, representing hitherto inaccessible information, should be of particular value.

The Halacaridae or Marine Mites

by Irwin M. Newell
University of Hawaii

Although the mites inhabiting the seas and the shores of the seas include representatives of the three major groups of Acari (Parasitiformes, Trombidiformes, Sarcoptiformes) one family alone is spoken of as the marine mites proper, namely the HALACARIDAE. This one family has been unusually successful in invading marine habitats and in adapting itself to the numerous niches present there, with the result that twenty-three generically distinct groups now exist. Some of these genera include forms which are phytophagous, others which are predaceous, and at least two have developed truly parasitic habits. The Halacaridae probably originated in the sea, but subsequently have invaded fresh-water habitats where an additional twelve genera or subgenera have developed.

Of the twenty-three marine genera, eleven are known to exist in the northeastern Pacific, and it seems unlikely that the number will increase markedly above this. Many species are still undescribed, however. One of the genera, *Thalassacarus*, is unknown except in this region.

Halacaridae have been found in virtually all marine habitats even down to abyssal depths. Nevertheless, the ecological distribution of most of the species is fairly restricted. Species of *Isobactrus* are usually confined to brackish waters in tide pools or estuaries; species of *Rhombognathus* are encountered only rarely below the tide zone, and there only in small numbers.

Ranging from 200μ to 800μ in body length, these mites are often extremely abundant, generally outnumbering all other intertidal Metazoa, except the Nemathelminthes and Copepoda. A pint of coralline algae from any point along the Pacific Coast may contain hundreds of individuals and as many as ten or a dozen species. Despite their small size, they are a conspicuous and nearly omnipresent element of the marine fauna, and well worth a few hours of the student's time.

Mites are best collected by washing algae, barnacle-encrusted rocks, coarse sand and other materials through successive sieves of about 1,000μ, 500μ, and 150μ mesh, using a strong spray of water. Sea water is best for washing if the mites are wanted alive, but if only fresh water is available under adequate pressure, this may be used and the mites transferred to dishes of sea water by means of a pipette. Intertidal mites will tolerate immersion in fresh water from one to three hours, but longer periods are usually fatal.

For cursory determinations of fresh specimens, it is possible in most cases to do an adequate job of clearing simply by squashing the mite between slide and coverslip, using about a 15 per cent glycerine-water mixture. This will permit storage of the specimen for a few weeks without special mounting procedures. Or they may be cleared in lactic acid and examined in a 15 per cent glycerine solution. Berlese fluid, or Hoyer's modification of it, is simple to use and satisfactory if permanent mounts are not desired. For research purposes the mites should be cleared with enzymes and mounted either in glycerine or Hyrax, following procedures outlined elsewhere (Newell, 1947).

Halacaridae pass through several stages in their development: larva (6-legs); protonymph (usually with a single pair of genital acetabula and femur IV undivided); deutonymph (2 pairs of acetabula and femur IV divided); and adult (genitalia well developed, three pairs of acetabula). The structures referred to in the following key are illustrated in figure 91A, numbers 1-5.

Key to Known Subfamilies and Genera of Marine Halacaridae from the Eastern North Pacific

1. Palpi short, closely appressed to sides of rostrum, either 3- or 4-segmented. Base of capitulum never with a pair of setae behind the insertions of the palpi. Middle piece of claw articulating with carpite, which in turn articulates with or is fused with end of tarsus.[6] Gut

[6]Although this is a difficult character to interpret, the presence of a carpite is the one character which separates all the Rhombognathinae from all the other species of Halacaridae. Oil immersion is often required.

caeca always green or greenish-black in living animals. Predominantly intertidal, and never found under conditions precluding algal growth . . . . RHOMBOGNATHINAE

1. Palpi longer, not especially closely appressed to sides of rostrum, 4-segmented. Base of capitulum usually (not always) with 1 or more pairs of setae behind the insertions of the palpi. Middle piece of claw articulating directly with end of tarsus; carpite absent. Gut caeca usually gray, brown, red, orange, yellow, or lavender. Often found at great depths, but also common intertidally 2

2. Palpi attached to capitulum laterally (fig. 91A, no. 4); bases of palpi separated by an interval much greater than the width of P-1, and clearly visible in ventral view 3

2. Palpi attached to capitulum dorsally, separated by an interval no greater than width of base of P-1; bases of palpi not visible in ventral view (except by transparency). 4

3. Larger forms, more than 300μ in length. Ocular plates well developed and readily visible in dorsal view (fig. 91A, no. 1). Base of capitulum usually (not always) with 1 or more pairs of setae posterior to base of palpi. Habitat varied, rarely arenicolous (sand-dwelling) . . . . HALACARINAE

3. Minute forms, adults less than 300μ in length. Ocular plates rudimentary, scarcely visible in dorsal view. Base of capitulum devoid of setae (fig. 91A, no. 4). Intertidal, arenicolous, and rarely found . . . . ACTACARINAE

4. Palpi 4-segmented. Rostrum about as long as palpi, and opposed to the downwardly directed palpi. Tarsus I essentially like IV in form . . . . LOHMANNELLINAE

4. Palpi 3-segmented. Rostrum much shorter than palpi. Tarsus I elaborately expanded and markedly different from IV . SIMOGNATHINAE

Rhombognathinae Viets, 1927

1. Capitulum readily visible in dorsal view, projecting well beyond end of body. Ocular plates with 2 pairs of setae. With 3 or more setae inserted near lateral margin between insertions of II and III, and clearly visible in dorsal view . . *Rhombognathus* Trouessart, 1888

1. Capitulum hidden in dorsal view in undistorted specimens. Ocular plates devoid of setae. With single pair of setae inserted at margins of body between II and III . . . . . *Isobactrus* Newell, 1947

Halacarinae Viets, 1927

1. Segment 4 (patella) of legs I-IV as long as, or very nearly as long

as, segments 3 and 5 (hitherto unrecorded from the eastern North Pacific) . . . . *Halacarus* Gosse, 1855

1. Segment 4 of legs I-IV only about 1/2 as long as segments 3 and 5 (fig. 91A, no. 2) 2

2. P-3 without a medial seta[7] . . *Copidognathus* Trouessart, 1888

2. P-3 with a medial seta 3

3. Rostrum (measured from base of P-1) not much longer than base of capitulum, and usually shorter 4

3. Rostrum much longer than base of capitulum, the distance from the tip of the rostrum to the base of P-1 more than 1¼ times the distance from the base of P-1 to the posterior margin of the capitulum . . . . *Agaue* Lohmann, 1889

4. Ocular plates each with a heavy, blunt, tail-like projection extending nearly to level of insertion of leg IV. Chelicerae with a heavy bidentate process at base of claw . . *Thalassacarus* Newell, 1949

4. Ocular plates simply rounded or sharply pointed posteriorly. Cheliceral claw without a heavy bidentate basal process 5

5. Combined length of P-3 and P-4 only about 1/2 as great as the length of P-2 (side view of dissected palpi). Capitulum short, never reaching to end of I-3. Leg I often with blunt, spike-shaped setae on segments 3, 4, and 5 (hitherto unrecorded from the eastern North Pacific) . . . . *Agauopsis* Viets, 1927

5. Combined length of P-3 and P-4 very nearly as great as the length of P-2 (side view of palpi). Capitulum longer, reaching well beyond end of I-3. Leg I with blunt, heavy setae only on segment 5 if present at all . . . . *Thalassarachna* Packard, 1871

Actacarinae Viets, 1939

A single genus, *Actacarus* Schulz, 1936, known from a single locality in the North Pacific: Dutch Harbor, Unalaska Island, Alaska..

Simognathinae Viets, 1927

A single genus, *Simognathus* Trouessart, 1889, known from Alaska and western Mexico (hitherto unrecorded).

Lohmannellinae Viets, 1927

A single genus, *Lohmannella* Trouessart, 1901, known from the Northeastern Pacific.

[7] In *Copidognathus pseudosetosus* Newell 1949, *C. styracifer* Newell 1951, and *C. imitator* Newell 1951, there is a small, sharp spine medially on P-3. But there is no alveolus, and it is not a seta.

CLASS PYCNOGONIDA

by Joel W. Hedgpeth
Scripps Institution of Oceanography, La Jolla

The Pycnogonida are a group of arthropods formerly considered an "appendix" to the Arachnida or a division of the old-fashioned but unnatural Arachnoidea. They are best considered a class of the Arthropoda, affiliated to the chelicerate stem. Although their phylogeny is obscure, pycnogonids constitute a characteristic element of the marine fauna. In some areas they are so common that they must be considered prominent members of the fauna of that locality. Hundreds of specimens of a single species have been taken in dredge hauls, especially in Arctic waters, and more than 10,000 specimens of one species were collected from buoys near the Golden Gate by a fouling survey in recent years. Yet so closely do these animals resemble the hydroids, which are their favorite haunts, that many shore collectors never notice them.

The following keys are intended to aid in the identification of those species which may be found in the intertidal zones from Oregon

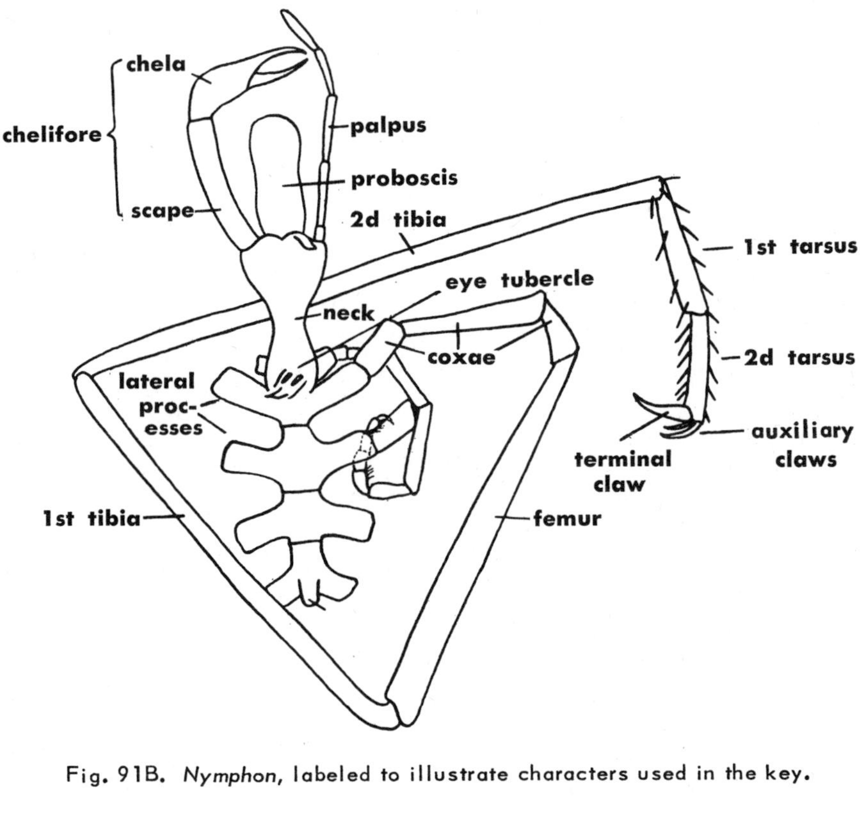

Fig. 91B. *Nymphon,* labeled to illustrate characters used in the key.

to Point Concepcion. As with so many other elements of the intertidal fauna, there is a conspicuous break at Point Concepcion, and many of the species common in the Los Angeles–San Diego area will not be found along the central coast. In order to convey an idea of the variation and classification of the Pycnogonida, the keys to the intertidal species are preceded by a key to the families.

The characters used in this key should be visible under the medium power of a dissecting microscope, although for many of the smaller species it is difficult to make out details because of the debris adhering to the surface of the animal. This is particularly true of the spiny forms, and in some cases it will be necessary to brush the animal off with a camel's hair brush, perhaps even to pick at it with a dissecting needle. Specimens are best killed for study purpose by placing them for a few minutes in fresh water, which causes them to die in an extended condition. They are strongly thigmotactic, and should be kept separately to prevent them from becoming inextricably tangled. For terms used in the keys, see figure 91B.

Key to the Families of Pycnogonids

1. Both chelifores and palpi present 2
1. Chelifores or palpi, or both, lacking or *greatly* reduced . . . 8

2. Chelifores and palpi both well developed 3
2. Chelifores or palpi, but not both, reduced 6

3. Palpi 17–20 jointed, first pair of legs with 12 or 13 secondary joints . . . . AMMOTHEIDAE (p. 205)
   Nymphonella, a Japanese form parasitic in venerid clams.
3. Palpi not more than 10-jointed; legs 8-jointed; usually with a terminal claw 4

4. Palpi 5-jointed, chelae well developed, or palpi 6–9 jointed, chelae small or rudimentary 5
4. Palpi 9–10 jointed; 5 or 6 pairs of legs . . COLOSSENDEIDAE
   Decolopoda, Dodecolopoda, antarctic genera.

5. Palpi never more than 5-jointed; chelae well developed, over-reaching proboscis; 4 or 5 pairs of legs . . . . . . NYMPHONIDAE
   Subtidal to moderate depths, a few abyssal. Rarely intertidal in northern latitudes.
5. Palpi 6–9 jointed; chelae small, chelifores usually shorter than proboscis . . . . AMMOTHEIDAE (p. 205)

6. Chelifores 2–3 jointed, chelae well developed; palpi 1–4 jointed . . . . .PALLENIDAE (p. 203)

6. Chelifores present, but chelae reduced to knobs; palpi 4–10 jointed 7

7. Chelifores 1–3 jointed, shorter than proboscis, chelae reduced or absent, palpi usually more than 7-jointed; body circular or oval in outline. . . . . AMMOTHEIDAE (p. 205)
7. Chelifores 1–2 jointed; palpi never more than 7-jointed; always compact, circular forms, never oval . . TANYSTYLIDAE (p. 209)

8. Chelifores or palpi lacking, but not both 9
8. Both chelifores and palpi lacking 12

9. Chelifores present, palpi lacking 10
9. Chelifores lacking, palpi present 11

10. Ovigers 10-jointed in both sexes . . . . PALLENIDAE (p. 203)
10. Ovigers less than 10-jointed; in ♂ only PHOXICHILIDIIDAE (p. 204)

11. Palpi 4–7 jointed; small forms . . . TANYSTYLIDAE (p. 209)
11. Palpi 8–10 jointed; chelifores sometimes persistent but usually deciduous; mostly huge deep water forms . . . COLOSSENDEIDAE

12. Body slender, legs about twice as long as body, auxiliary claws present. . . . . ENDEIDAE
    Not represented on California coast.
12. Body stout, legs short, not much longer than body; auxiliary claws absent or deciduous; 4-5 pairs of legs . PYCNOGONIDAE (p. 209)

Keys to Pycnogonids of the Central California Intertidal

Pallenidae

1. Trunk compact, legs comparatively short and stout 2
1. Trunk extended, legs long and slender 3

2. Trunk and legs spinose; propodus not subchelate . *Pseudopallene*
    A boreal genus, not represented south of British Columbia.
2. Trunk and legs not conspicuously spinose; propodus with large central spine opposable to terminal spine, forming a subchelate process (fig. 92, *b*) . . . . *Decachela discata*
    The type locality for this curious little species is Monterey. It has also been found off the western coast of Hokkaido.

3. With a small, 4-jointed palpus. . . . . *Oropallene*
    O. palpida Hilton is reported from Santa Monica, but the genus is not yet known north of that locality.

3. Without palpus . . . . *Callipallene*

   Also a southern genus on the California coast; *C. californiensis* (Hall) from Laguna Beach is the only intertidal species.

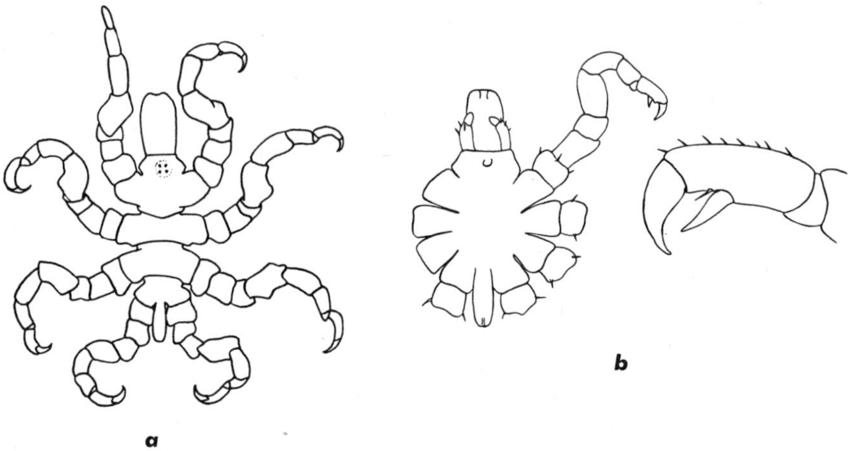

Fig. 92. PYCNOGONIDAE and PALLENIDAE. *a, Pycnogonum stearnsi; b, Decachela discata* and claw, from the holotype (Hedgpeth, original).

Phoxichilidiidae

1. Cephalic segment extended forward as a conspicuous neck, overhanging insertion of proboscis; auxiliary claws minute (fig. 93, *d*) . . . . *Anoplodactylus erectus*

   This genus is characteristic of subtropical waters; the only California intertidal species, *A. erectus*, is common in the Balboa Bay region and has been reported as far north as Monterey.

1. Cephalic segment not extended forward 2

2. Lateral processes separated by at least 1/2 their own diameter; auxiliary claws small but well developed . . . . *Phoxichilidium femoratum*

   This is the only true intertidal species of this genus to be found in central California. It is also a widely distributed boreal species, occurring on both sides of the North Atlantic and in Alaskan waters. A closely similar species, *Phoxichilidium quadridentatum* Hilton (fig. 93, *c*), occurs in tremendous numbers on buoys off the Golden Gate, but has not yet been found on the shore. In *Ph. quadridentatum* there are 4 spines at the heel of the propodus, 2 large ones side by side and a pair of smaller ones; in *Ph. femoratum* the spines are placed in a single row and there may be 4–5.

2. Lateral processes separated by less than their own diameter (trunk usually circular); auxiliary claws minute, inconspicuous . *Halosoma*

   Halosoma viridintestinale Cole (fig. 93, *a*) the genotype, is common in Tomales Bay where it is the most abundant and characteristic pycnogonid of shallow water, and at Monterey. It is a small, delicate form, conspicuous by virtue of its bright green intestines, which branch out to the legs. Another species, described from Pacific Grove, *H. compactum* Hilton (fig. 93, *b*), is very similar but has a much broader and blunter eye tubercle and abdomen and a stubbier propodus. There are no other species on the Pacific Coast.

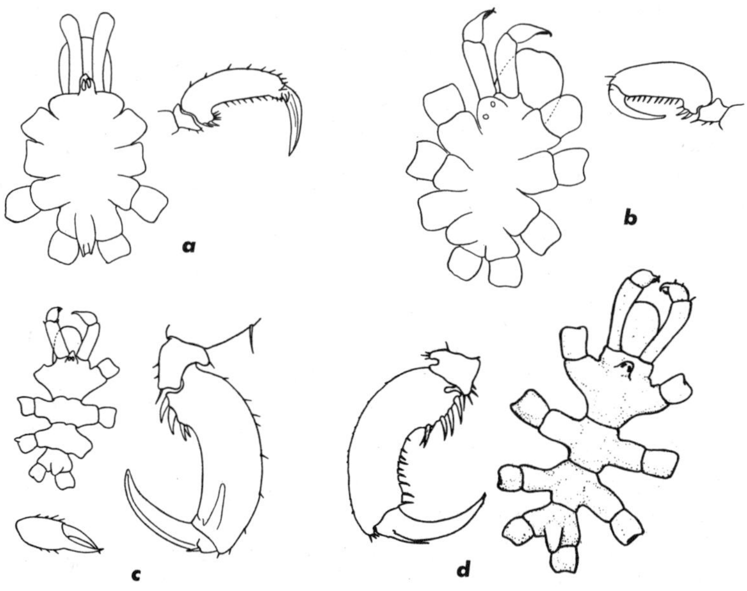

Fig. 93. PHOXICHILIDIDAE. *a, Halosoma viridintestinale* and claw; *b, Halosoma compactum* and claw, from the holotype; c, *Phoxichilidium quadridentatum*, chela, and claw; *d, Phoxichilidium femoratum* and claw (*a-c*, Hedgpeth original; *d*, body, after Hedgpeth, 1948; *d*, claw, after Lebour).

Ammotheidae

1. Proboscis not 2-jointed 2
1. Proboscis 2-jointed . . . . *Eurycyde*

   This genus is represented on the southern California coast by *E. spinosa* Hilton.

2. Chelifores 1-jointed (may be reduced to an almost imperceptible stump), lateral processes well separated 3
2. Chelifores 2–3-jointed; trunk usually compact, disc-shaped, or lateral processes narrowly separated 4
3. Chelifore an easily recognized, papillate process; eye tubercle and eyes present; legs long, slender; a medium-sized (18–20 mm. extent) form with conspicuous brownish purple bands on the legs (fig. 94, *j*) . . . . *Lecythorhynchus marginatus*

   This common species, living among hydroids and in sheltered crevices, is one of the characteristic species of the central California intertidal.
3. Chelifores reduced to almost imperceptible stumps; eye tubercle and eyes absent; legs short; a minute form (2-3 mm. extent) living beneath the surface of coarse sand in the sheltered coves (fig. 95) . . . . *Rhynchothorax philopsammum*

   So far known only from the inner side of Tomales Point, where it occurs several inches beneath the surface of the sand in association with several other forms, including harpacticoids, small holothurians and isopods.
4. Chelifores 3-jointed 5
4. Chelifores 2-jointed . . . . . *Achelia* (=*Ammothea* auct.) 8

   Ammothea, type of the family, is an Antarctic genus
5. Small species, without prominent spinose tubercles on legs . . 6
5. Conspicuous (trunk, 10 mm.) form with prominent spinose tubercles, especially on tibial joints of legs (fig. 94, *f*). From Marin County to Laguna Beach . . . . *Nymphopsis spinosissima*
6. Trunk rather delicate, with lateral processes narrowly separated and long, pointed dorsal tubercles along mid-line; chelifores with blunt, spurlike processes in addition to spines 7
6. Trunk compact, circular; chelifores rather thick, without spurs or processes; dorsal trunk tubercles rounded (fig. 94, *c*). Marin County to Laguna Beach . . . . *Ammothella tuberculata*
7. Propodus about 4 times as long as its dorsoventral width without large basal spines at "heel" but with regular series of large spines along "sole", auxiliary claw 3/4 as long as terminal (fig. 94, *i*) . . . . *Ammothella setosa*

Fig. 94. AMMOTHEIDAE. *a*, *Achelia gracilipes*; *b*, *Achelia nudiuscula*, male; *c*, *Ammothella tuberculata*; *d*, *Achelia spinoseta* with palp and claw at right; *e*, *Achelia simplissima* with palp and claw at left; *f*, *Nymphopsis spinosissima*; *g*, *Achelia chelata* viewed from front; *h*, same, from above; *i*, *Ammothella setosa* with claw, from the holotype; *j*, *Lecythorhynchus marginatus*; *k*, *Achelia echinata* with claw; *l*, *Ammothella menziesi* with claw, from the holotype; (b, from Hall 1913; *g*, *h*, from Hedgpeth, 1940; *d*, *e*, *i*, *k*, *l*, Hedgpeth, original).

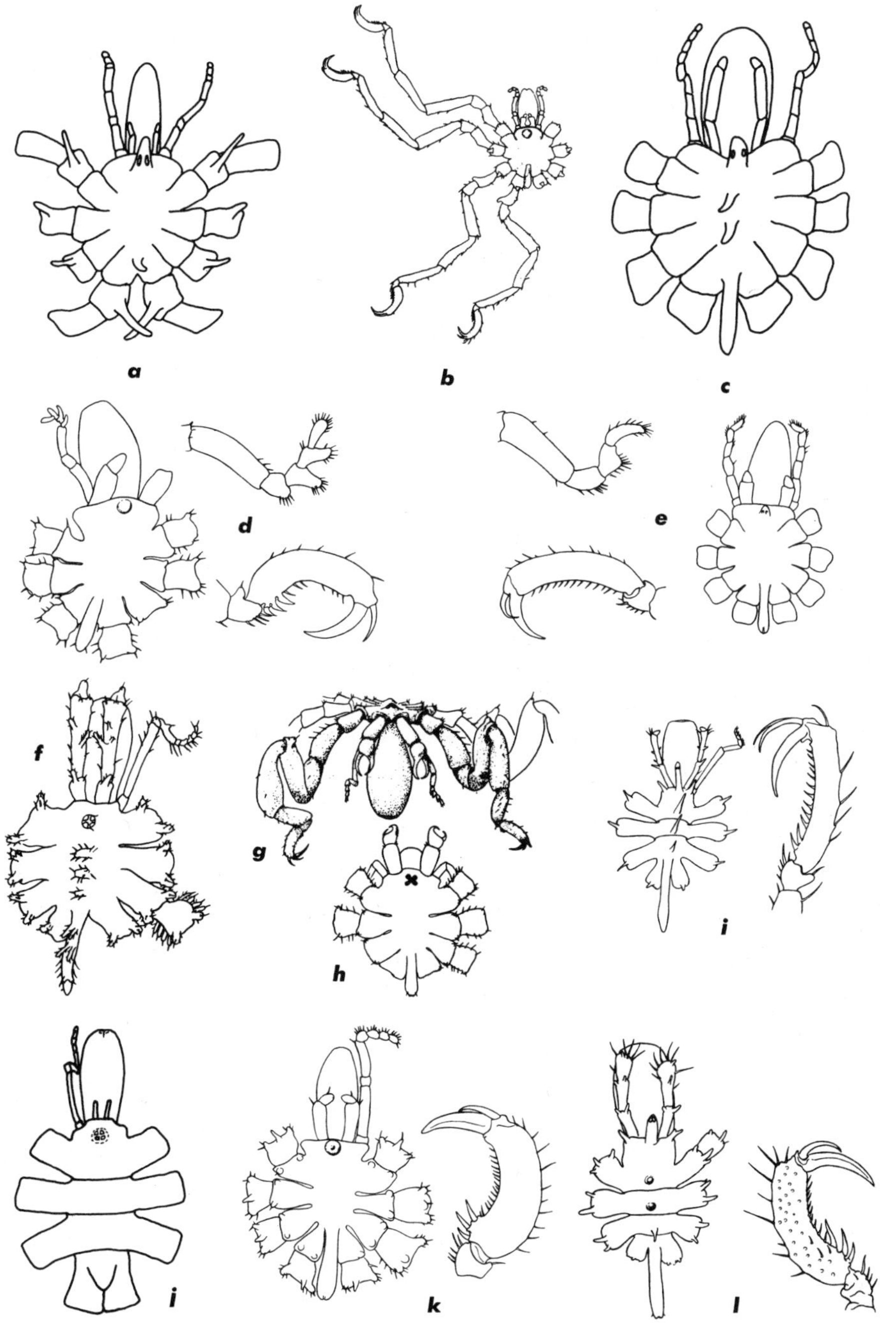

Fig. 94. AMMOTHEIDAE. (Legend on facing page)

7. Propodus about 3 times as long as wide, heel well developed, with 3 large basal spines; auxiliary claw 1/2 as long as terminal claw (fig. 94, *l*). Dillon Beach . . . . *Ammothella menziesi*

   A common species from Southern California, *A. bi-unguiculata* (Dohrn), distinguished by the absence of a terminal claw on the legs, was originally described from the Bay of Naples, and is also found in Hawaii, Japan, and Western Australia.

8. Chela of adult not chelate 9
8. Chela of adult chelate (fig. 94, *g*, *h*). On *Bugula*, Pescadero; Moss Beach, San Mateo County . . . . *Achelia chelata*

9. Trunk and legs conspicuously spinose, with small spinose tubercles on first coxae 10
9. Not so spiny, without spiny processes or knobs on first coxae, but some species with fingerlike spurs 11

10. Terminal segments of palpi with ventral lobes, i.e., "pectinate" (fig. 94, *d*). Marin County; Moss Beach, San Mateo County . . . . *Achelia spinoseta*
10. Palpi without ventral lobes, but subcylindrical (fig. 94, *k*). Mile Rock, San Francisco . . . . *Achelia echinata*

    This species, or closely related varieties, is reported from the Atlantic and Mediterranean coasts of Europe and the Northwestern Pacific.

11. First coxae with fingerlike dorsal processes; propodus with basal spines; palpi 8-jointed 12
11. First coxae without such processes; no large basal spines on propodus; palpi 7-jointed (fig. 94, *e*). Dillon Beach . . . . *Achelia simplissima*

12. Processes on coxae less than 1/2 as long as the joint; proboscis broadly elliptical (fig. 94, *b*). San Francisco Bay. . . . .*Achelia nudiuscula*

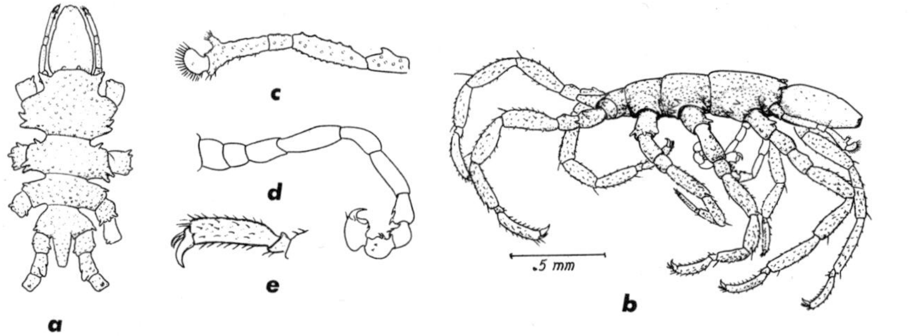

Fig. 95. AMMOTHEIDAE (cont.): *Rhynchothorax philopsammum* (female). *a*, trunk in dorsal view; *b*, whole animal from right; *c*, palp; *d*, ovigerous leg; *e*, tarsus, propodus, and terminal claw (from Hedgpeth, 1951).

12. Processes 3/4 as long as joint; proboscis narrowly elliptical (fig. 94, *a*). Alaska to San Francisco . . . . . . *Achelia gracilipes*

Tanystylidae

1. Proboscis tapered to subconical point 2
1. Proboscis rounded; trunk very compact. Washington to Pacific Grove . . . . *Tanystylum occidentalis*

2. With prominent basal spines on heel of propodus 3
2. Without prominent basal spines on heel of propodus; small delicate form. Central California to San Diego . . *Tanystylum intermedium*
   Characteristically a southern California species.

3. Basal spines of propodus 2, angular; abdomen longer than last pair of lateral processes (fig. 96, *b*) . . . *Tanystylum duospinum*
3. Three straight or slightly curved basal spines; abdomen about as long as last lateral processes (fig. 96, *a*). Conspicuous brown knobby form with white articulations. Marin County to Pacific Grove, especially on *Aglaophenia* . . . . *Tanystylum californicum*
   This is one of the characteristic species of the central California intertidal.

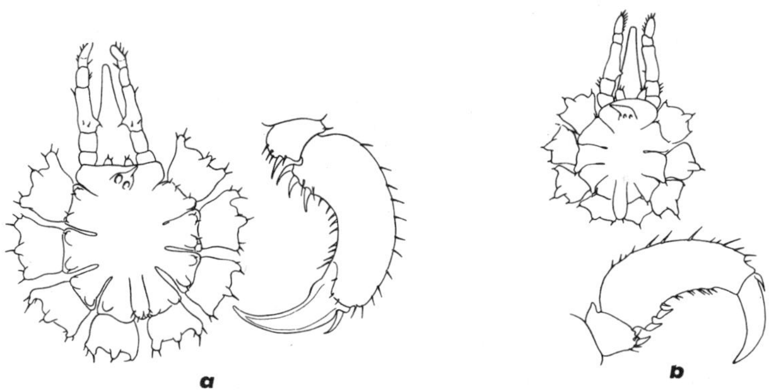

Fig. 96. TANYSTYLIDAE. *a, Tanystylum californicum* and claw; *b, Tanystylum duospinum* and claw, from the holotype (Hedgpeth, original).

Pycnogonidae

1. Body granular, dorsal tubercles inconspicuous or absent (fig. 92, *a*). The most commonly collected intertidal species, often on *Anthopleura, Metridium,* or *Aglaophenia.* Body of female 6 mm. long, male smaller . . . . *Pycnogonum stearnsi*
1. Body reticulated; prominent dorsal trunk tubercles which are higher than eye tubercle; male 6-7 mm. long. Subtidal to intertidal, wharf

piling and anemones and among hydroids. Marin County to Monterey . . . . *Pycnogonum rickettsi*

List of Pycnogonids

Achelia chelata (Hilton, 1939) (=*Ammothea euchelata* Hedgpeth, 1940)
Achelia echinata Hodge, 1864
Achelia gracilipes (Cole, 1904)
Achelia nudiuscula (Hall, 1913)
Achelia simplissima (Hilton, 1939)
Achelia spinoseta (Hilton, 1939)
Ammothella bi-unguiculata (Dohrn, 1881)
Ammothella menziesi Hedgpeth, 1951
Ammothella setosa Hilton, 1942
Ammothella tuberculata Cole, 1904
Anoplodactylus erectus Cole, 1904
Callipallene californiensis (Hall, 1913)
Decachela discata Hilton, 1939
Eurycyde spinosa Hilton, 1916
Halosoma viridintestinale Cole, 1904
Halosoma compactum (Hilton, 1939)
Lecythorhynchus marginatus Cole, 1904
Nymphopsis spinosissima (Hall, 1912)
Phoxichilidium femoratum (Rathke, 1799)
Phoxichilidium quadridentatum Hilton, 1942
Pycnogonum rickettsi Schmitt, 1934
Pycnogonum stearnsi Ives, 1892
Rhynchothorax philopsammum Hedgpeth, 1951
Tanystylum californicum Hilton, 1939
Tanystylum duospinum Hilton, 1939
Tanystylum intermedium Cole, 1904
Tanystylum occidentalis (Cole, 1904)

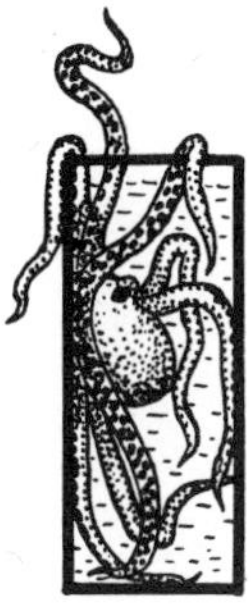

Phylum Mollusca

Among all the invertebrates, the Mollusca rank second only to the Arthropoda as an interesting and diverse group. They have adapted themselves to almost all habitats, marine, fresh-water, and terrestrial, from the 35,000 foot depths of the Philippine Trench to mountain masses 15,000 feet above the sea, from the tropics to the Arctic, from the surface of the open oceans to barren deserts. Since earliest times, man has used them for food, as witness the great prehistoric shell heaps (kitchen middens) in all parts of the world, for tools and weapons, for money and jewelry. Their shells have appealed to lovers of beauty as much as to scientists and to the great and indefatigable fraternity of conchologists. References to molluscs pervade our legends, our literature, and our everyday speech; they provide such symbols as the snail for slowness, the clam for silence, the pearl for virtue, and the octopus for greed; they have been the livelihood of the poor, and have dyed the robes of kings.

Molluscs apparently arose from a stem that also produced the annelids, as evidenced by the extremely similar pattern of "spiral cleavage" in the eggs of the more primitive members of these phyla. In each group a larva of the trochophore type is produced, after which the development takes its characteristic course. The segmented annelids develop an elongated body, an extensive coelom, and a muscular body wall provided with setae. But the molluscs reduce the coelom to a small pericardium and associated structures and develop a compact body, divisible into head, foot, and visceral mass, the latter covered by a mantle which often secretes a shell, and which overhangs the sides and rear of the body to form a pallial chamber or mantle cavity. In this space are found respiratory organs of unique form, known as ctenidia (singular, ctenidium), which best show the primitive aspect in the lower gastropods (order Aspidobranchia); here each ctenidium has the form of a long axis bearing two lateral rows of flat plates (filaments), set closely together. On the neighboring faces of these filaments, cilia move water to create a respiratory current. In different groups of molluscs we may find the ctenidia variously modified, especially for feeding in the lamellibranchs, or lost altogether as in nudibranchs and pulmonates. The study of the ctenidia and associated organs provides one of the most illuminating means of appreciating the adaptations of molluscs to diverse ecological situations. The most important work in this field is that of Yonge, whose paper on the pallial organs of the gastropods and other molluscs (1947) should be read by any serious student of the Mollusca.

Molluscs apparently arose as creeping types, probably living on hard surfaces and scraping up their food by means of a unique organ, the radula, which is found in all classes except the lamellibranchs. The latter have extensively modified their ctenidia for the filtration of microscopic food material from the water and have lost all semblance of a head. Further details are discussed in relation to the various classes and orders. There is no single good, modern, comprehensive work on Mollusca in general, although the literature is vast. Pelseneer (1906), though rather outdated, is still very useful. There are numerous up-to-date works on conchology, listed in the bibliography, as well as many excellent works on individual species and groups. The check list of Smith and Gordon (1948) is exceedingly useful to the general collector in this area, and the charmingly written *West Coast Shells* by Keep and Baily (1935) is an excellent guide for the beginner.

CLASS SCAPHOPODA

The small class of the scaphopods (tooth shells) is not found in the local intertidal, but *Dentalium* (fig. 97) occurs abundantly in soft bottom at a depth of several fathoms in the Puget Sound region. *Dentalium* possesses a tubular tapering shell open at both ends, an extensible lobed foot for burrowing, a small radula, a mass of slender capitate oral tentacles (captacula) used for feeding, and powerful bands of cilia within the mantle cavity, used for maintaining a water current for respiration and possibly for feeding. There are no ctenidia, but the rectum apparently serves as a water lung, periodically taking in and expelling water.

CLASS CEPHALOPODA

The cephalopods are represented in the local intertidal by one species of octopus, probably *Octopus apollyon* (Berry, 1912), a form that may or may not be identical with *O. hongkongensis.* Small individuals are frequently found under boulders and in crevices. In addition, larger octopi are occasionally encountered intertidally or are washed in; these are probably of other species. Along the southern California shores the two-spotted octopus population has recently been found to comprise two very similar species, one the *Octopus bimaculatus* Verrill, 1883, and the other, newly recognized, *Octopus bimaculoides* Pickford and McConnaughey, 1949. Interesting accounts of the reproduction of octopods are given by Fox (1938) and Fisher (1923, 1925).

Although rarely encountered alive in the intertidal, the squid, *Loligo opalescens* Berry, 1911, is extremely abundant in coastal waters and is fished commercially off Monterey, the annual catch being more than 5,000 tons. Occasionally vast numbers of the gelatinous, fingerlike, egg masses are washed ashore. Squid of other genera are occasionally brought in.

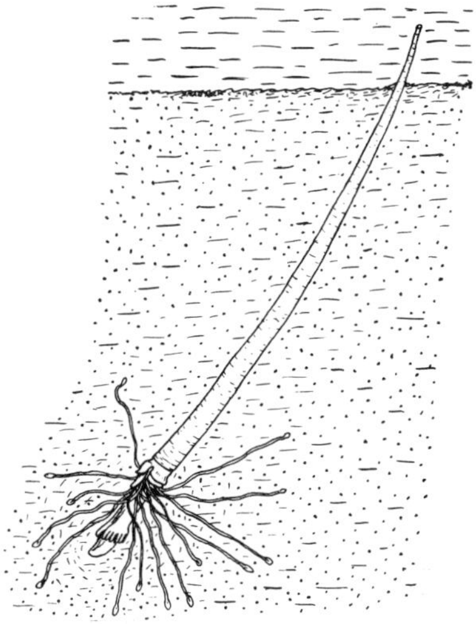

Fig. 97. Scaphopoda, *Dentalium* sketched at Friday Harbor in act of feeding, buried in silt against side of aquarium. The fringed foot serves in burrowing and a mass of captacula (the majority not shown) extend themselves actively into the loosened soil about the anterior end of the body.

CLASS AMPHINEURA

The familiar chitons, order Polyplacophora or Loricata, are common and interesting members of the intertidal community. The various species show distinct ecological preferences and furnish excellent material for studies of intertidal zonation. Chitons are all marine and are restricted to hard surfaces, to which they are well adapted for clinging by the broad foot. The low form of the body makes them capable of withstanding wave shock; the flexible edge of the mantle allows them to fit very closely the contours of rough surfaces, and so to lessen the rate of desiccation when exposed by the tides. The jointed shell permits many chitons to roll up for protection if dislodged. Chitons in general are not well adapted to cope with silt and suspended matter which may tend to clog their ctenidia. This fact, together with their high degree of adaptation to rock surfaces, suggests why they are most successful on rocky coasts with clear water. Chitons as a group are most abundant in the intertidal or in shallow water, although a few occur in very deep water.

In contrast to the Polyplacophora, the members of the shell-less order Aplacophora or Solenogastres are found only at depths greater than 90–100 feet, generally in oozy bottoms, or creeping over corals and hydroids, upon which they feed. They are small and wormlike, with a mantle beset with spicules which encases the entire body. The foot is reduced. A radula is present in most; absent in some. There is a small posterior mantle cavity, which in some members of the group contains a pair of ctenidia. The shell is entirely lacking in the adult, but in some, at least, the larva is said to develop a series of seven calcareous plates made up of close-set spicules. Since solenogastres are rarely encountered, we shall confine our attention to the abundant chitons.

The characters used in the identification of chitons have chiefly to do with the eight-valved shell (fig. 98), the width and ornamentation of

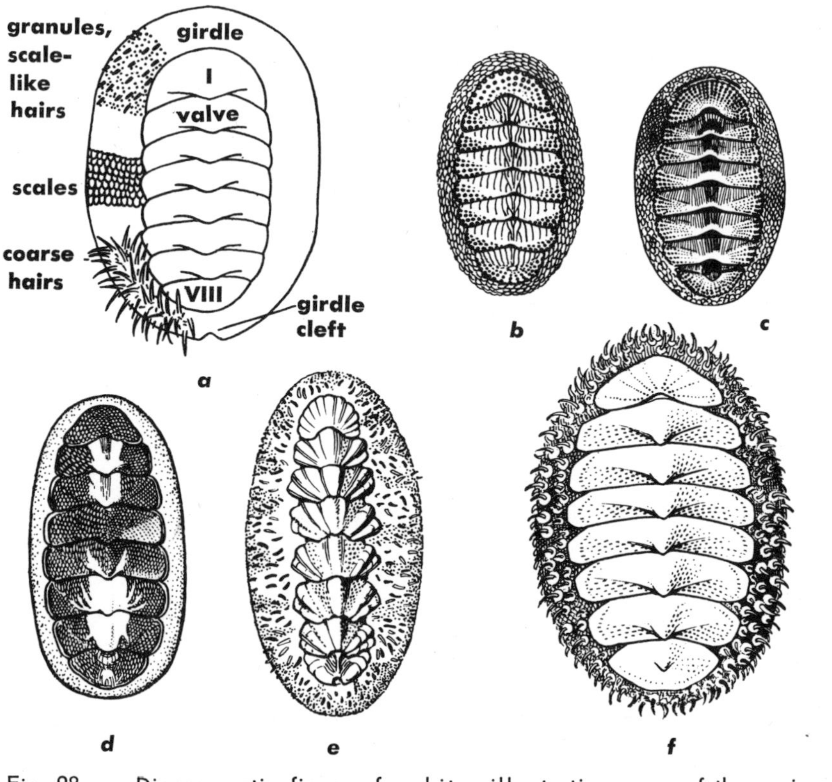

Fig. 98. *a*, Diagrammatic figure of a chiton illustrating some of the various types of girdle ornamentation.; *b*, *Ischnochiton mertensii*; *c*, *Ischnochiton cooperi*; *d*, *Cyanoplax dentiens*;, *e*, *Nuttallina californica*; *f*, *Mopalia lignosa*.

the muscular integumental area, known as the girdle, between the shell and the margin, and with the proportions of the body.

Each valve consists typically of two layers: a basal and more extensive articulamentum which extends under the other plates anteriorly and is more or less covered by the girdle laterally; and the upper tegmentum, often sculptured, consisting of the exposed part of the valve. In *Cryptochiton* the valves are entirely covered by the girdle, and the tegmentum is lacking. The valves overlap posteriorly; the part of a valve underlying the one in front is also often designated the articulamentum, and that part covered by the girdle, the insertion plate. The insertion plates may bear notches which are usually said to be of systematic value, a statement that does not make clear their value, if any, to the animal. The girdle varies markedly in width and proportions. It may be rough or smooth, and variously beset with spicules in the form of granules, short bristlelike rods, spines, or imbricated scales. Hairs of diverse form, stiff or flexible, simple or branched, coarse or minute, may be present also. In *Cryptochiton* the girdle covers the shell completely except in very young animals, a condition approached in *Katharina*, which has only a small area of each valve exposed. The edges of the mantle (girdle) overhang the body of the animal on all sides. The head bears simple sensory lappets but no definite tentacles, and, as might be expected from the fact that it cannot be protruded from beneath the girdle, the head lacks eyes. Light sensitivity is often present, however, the receptors consisting of tiny eyelike organs, esthetes, set in pits in the tegmentum of the valves and best seen in young, uneroded individuals. The space between the girdle and the margin of the foot on either side of the body houses numerous ctenidia, from several to many pairs. These are provided with cilia that maintain a respiratory flow of water as described by Yonge (1939). Water enters anteriorly beneath the mantle edge, which can be lifted at suitable points. The water streams converge behind the anus and flow out as a single posterior stream. This point may be marked by a distinct girdle cleft, especially noticeable in *Mopalia hindsii*.

Although the commoner chitons are fairly easy to recognize, the identification of some members of the group presents difficulties because of variation in color, ornamentation, and erosion of the valves. In collecting, the use of "chiton sticks" is advisable to prevent the specimens from curling up. These are simply strips of wood or plastic a few inches or more in length and of various widths to accommodate chitons of different sizes. To one end of each stick is attached a strip of cloth tape, up to a few feet long. As each chiton is collected, it is placed with its foot against the stick and bound down firmly by a couple of turns of the wet tape, secured by an overhand knot. Care must be taken to flatten the animal without crushing the sometimes delicate valves. By using both sides of each stick a number of chitons can easily be accommodated. Before preservation, the chitons may be killed and stiffened by immersing them, still bound to the stick, in hot water (60° C or 140° F), the time depending upon the size of the specimens. After killing (not cooking) the chitons, they may be preserved in alcohol (formalin corrodes the valves and destroys the spicules). Preservation by drying is also possible but not very satisfactory. The specimens should be soaked in alcohol or fresh water, the foot and viscera removed, and the shell and mantle rebound to a flat surface for drying. One suitable method of preserving chitons for general display, a method that preserves colors well and renders the specimen indestructible, is embedding them in plastic. A general account of the procedure has been published by Ward's Inc. (1950).

Key to the Chitons

1. Girdle completely covering valves, thick, red, gritty; large (6-12 in.) . . . . *Cryptochiton stelleri*
1. Girdle 2–3 times as wide as exposed part of valves; girdle smooth and black . . . . *Katharina tunicata*
1. Exposed part of valves at least as wide as girdle 2

2. Valves marked with conspicuous zigzag lines of color; girdle smooth . . . . *Tonicella lineata*
2. No such lines; girdle otherwise 3

3. Girdle covered with hairs 4
3. Girdle covered with scales like coat of mail (scales minute in *Ischnochiton fallax*) 9
3. Girdle with minute granules, short bristles, or minute papillae . 16

4. Hairs coarse and stiff 5
4. Hairs lighter, flexible 6

5. Oval in shape; girdle coarsely mossy, narrowed at ends; valves often corroded, with dorsal ridge, ornamented with nodules in lines . . . . *Mopalia muscosa*
5. Nearly circular; much elongated in front, with scattered hairs; valves short and broad, without nodules . . . . *Placiphorella velata*

6. Girdle conspicuously cleft posteriorly. 7
6. Girdle not conspicuously cleft. 8

7. Girdle clothed with curling, straplike brown hairs which bear, at least near their bases, a bunch of minute, white, acute spines . . . . *Mopalia ciliata*

7. Hairs of girdle not bearing white acute spines; girdle clothed with flexible, threadlike hairs . . . . *Mopalia hindsii*

8. Valves sculptured, central areas with longitudinal beaded lirae, lateral areas with 5–7 radiating rows of distinct, clear-cut tubercles; color red, orange, olive-ashen or yellow; rear valve generally black or very dark; girdle sparsely clothed with short hyaline hairs . . . . *Chaetopleura gemma*

8. Valves smooth, greenish-gray with brown lines radiating from apices; girdle hairs thick at base, slender and incurved distally, each rising in a light spot; underside of girdle orange; (fig. 98, *f*) . . . . *Mopalia lignosa*

9. Central areas of valves set with raised riblets; lateral areas raised and sculptured with radiating rows of nodules, radiating lirae, or strong ribs 10

9. Central areas smooth, lacking raised riblets (but may have fine threads suggesting riblets); lateral areas slightly raised, but lacking nodules and ribs 13

10. Anterior valve with 7–9 strong ribs with or without shallow median groove 11

10. Anterior valve lacking 7–9 strong ribs 12

11. Anterior valve with 7 very strong ribs, each divided by shallow median groove; lateral area greatly elevated, not split by median sulcus . . . . *Callistochiton crassicostatus*

11. Anterior valve with up to 11 strong ribs, the outer 2 joined; ribs of lateral areas split by median sulcus . . *Callistochiton palmulatus*

12. Red to dull brown; occasionally speckled and blotched with white; nodules and ribs of valves and scales of girdle conspicuous (fig. 98, *b*) . . . . *Ischnochiton mertensii*

12. Olivaceous to gray; nodules, ribs, and scales less conspicuous, (fig. 98, *c*). . . . . *Ischnochiton cooperi*

12. Lateral areas with 3–4 granulose radiating lirae, crossed with brown and white lines giving wavy appearance; ribbing of central area not parallel to dorsal ridge in lateral-most part; color greenish . . . . *Ischnochiton sinudentatus*

13. Body elongated, considerably more than twice as long as wide . 14

13. Body about twice as long as wide 15

14. Valves roseate in general coloration; central areas pitted; girdle covered with minute scales . . . . . . *Ischnochiton fallax*

14. Valves gray or eroded almost to white with pinkish mottlings; central areas lacking pitting; scales of girdle large; common . . . . *Ischnochiton heathiana*

15. Central areas with fine longitudinal threads paralleling dorsal ridge; lateral areas slightly raised with fine radial striations and concentric rings; color uniform olive to slaty or cobalt blue . . . . *Ischnochiton regularis*
15. Central areas lacking longitudinal threads; lateral areas scarcely defined; entire surface minutely granulated; color olivaceous, radially streaked with brown . . . . *Ischnochiton radians*

16. Girdle covered with short bristles; valves nearly as long as wide; girdle broad; (figure. 98, *e*) . . . . . . *Nuttallina californica*
16. Girdle narrow, ornamented otherwise; valves short and broad. . 17

17. Girdle minutely papillose; body elongate; valves slightly beaked, covered with very fine granulations; color extremely variable; anterior valve with 8 slits . . . . *Cyanoplax dentiens*
17. Girdle finely granular; body broad oval when adult; valves broad and without beak, sculpture granular with occasional heavier warts; olive green unless eroded; anterior valve with 10–11 slits . . . . *Cyanoplax hartwegii*

List of Amphineura

Order Polyplacophora

**Basiliochiton heathii* (Pilsbry, 1898)
Callistochiton crassicostatus Pilsbry, 1892
Callistochiton palmulatus Carpenter in Pilsbry, 1892
**Callistochiton palmulatus mirabilis* Pilsbry, 1892
Chaetopleura gemma Carpenter in Pilsbry, 1892
Cryptochiton stelleri (Middendorff, 1846) (Scarce in Monterey area; more common farther north)
Cyanoplax dentiens (Gould, 1846) [This has been widely referred to as *C. raymondi* (Pilsbry, 1894). See Berry, 1948.]
**Cyanoplax fackenthallae* (Berry, 1919)
Cyanoplax hartwegii (Carpenter, 1855)
**Ischnochiton berryi* Dall, 1919
Ischnochiton cooperi (Carpenter in Pilsbry, 1892)
Ischnochiton fallax Carpenter in Pilsbry, 1892
Ischnochiton heathiana (Berry, 1946) (Common under boulders, sometimes even when these are bedded in sand. This species has for many years been miscalled *I. magdalenensis*. In its mantle folds are frequently found large numbers of a small commensal snail, *Vitrinella oldroydi* Bartsch, not encountered elsewhere.)
Ischnochiton mertensii (Middendorff, 1846)

Ischnochiton radians Carpenter in Pilsbry, 1892
Ischnochiton regularis (Carpenter, 1855)
Ischnochiton sinudentatus Carpenter in Pilsbry, 1892
Katharina tunicata (Wood, 1815) (The "black chiton," found with *Nuttallina* on exposed mid-tidal ledges)
**Leptochiton rugatus* Carpenter in Pilsbry, 1892
Mopalia ciliata (Sowerby, 1840)
Mopalia hindsii (Reeve, 1847) (Tolerates silt better than most chitons, and is characteristic of rocks and pilings in bays or estuaries)
Mopalia lignosa (Gould, 1846)
Mopalia muscosa (Gould, 1846)
Nuttallina californica (Reeve, 1847) (This is the commonest chiton on exposed coralline and mussel-covered rocks. In soft stone occupies small pits.)
**Nuttallina thomasi* Pilsbry, 1898 (A rare form, known from Monterey and areas somewhat to the south)
Placiphorella velata Carpenter in Dall, 1878
Tonicella lineata (Wood, 1815) (Abundant on coralline-encrusted, shaded rocks, well camouflaged by its color and broken pattern)

CLASS LAMELLIBRANCHIA (PELECYPODA)

The class Lamellibranchia consists of basically bilaterally symmetrical molluscs in which there has occurred a downgrowth of the mantle on each side, completely encasing the head, foot, and visceral mass. The mantle secretes a shell in the form of two lateral valves, hinged dorsally. With the retreat of the body from direct contact with the substratum, or perhaps making this retreat possible, there has developed a unique mode of feeding. Primitively, as in many of the Protobranchia, elongated ciliated palps are extended from within the shell, and by ciliary activity sweep up detritus and convey it to the mouth. With this indirect method of feeding has come a loss of the radula and other structures of the head. The two ctenidia of the primitive bivalve, as seen today in *Nucula*, consist each of a central axis, bearing on either side a series of flattened filaments (fig. 100, *a*) between which cilia maintain the respiratory flow of water. This water stream enters the mantle cavity ventrally or anteriorly, passes the gills, and is discharged posteriorly between the mantle edges. The next step in the evolution of the ctenidia as feeding organs was the development of forwardly directed ciliary currents along each ctenidium, and a transfer of the functioning of the palps from searching outside the shell to collecting food from the ctenidia. This accomplished, lamellibranchs became complete introverts, using their ctenidia both for respiration and for food collecting, whereas the palps retained their basic function, that of gathering and sorting food (now taking it from the ctenidia) and transferring it to the mouth. No understanding of lamellibranchs is possible without reference to this mode of feeding, around which their structure and way of life have evolved. Ctenidial food collecting, though successful in its way, has imposed certain limitations; thus no lamellibranch can lead a terrestrial existence, and the enclosure of the body within mantle and shell almost entirely precludes a really

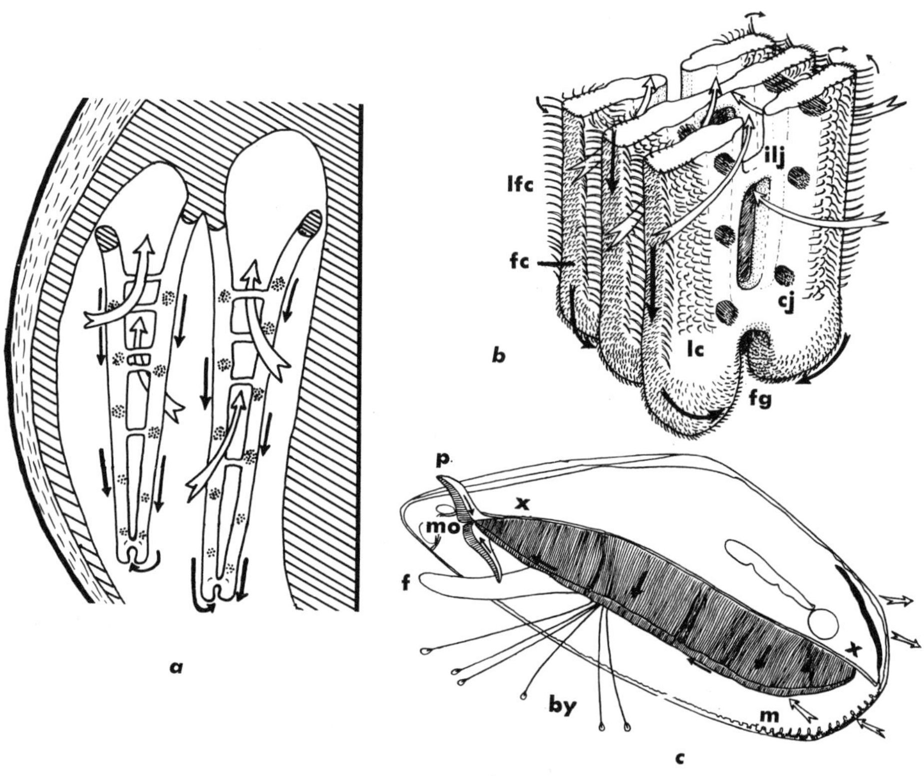

Fig. 99. **Filibranchia.** *a*, schematic cross section of one ctenidium, consisting of two demibranchs, suspended between foot (at right) and mantle; *b*, representation of part of bottom of a demibranch, including food groove, semischematic, cilia except for frontals grossly exaggerated in size; c, *Mytilus edulis* with left valve removed and left mantle cut along the line X-X, left ctenidum and left-hand pair of palps shown. Open arrows show course of water flow; heavy black arrows show general course of food transport on ctenidia and palps. f, foot; by, byssal threads; fc, frontal cilia; fg, food groove; cj, ciliary junction; ilj, interlamellar junction; lc, lateral cilia; lfc, laterofrontal cilia; mo, region of mouth; m, mantle edge; p, palp (b, c, modified after Orton).

active life. On the other hand, the ability to collect from the water has ensured a steady source of food and made possible the burrowing of the animal into protected situations and the general security of a sheltered, if monotonous, existence.

Lamellibranchs furnish a splendid example of evolutionary diversication and adaptive radiation, operating, it is true, within certain limits. Various members of the class can swim, climb, creep, cement themselves down, hang themselves up by byssal threads; burrow in sand, mud, shell, wood, or stone; and feed either upon suspended material or on surface detritus. A few are carnivorous; some are commensal or even parasitic; others may farm algal cells in their tissues, make nests, or brood their young. In size they vary from the giant tridacnas of the Indo-Pacific to forms not larger than coarse sand grains; their range extends from the floors of the deepest seas to high mountain brooks.

Among lamellibranchs, those features which are of taxonomic importance are usually subject to adaptive modification. This is of value to the student in that closely related groups such as families or suborders often show a certain uniformity in way of life. Thus the Pholadidae and Tellinacea each has its own distinctive morphology and characteristic ecological distribution. On the other hand, similar ways of life may produce parallelisms in structure and adaptations, such as we see among such distantly related genera as *Chama, Ostrea,* and *Hinnites,* or between *Zirfaea* and *Panope.* Certain structures such as the hinge, being less subject to environmental pressures, often permit the recognition of affinities despite outward dissimilarity. Thus the ligament and cardinal teeth in the left valves of *Tivela, Mactra,* and *Schizothaerus* readily demonstrate the affinity of the latter two genera, whereas in outward form and habits, *Mactra* and *Tivela* seem most similar. Proper understanding of the taxonomic and ecological position of any lamellibranch must be based upon a knowledge of the animal as a whole

The orders of lamellibranchs are based upon the form of the ctenidia, which are always a single pair, one ctenidium lying on either side of the body within the upper and rear part of the pallial (mantle) cavity.

1. Order Protobranchia. The ctenidia retain the primitive form, not unlike that seen in the lower gastropods (suborder Aspidobranchia). Each ctenidium consists of an axis bearing a double series of short, platelike filaments (fig. 100, *a*) which are neither long nor reflexed. Ctenidia are used for food collecting in the family SOLEMYIDAE, but in the NUCULIDAE and NUCULANIDAE the palps search for food outside the shell (fig. 100, *b, c*). Protobranchs are further distinguished by a flat rather than a wedgelike foot, sometimes lobed at its margins and very effective in burrowing. This order is not found in the local intertidal and is not covered in the key. Good general accounts may be found in Drew (1899, 1900, 1901) and Yonge (1939).

2. Order Filibranchia. The ctenidia have elongated filaments which are reflexed so that each ctenidium has in cross section somewhat the form of a tall narrow W from the center peak of which it is suspended (fig. 99, *a*). Adjacent filaments are united by ciliary junctions (patches of interlocked cilia) to form lamellae. Each half of the ctenidium is called a demibranch; each demibranch has a descending lamella nearest the axis and an ascending lamella away from the axis. The two lamellae of each demibranch are joined by bridges of tissue (interlamellar junctions). The filibranch type of ctenidium may be seen almost diagrammatically in *Mytilus*. Between adjacent filaments are lateral cilia (fig. 99, *b*) which drive water from the mantle cavity into the interlamellar spaces and so upward into the epibranchial space. Laterofrontal cilia on each filament aid in catching food particles, while frontal cilia on the face of each filament beat toward the ventral margin, sweeping thin sheets of mucus with entrapped food material down to the lower extremity of each demibranch. Here usually occur food grooves, in which cilia transport food forward to the palps, one pair of which lies on each side of the mouth (fig. 99, *c*). Filibranchs include the families of mussels and and date-stone borers (MYTILIDAE), scallops (PECTINIDAE), and rock jingles (ANOMIIDAE). These are further characterized by a secreted holdfast of threads, the byssus, and by a widely open mantle without definite siphons (fig. 99, *c*). There is a tendency for one adductor muscle (the posterior) to be larger than the other or for only this one muscle to remain.
3. Order Eulamellibranchia. In this order the ctenidia are more solidly constructed, adjacent filaments being united by vascularized bridges of tissue into firm lamellae. The general form of the ctenidia, although subject to variation, is much the same as in the Filibranchia. The mantle edges are usually united at one or two points (sutures) or broadly fused, leaving the apertures of incurrent and excurrent (anal) siphons, the pedal aperture, and sometimes a fourth pallial aperture serving for expulsion of rejected food material from the ctenidia (pseudofeces) or the relief of pressure when the foot is withdrawn. The siphons may be indistinct, distinct and short, long and separate, or fused into a "neck" with inhalent and exhalent apertures at its tip as in the deep burrowers (fig. 103). There are usually two adductor muscles, the oysters and certain others being exceptions. The order includes the great majority of all bivalve families.
4. Order Septibranchia. This is a little-known group inhabiting deep water. The ctenidia are lacking or very reduced, their place being taken by a perforated horizontal muscular septum, which serves to pump water for respiration and food getting (fig. 100, *d*, *e*). Septi-

branchs are carnivorous, feeding upon small crustaceans, annelids, and the like, either dead or alive, which are sucked in by muscular contraction of the septum. For an account of the feeding and respiratory movements of septibranchs, see Yonge (1928).

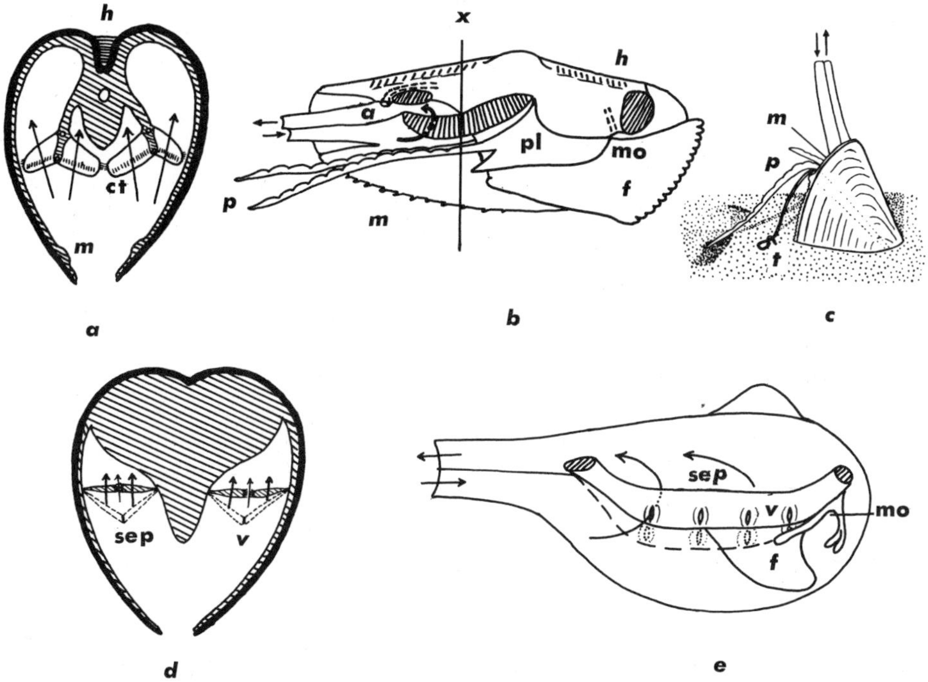

Fig. 100. *a-c*, **Protobranchia:** *a*, diagrammatic cross section of *Nucula* in region of ctenidia, showing simple aspidobranchlike form of filaments; *b*, *Yoldia*, right valve and mantle removed; *c*, *Yoldia limatula*, feeding with palp proboscides. *d-e*, **Septibranchia:** *d*, diagrammatic section of *Cuspidaria*, showing septa in raised position passing water through ciliated valves, and (dotted) in lowered position before start of massive lifting of water; e, side view of *Cuspidaria* showing hammocklike septum in raised and lowered positions. Arrows show course of water movement. a, anus ct, ctenidium; f, foot; h, hinge; m, mantle edge; mo, mouth; sep, muscular septum; t, unpaired mantle tentacle; v, ciliated valves of septum; x, region in *b* corresponding to cross section *a* (*a*, *d*, e, modified from Yonge; *b*, c, after Drew, from Yonge).

The classification of bivalves into families is on the basis of characters such as the nature of the siphons, degree of closure of the mantle, variation in form of the gills, foot, and adductor muscles, and the character of the shell, including its texture, composition, general form, ligament, and dentition. Some bivalves lie upon or are attached by one or the other valve (the right in *Pecten, Hinnites, Pseudochama, Pododesmus*; the left in *Ostrea, Chama*, and *Macoma nasuta*). In these the morphological right and left valves become more or less dissimilar, but in most bivalves the two valves are nearly alike. Growth in each valve proceeds concentrically (not necessarily equally) from the first-formed shell of the young animal. This point is termed the beak (fig. 101, *b*). The highest or most prominent point of each valve, near or coinciding with the beak, is the umbo (plural umbones).

The outer surface of the valves may be covered with a horny layer (periostracum), often worn away in older animals. The main outer layer of the shell beneath the periostracum is the prismatic layer; this is laid down concentrically during growth and may be variously sculptured into growth rings, concentric ridges or lamellae, or into fine radial striations (striae) or coarser ribs. The lining of the shell, by the addition of which the shell is increased in thickness, is the nacreous layer. The texture of shells may be described as nacreous (pearly), porcelaneous, or chalky.

The valves are joined dorsally at the hinge, where there is a horny elastic ligament serving to hold the valves open; closure must be by the pull of adductor muscles. The ligament may be entirely internal (in which position it is often called a resilium) and hidden from view, or it may be wholly or partly external and visible. The internal resilium is often supported by calcareous shelves or plates (chondrophores; fig. 101, *a*). In most bivalves the hinge is strengthened by interlocking hinge teeth, those below the umbones being called cardinal teeth, and those lying anterior or posterior to the umbonal region are called (not too logically) lateral teeth (fig. 101, *b*). In certain boring clams (pholads and teredos) there may be a calcareous bar or myophore within each valve below the hinge, serving as a point of attachment of muscles (fig. 102, *f*).

The inner surface of the shell bears adductor muscle scars. If two muscles are present the form is dimyarian; muscles roughly equal, isomyarian (*Mya*); muscles markedly unequal, anisomyarian (*Mytilus*); with single muscle, monomyarian (*Pecten, Ostrea*). Near the inner border of

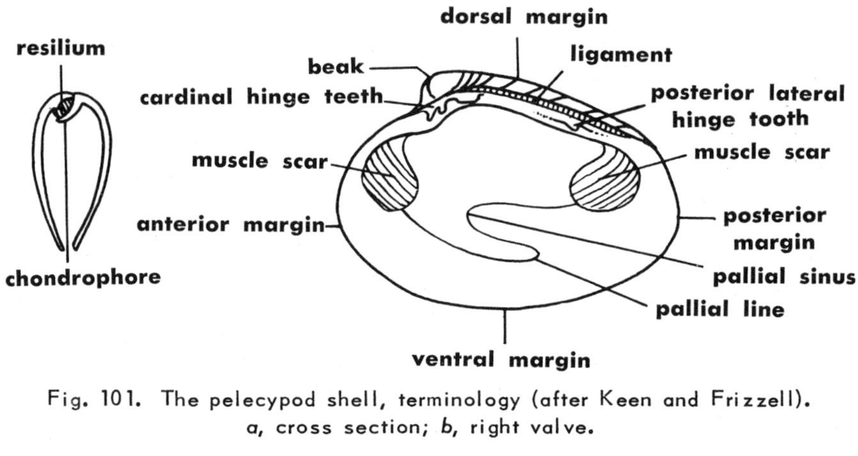

Fig. 101. The pelecypod shell, terminology (after Keen and Frizzell). *a*, cross section; *b*, right valve.

the shell, connecting the muscle impressions, runs the pallial line (fig. 101, *b*), marking the line of fusion of the mantle to the shell. Posteriorly, the pallial line may be indented about the pallial sinus, which marks the position of the siphonal retractor muscles in life.

When we examine a "dead" shell, certain inferences as to the nature of the soft parts can be drawn. The size and position of muscles are obvious. A large pallial sinus, especially if combined with a posterior gape or unclosable rear of the shell, indicates large retractable siphons. Sometimes it is difficult to determine the front or rear end of a shell; the umbo is not, as is often supposed, a reliable guide, since it may be anterior, central, or posterior. But the ligament when present lies to the rear of the umbo, and the pallial sinus if present is always posterior.

The problem of identifying bivalve molluscs varies somewhat with the needs and with the background of the individual collector. Paleontologists by necessity and most conchologists by preference deal with the hard and readily preserved shells. The general biologist is equally interested in the soft parts of the animal, especially foot, mantle, ctenidia, siphons, and adductor muscles, since it is the sum total of functional parts, including shell, which determines an animal's place in nature and in the taxonomic hierarchy. Once an understanding of the interrelationships between the various parts of a bivalve and between the animal and its environment is gained, it becomes possible to assign a specimen fairly accurately to its taxonomic or ecological niche either by means of the shell or by the soft parts. In practice a combination of these features is used, since the marks of such soft parts as muscles and mantle edge are revealed within the empty shell, and the environmental adaptations for boring, attachment, and the like are often clearly stamped upon the shell's exterior.

In the key that follows, the attempt is made to key out the families of the common local intertidal bivalves. This does not necessarily lead to the most efficient identification of genera and species, but after a little use this method should give a better understanding of the adaptive types and natural groups of bivalves, indispensable to those who wish to understand the class. For identification of genera by means of hard parts, the excellent empirical key of Keen and Frizzel is readily available to the serious worker, and the needs of the beginner can be pleasantly met by Keep and Baily, *West Coast Shells*. The recent California Fish Bulletin no. 90, *Common Marine Bivalves of California*, by John E. Fitch (1953) should also prove very useful. The present manual is intended for general zoölogists and for those entering the field, and since we do not wish to duplicate existing guides, we have formulated a key which, though far from complete or free of ambiguity, is essentially biological in its approach, with the aim of basing factual identification upon an understanding of the animals involved.

The orders Protobranchia and Septibranchia, not being encountered in the local intertidal, are omitted, as are the numerous fresh-water members of the Eulamellibranchia (for the latter, see Ingram, 1948).

Key to Marine Lamellibranchia

(Filibranchia and Eulamellibranchia)

1. Animal firmly cemented to hard substrate by one valve; valves distinctly different, one or both rough and irregular 20
1. Animal not cemented down; may be free, burrowing, attached by a byssus, nestling, ensconced in a boring made by itself, or overgrown by sponges or ascidians; valves may be irregular but not markedly unlike 2

2. Shell brown to black, usually tapered either anteriorly or posteriorly; umbo well forward; no hinge teeth; animal either attached by byssus, firmly nestled, or boring in rock (fig. 102, *a-e*) . . . . Family MYTILIDAE 3
2. Shell white, with filelike boring teeth on anterior surfaces; internally with a myophore (calcareous process for muscle attachment); all are borers into wood, stone, clay, firm mud, or shell (fig. 102, *f-k*) . . . . Families PHOLADIDAE and TEREDIDAE 10
2. Shell fanlike, with radiating ribs; flat expansions ("ears") on either side of umbo; animal rests on right valve; may swim by flapping valves; eyes present in mantle edges; free living, or attached by byssus, or cemented by one valve (scallops) . . . . Family PECTINIDAE 15
2. Shell elongate with both ends gaping, generally with shining periostracum; umbo anterior or in anterior 1/3; foot powerful, adapted for rapid burrowing in sand; siphons short (razor shells) . . . . Family SOLENIDAE 17
2. Not assignable to any of above families 25

3. Boring in rock 4
3. Attached by byssus to rocks or piles, or half-buried in debris or firm mud 6

4. Front of shell rounded, almost circular in cross section; posterior end tapers markedly; secreted calcareous material may partly cover shell and form a sleeve lining the opening of the burrow . . . . *Lithophaga plumula*
4. Shell not circular in cross section; generally with oblique ridge running posteriorly from umbo; shell appears slightly arched or bent downward at ends . . . . *Botula* 5

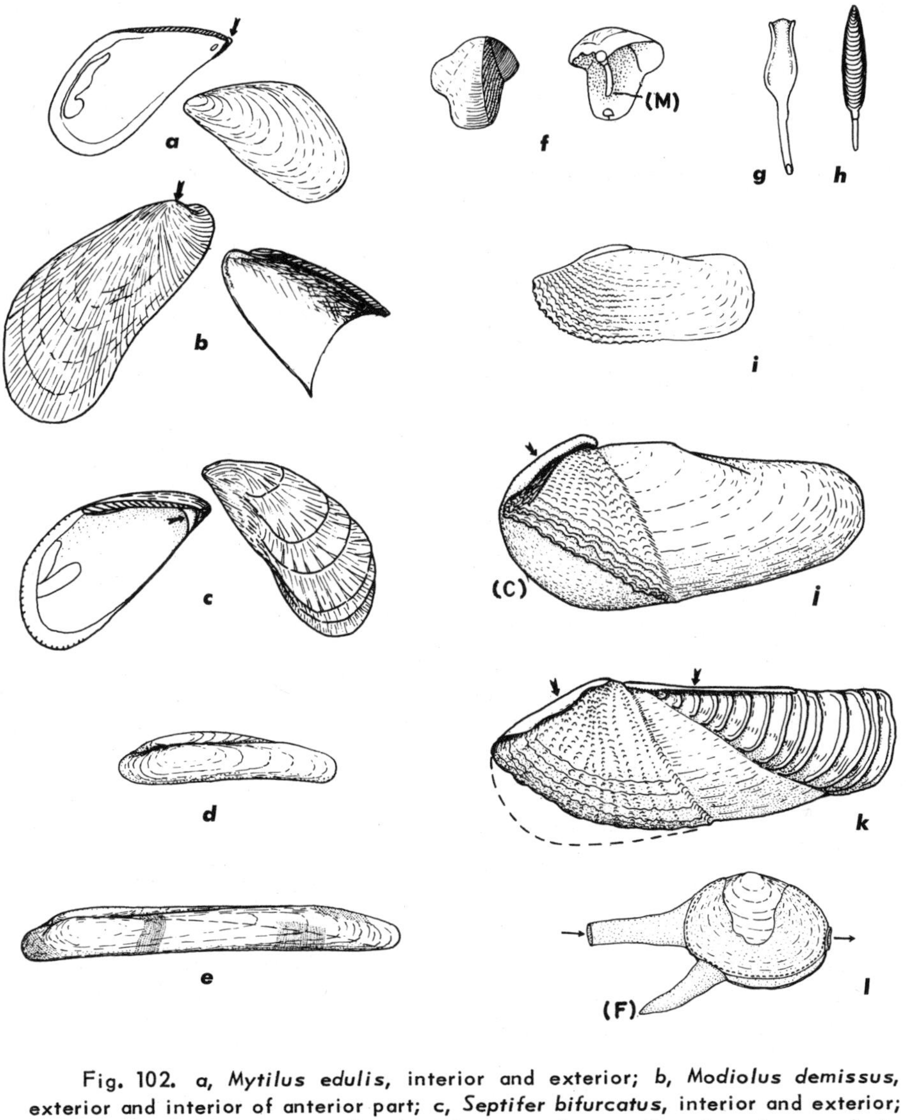

Fig. 102. *a, Mytilus edulis*, interior and exterior; *b, Modiolus demissus*, exterior and interior of anterior part; c, *Septifer bifurcatus*, interior and exterior; *d, Botula californiensis*; e, *Botula falcata*, surface sculpturing shown only in part; *f, Teredo navalis*, exterior and interior of valves, with myophore (m) (after Miller, 1924); *g*, pallet of *Teredo*; *h*, pallet of *Bankia* (*g* and *h* after Keen and Frizzell); *i, Zirfaea pilsbryi*; *j, Pholadidea penita*, mature individual with callum (c); *k, Parapholas californica*, immature individual lacking callum, extent of which is outlined; *l, Kellia suborbicularis*, living, showing anterior inhalent siphon, foot (f), and thin extension of mantle over shell. Scales various.

5. Periostracum not sculptured except for growth lines; shell moderately long, resembling a date stone (fig. 102, *d*) . *Botula californiensis*
5. With fine transverse sculpturing over most of periostracum; shell much longer, proportionately, than a date stone (fig. 102, *e*) . . . . *Botula falcata*

6. A shelf or deck internally across anterior end of valves (fig. 102, *c*) . . . . *Septifer bifurcatus*
6. No such shelf within valves 7

7. Beaks terminal, at anterior end of shell (fig. 102, *a*) . . *Mytilus* 8
7. Beaks not terminal, anterior end of shell projects beyond beaks (fig. 102, *b*) . . . . *Modiolus* (=*Volsella*) 9

8. Length rarely more than 3 inches; valves black, smooth, and fairly thin; posterodorsal edge sharp; restricted to bays, estuaries, and sheltered situations (the bay mussel) . . . . . *Mytilus edulis*
8. Larger and coarser; valves somewhat irregularly radially ribbed; growth lines prominent; shell heavy, with blunt posterodorsal margin; forms extensive mid-tidal beds on exposed rocks, but also found in bays (the coast mussel) . . . . . . *Mytilus californianus*

9. With prominent radial ribbing; anterior end narrowed; characteristically in bays or brackish waters (ribbed mussel) (fig. 102, *b*) . . . . *Modiolus demissus*
9. Without ribs; width nearly uniform; anterior part dark and polished, posterior yellowish and hairy . . . . *Modiolus* spp.

10. Valves very reduced (fig. 102, *f*); body elongated and wormlike; boring in wood; pallets present at siphonal end (shipworms) . . . . Family TEREDIDAE 11
10. Valves somewhat reduced or not reduced; not boring in wood . . . . Family PHOLADIDAE 12

11. Pallets paddlelike, with concave ends and sides extending back as sharp points (fig. 102, *g*) . . . . *Teredo navalis*
11. Pallets appear featherlike, with 15–30 striations (fig. 102, *h*) . . . . *Bankia setacea*

12. Sculpture coarse, in radiating rows over anterior 1/2 of shell, not sharply limited posteriorly; lacking conspicuous accessory plates over hinge; shell not covering siphons, not tapered posteriorly; anterior gape not closed by callum (rounded overgrowth of shell) boring in sticky mud or clay (fig. 102, *i*) . . . . *Zirfaea pilsbryi*
12. Sculpture finer, the sculptured area sharply limited at rear by diagonal groove; accessory plates present over hinge; shell covers siphons when retracted; anterior gape in mature (but *not* in young) specimens closed by a callum (fig. 102, *j*); boring in stone or shell 13

13. Dorsal shell margin doubled back both anterior and posterior to umbo (fig. 102, *k*) . . . . *Parapholas californica*
13. Dorsal margin doubled back only anterior to umbo (fig. 102, *j*) . . . . *Pholadidea* 14

14. Large (2–3 in.); boring in shale . . . . . . *Pholadidea penita*
14. Small; boring in red abalone shells . . . . . *Pholadidea parva*

15. Free-living; well over 1 in. in diameter; general color pinkish . . . . *Pecten* 16
15. Attached by byssus, sometimes detached; 1 in. or less in diameter; color yellowish . . . . Young of *Hinnites* (see 23)

16. Margins nearly smooth; ribs of left (upper) valve smooth and nearly equal . . . . *Pecten hindsii*
16. Margins rough; main ribs of upper valve rough or spiny and alternating with smaller ribs . . . . *Pecten hericius*

17. Valves moderately long, somewhat flattened, anterior end rounded; umbo nearly 1/3 of length back from anterior end; diagonal rib running internally from umbo; in clean sand (jackknife or short razor clam). This genus superficially resembles *Tagelus* (38) . . . *Siliqua* 18
17. Valves markedly elongate, anterior end squared; umbo at anterior end (razor clams) 19

18. Shell opaque, with periostracum; posterior end rounded . . . . *Siliqua patula*
18. Shell translucent, without periostracum; posterior end truncate . . . . *Siliqua lucida*

19. One cardinal tooth in each valve . . . . *Solen sicarius*
19. Four cardinal teeth in one valve, 2 or 3 in other . *Ensis californicus*

20. Upper valve flattened, subcircular, obscurely radially striated, thin, translucent and pearly; lower (attached) valve somewhat reduced, with hole or deep notch near hinge (family ANOMIIDAE, rock-jingles) . . . . *Pododesmus macroschisma*
20. Upper valve heavy, roughened; lower valve often somewhat cuplike, without a hole 21

21. Form of shell massive, solid, without much tendency for the margins of the valves to project; beaks spiraled (but may be obscured externally) . . . . Family CHAMIDAE 22
21. Form more or less flattened, with the edges of the shell projecting; elongated to circular; beaks not spiraled 23

22. Attached by left valve; beaks spiral to right (clockwise) as viewed from free side; free valve often with numerous flattened spines; shell white to pinkish; beneath rocks . . . . . . *Chama pellucida*
22. Attached by right valve; beaks spiral to left; usually on sides or

tops of rocks in mid-tidal, overgrown with seaweeds and resembling a nubbin of rock . . . . *Pseudochama exogyra*

23. Form circular unless distorted; up to several inches in diameter; surface of upper valve with radiating ribs, and wings at side of umbo; lower valve shows the regular shell of the younger stage before attachment, the rest of lower valve being cemented to substrate, and hence irregular; mantle edges with row of eyes (the adult rock scallop, family PECTINIDAE) . . . *Hinnites multirugosus*
23. Form elongate to subcircular; surface of upper valve wavy, laminated, or irregularly plaited; no wings at sides of umbo; no eyes in mantle edges (family OSTREIDAE) . . . . *Ostrea* 24

24. Small; native and generally distributed (western or Olympian oyster) . . . . *Ostrea lurida*
24. Large; introduced and of local occurrence (Japanese or Pacific oyster). . . . . *Ostrea gigas*

25. Shell with externally distinct ligament 35
25. Shell without externally distinct ligament. 26

26. Triangular resilium (internal ligament) supported by vertical chondrophore in each valve; an Λ-shaped cardinal tooth in left valve anterior to ligament . . . . . . Family MACTRIDAE 27
26. Resilium supported by a horizontal shelf (chondrophore) arising in the left valve; no cardinal teeth . . . . Family MYIDAE 29
26. Hinge lacking teeth and chondrophore, but with internal ligament; ventral surface of ligament covered with oblong shelly plate (lithodesma); shell nacreous, elongated or inflated, often deformed; free, nestling, or commensal in ascidians or sponges . . . . Family LYONSIIDAE 31
26. Hinge with cardinal and lateral teeth, as well as internal ligament, but no chondrophore; shell short, sometimes inflated; nestling, or commensal on crustaceans or annelids; brood the young in epibranchial chamber . . . Family ERYCINIDAE (LEPTONIDAE) 33

27. Large; valves gaping posteriorly; siphons fused and very long; a deep burrower in muddy sand (the horse-neck clam) . . . . *Schizothaerus nuttallii*
27. Valves with small or no posterior gape 28

28. Chondrophore set off from resilium by shelly layer . . . . *Mactra californica*
28. Chondrophore not set off from resilium by shelly layer . *Spisula*

29. Shell somewhat flattened with faint radial striations on rear, oval in outline; siphons very short, ordinarily opening into burrows of *Urechis* or *Callianassa* (fig. 103, *c*) . . . *Cryptomya californica*
29. Siphons elongated; deep burrowers in muddy sand or in rock . . 30

30. Boring in soft rock; valves thick and heavy, narrowed and gaping posteriorly; umbo to rear of center; 4 scales near tip of siphon close siphons when withdrawn . . . . *Platyodon cancellatus*
30. Burrowing in mud or sand; valves rather thin and chalky, with posterior gape; umbo central (soft-shell or long-neck clam) . . . . *Mya arenaria*
30. With shell resembling that of *Mya arenaria*, but appearing as if rear fourth had been chopped off; gaping posteriorly more than *Platyodon*; umbo to rear of center . . . . *Mya truncata*

31. Umbones bent forward so that beaks appear spiraled and are visible in lateral view of shell; shell inflated, higher than long, very thin, with nacreous lining; embedded in compound tunicates . . . . *Mytilimeria nuttallii*
31. Beaks not so twisted 32

32. Periostracum, if present, with radial striations, thin and often a characteristic silvery gray; shell elongated, thin and pearly . . . . *Lyonsia californica*
32. Periostracum thick; shell thicker, shorter, less pearly, tending to irregularity. . . . . *Entodesma saxicola*

33. Shell small, white, somewhat triangular; attached to *Upogebia*, *Blepharipoda*, *Aphrodita*, and other animals . . . . *Pseudopythina rugifera*
33. Shell tiny, reddish, with umbo set to rear; abundant in byssal threads of mussels, among barnacles, and in kelp holdfasts . . . . *Lasaea cistula*
33. Shell of medium size, inflated, with umbo central; greenish periostracum; inhalent siphon anterior (fig. 102, *l*); mantle can be extended to cover entire shell; nestling in holdfasts, crevices, and holes of rock-boring clams . . . . *Kellia* 34

34. Shell nearly spherical . . . . *Kellia suborbicularis*
34. Shell somewhat elongated . . . . *Kellia laperousii*

35. Deep burrowers; siphons long and *separate* (fig. 103, *d*); shells of most are laterally flattened, with umbo central or behind the middle; a few show elongation of the shell reminiscent of short razor clams . . . . Suborder Tellinacea, including families TELLINIDAE, SEMELIDAE, SANGUINOLARIIDAE, and DONACIDAE 36
35. Shallow burrowers or nestlers; siphons short or absent; shells usually stout, not markedly flattened or elongated, smooth or sculptured either radially or concentrically (cockles and similar forms) . . . . Families VENERIDAE, CARDIIDAE, CARDITIDAE, and LUCINIDAE 45
35. Nestlers, deep burrowers, or borers; siphons fused, sometimes quite

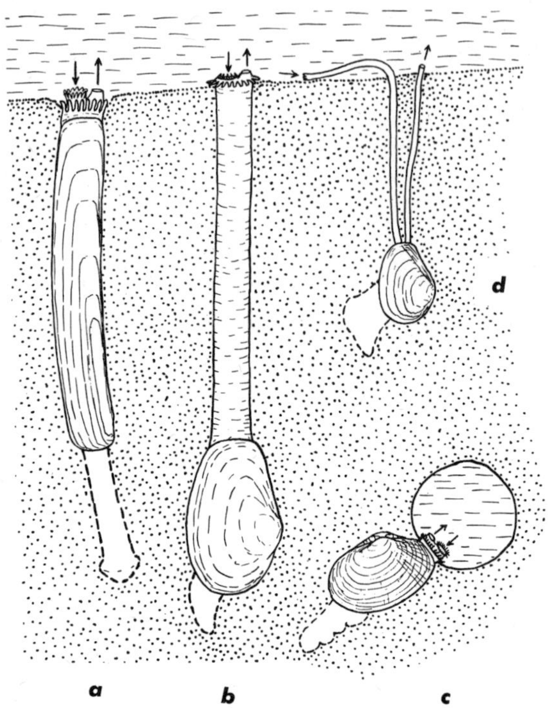

Fig. 103. Forms of lamellibranch siphons and mode of burrowing. *a*, SOLENIDAE, e.g., *Ensis*, short siphons but can burrow actively if disturbed; *b*, MYIDAE, *Mya arenaria*, long fused siphons, lies deeply buried, retracts siphons if disturbed; c, MYIDAE, *Cryptomya californica*, lies deeply but utilizes *Callianassa* or *Urechis* burrow instead of long siphons; *d*, TELLINIDAE, e.g., *Macoma irus*, burrows fairly deeply and actively, has long separate siphons, and feeds upon detritus rather than upon suspended material (after Yonge, 1949).

long; shells white, variously pholadlike, elongated, distorted, or quadrangular and reduced . . . . Families PETRICOLIDAE and SAXICAVIDAE 54

36. Ligament largely or wholly external 37
36. Ligament partly internal, in diagonal, spoon-shaped concavity of the hinge; shells subcircular or ovate . . . Family SEMELIDAE 44

37. Shell elongated, somewhat flattened, oval or rectilinear; some resembling a short razor clam . . Family SANGUINOLARIIDAE 38
37. Shell flattened, oval to rounded-triangular in side view; if elongated, tends to taper posteriorly . . . . . Family TELLINIDAE 39
37. Shell elongated anteriorly, so that umbo is well to rear; posterior slope steep and flat; general form wedge-shaped . . . . . Family DONACIDAE (Not represented locally, but the bean clam, *Donax gouldii*, is extremely abundant on southern California beaches)

38. Large; oval in outline with a heavy, dark brown, shining periostracum; ligament bulging; right valve much flatter than left; in mud; rare . . . . *Sanguinolaria nuttallii*
38. Oval or somewhat elongated; shell smooth, light colored, with radiating pinkish bands; in coarse sand of low intertidal or washed up . . . . *Gari californica*
38. Elongated, more so than *Siliqua* (17) with which it might be confused, but with umbo central; gaping at both ends; inhabits tubular vertical burrows in muddy sand with paired openings at surface for the two siphons . . . . *Tagelus californianus*

39. Shell chalky or with a dull porcelaneous surface; more or less oval in outline . . . . *Macoma* 40
39. Shell with a shining outer surface; inclined to a triangular outline . . . . *Tellina* 43

40. Inequivalve, with one valve flatter than the other or with valves bent to right at rear; usually lie on side 41
40. Essentially equivalve; tend to lie vertically in substrate . . 42

41. Ligament slender; valves bent to right at rear; not more than 2 in. long (bent-nose clam) . . . . *Macoma nasuta*
41. Ligament short, thick, and bulging; left valve flatter than right, but valves usually not bent at rear; shell high, somewhat angulated at rear, with diagonal ridge extending from umbo; posterior slope steep, usually elevated into a keel behind ligament; commonly greater than 2 in. long . . . . *Macoma secta*

42. Small (less than 1 in.); shell flattened; outline oval; periostracum thin . . . . *Macoma inconspicua*
42. Up to 2 in. long; shell somewhat inflated; outline oval, pointed at rear; periostracum prominent . . . . *Macoma irus*

43. Shell small, thin, ovate; umbo posterior; salmon-pink within . . . . *Tellina salmonea*
43. Shell as above, but white, with umbo more or less central . . . . *Tellina buttoni*
43. Shell of medium size; elongated; somewhat pointed posteriorly; umbo behind center; pale yellowish, shining . . . *Tellina bodegensis*

44. Shell thin, ovate, slightly pointed and gaping at rear; often irregular . . . . *Cumingia californica*
44. Shell subcircular (in shape much like *Lucina*), with slightly elevated beak; fairly thick; often with reddish coloration in radiating lines or at inner edges; muscle scars rounded; pallial sinus wide and deep . . . . *Semele*

45. With heavy radial ribbing and brown markings 46

45. Radial markings fine or lacking; concentric growth rings or sculpturing more or less obvious 47

46. Large; stout and nearly circular in profile; height including beaks about equal to length (family CARDIIDAE). *Cardium* (*Clinocardium*) *nuttallii*
46. Small (rarely more than 5/16 inch); posteriorly elongated and subquadrate (family CARDITIDAE) . . . . . . *Glans carpenteri*

47. Medium sized; subcircular in outline (very similar to *Semele*); fine concentric sculpturing; white; narrow impression (lunule) anterior to umbones; posterior muscle scar elongated; no pallial sinus (family LUCINIDAE; several species and subgenera) . . . . "*Lucina*" (= *Phacoides*)
47. Not referable to above groups (here we are dealing only with common central California forms) . . . . . Family VENERIDAE 48

48. Minute; with delicate concentric sculpture 49
48. Larger; sculpturing varies. 50

49. Shell slightly elongated or oval; general color yellowish; inner side of ventral margin smooth; abundant in coarse sand . . . . *Transennella tantilla*
49. Shell no longer than high, slightly triangular; general color white with purple tinge; inner side of ventral margin finely crenulated; introduced from East coast and abundant in San Francisco Bay and elsewhere . . . . *Gemma gemma*

50. Sculpture decidedly concentric and raised into lamellae, especially on rear; without radial sculpture 51
50. Sculpture both radial and concentric 52
50. Essentially unsculptured; shell large, heavy, somewhat elongated; umbo central; ligament bulging; exterior appears varnished; in sand of exposed beaches (Pismo clam) . . . . . . *Tivela stultorum*

51. Shell short, almost globose, of medium size; not gaping posteriorly; concentric lamellae very pronounced; often irregular; nestling among rocks or in borer holes . . . . *Irus lamellifer*
51. Shell large; elongated or oval; gaping posteriorly; concentric lines raised into lamellae posteriorly; burrowing in sand and mud (Washington clam) . . . . *Saxidomus nuttalli*

52. Large, oval in outline, and somewhat flattened; concentric raised lines thin and sharp, widely spaced, more prominent than the fine radial striations; inner ventral margins smooth . . . . *Protothaca tenerrima*
52. Medium sized; oval to subcircular in outline; stouter than *P. tenerrima*; radial sculpture more pronounced than concentric; inner ventral margins slightly roughened 53

53. Curvature of anterior margin not much sharper than that of posterior; subcircular in side view; siphons fused for whole length (rock cockle) . . . . *Protothaca staminea*
53. Curvature of anterior margin noticeably sharper than that of posterior; ovate in side view; often with purple coloration inside valves and an external diagonal dark stripe or broken pattern; siphons separated at tips (the introduced "Philippine cockle"; very common in San Francisco Bay). . . . . *Tapes semidecussata*

54. Nestling or boring; valves heavily hinged, with 2 cardinal teeth in each valve (a 3d in left is obscure) Family PETRICOLIDAE 55
54. Nestling or burrowing; valves weakly hinged, cardinal teeth 1 in each valve in adult, or lost . . . . Family SAXICAVIDAE 56

55. Resembling an elongated pholad, with radial rows of teeth anteriorly and radial striations posteriorly (but lacks myophore, accessory plates, and other features of pholads); shell gaping; bores in clay, peat, and soft rock. . . . . *Petricola denticulata*
55. Shell very heavy, broad, ovate; not gaping; nestles in borer holes; usually much eroded and distorted; tips of siphons bright purple . . . . *Petricola carditoides*

56. The largest bivalve on this coast, body much too big to retract within shell; valves quadrangular, gaping widely; burrows deeply in muddy sand (the geoduck, pronounced gooeyduck) . . . . *Panope generosa*
56. Small nestling clams with red-tipped siphons (often confused with those of the ascidian *Styela truncata*); 2 small teeth in each valve of young, teeth lost in adult; shells wedge-shaped, tapering posteriorly; chalky and usually distorted; nestling in holdfasts, sponges, and borer holes . . . . *Saxicava* spp.

List of Lamellibranchia

Order Filibranchia

ANOMIIDAE

Pododesmus macroschisma (Deshayes, 1839)

MYTILIDAE

Botula californiensis (Philippi, 1847)

Botula falcata (Gould, 1851)

Lithophaga plumula Hanley, 1844

Modiolus demissus Dillwyn, 1817 (*Volsella* may have priority over the more widely used *Modiolus*)

**Modiolus rectus* Conrad, 1837
Modiolus spp.
Mytilus edulis Linnaeus, 1758
Mytilus californianus Conrad, 1837
Septifer bifurcatus (Conrad, 1837)

PECTINIDAE
Hinnites multirugosus (Gale, 1928) (Has often been called *H. giganteus* Gray)
Pecten hericius Gould, 1850 (The pectens listed here are common farther north. Several other species are about equally common in Monterey Bay, all subtidal and seldom seen)
Pecten hindsii Carpenter, 1864

Order Eulamellibranchia

OSTREIDAE
Ostrea gigas Thunberg, 1793
Ostrea lurida Carpenter, 1863

CHAMIDAE
Chama pellucida (Broderip, 1834)
Pseudochama exogyra (Conrad, 1837)

TELLINIDAE
**Macoma calcarea* Gmelin, 1792 (Somewhat resembles a small *M. nasuta*; Elkhorn Slough, rare)
Macoma inconspicua (Broderip and Sowerby, 1829) (Common in San Francisco Bay; has usually been called *M. balthica* Linnaeus)
Macoma irus (Hanley, 1845) (Has commonly been identified as *M. inquinata* Deshayes.)
Macoma nasuta (Conrad, 1837)
Macoma secta (Conrad, 1837)
Tellina bodegensis Hinds, 1844
Tellina buttoni Dall, 1900
Tellina salmonea (Carpenter, 1864)

SEMELIDAE
Cumingia californica Conrad, 1837
**Semele rupicola* Dall, 1915
**Semele rubropicta* Dall, 1871

SANGUINOLARIIDAE
Gari californica (Conrad, 1837)
Sanguinolaria nuttallii Conrad, 1837
Tagelus californianus Conrad, 1837

DONACIDAE
Donax gouldii Dall, 1919 (southern California)

SOLENIDAE
Ensis californicus Dall, 1899
Siliqua lucida (Conrad, 1837)

Siliqua patula (Dixon, 1788)
Solen sicarius Gould, 1850

MACTRIDAE
Mactra californica Conrad, 1837
Schizothaerus nuttallii (Conrad, 1837)
Spisula spp.

CARDIIDAE
Cardium (Clinocardium) nuttallii Conrad, 1837 [Often incorrectly called *C. corbis* (Martyn)]

CARDITIDAE
Glans carpenteri Lamy, 1922 [= *Cardita subquadrata* (Carpenter)]

VENERIDAE
Gemma gemma (Totten, 1834)
Irus lamellifer (Conrad, 1837)
Protothaca staminea (Conrad, 1837)
Protothaca tenerrima (Carpenter, 1865)
Saxidomus nuttalli Conrad, 1837
Tapes semidecussata (Reeve, 1864)
Tivela stultorum (Mawe, 1823)
Transennella tantilla (Gould, 1852)

LUCINIDAE
"*Lucina*" (= *Phacoides*) spp. (*L. californica* is the form usually encountered intertidally.)

LYONSIIDAE
**Entodesma inflatum* (Conrad, 1837)
Entodesma saxicola (Baird, 1863)
Lyonsia californica Conrad, 1837
Lyonsia sp.
Mytilimeria nuttallii Conrad, 1837

ERYCINIDAE (LEPTONIDAE)
Kellia laperousii (Deshayes, 1839)
Kellia suborbicularis (Montagu, 1804)
Lasaea cistula Keen, 1938 (Has commonly been called *L. rubra*.)
Pseudopythina rugifera (Carpenter, 1864)

MYIDAE
Cryptomya californica (Conrad, 1837)
Mya arenaria Linnaeus, 1758
Mya truncata Linnaeus, 1758
Platyodon cancellatus (Conrad, 1837)

PETRICOLIDAE
Petricola carditoides Conrad, 1837
Petricola denticulata Sowerby, 1834

SAXICAVIDAE [*Hiatella* (-idae) appears to have priority over *Saxicava* (-idae), but the latter name is in far more general use in this country.]

Panope generosa Gould, 1850
**Saxicava arctica* Linnaeus, 1771 (?) (The common form in the Monterey area identified as *S. arctica* is a nestler, there being no intertidal shales for it to bore into as the species is reported to do in Europe.)
**Saxicava pholadis* Linnaeus, 1771 (This is possibly more common as a borer in intertidal shales further north.)

PHOLADIDAE
Parapholas californica (Conrad, 1837)
Pholadidea parva (Tryon, 1865)
Pholadidea penita (Conrad, 1837)
Zirfaea pilsbryi Lowe, 1931 (Formerly identified as *Z. gabbi* Tryon, 1863)

TEREDIDAE
Bankia setacea (Tryon, 1863)
**Teredo diegensis* Bartsch, 1916
Teredo navalis (Linnaeus, 1758)

CLASS GASTROPODA

Gastropods form the largest and most successful group of molluscs, exhibiting enormous diversity in form and habitat. No other invertebrate group offers better material for studies in evolution, distribution, ecology, comparative physiology, and functional adaptations of morphology. Because of their convenient sizes and often beautiful shells they have long been favorite objects for the collector.

In body form, gastropods adhere fairly closely to the generalized molluscan pattern (fig. 104), having a solelike foot for creeping, a head with the characteristic scraping radula and usually with tentacles and eyes, a compact visceral mass containing digestive and reproductive organs, and a mantle which secretes the shell and provides in the pallial cavity a shelter for the respiratory organs. The latter are primitively ctenidia, but may be replaced by adaptive gills or other arrangements.

The hallmark of the gastropods, which sets them sharply apart from

---

Fig. 104. *a-b*, a dextrally coiled gastropod shell (*Buccinum*) showing structures of shell referred to in the key (after Oldroyd); *c*, *Calliostoma*, an aspidobranch (after Eales); *d*, a pectinibranch (after Eales); *e-f*, *Helix*, a terrestrial pulmonate (after Selenka): *e*, with the shell in place; *f*, shell removed; mantle chamber ("lung") opened by cutting mantle fold anteriorly and turning it back.

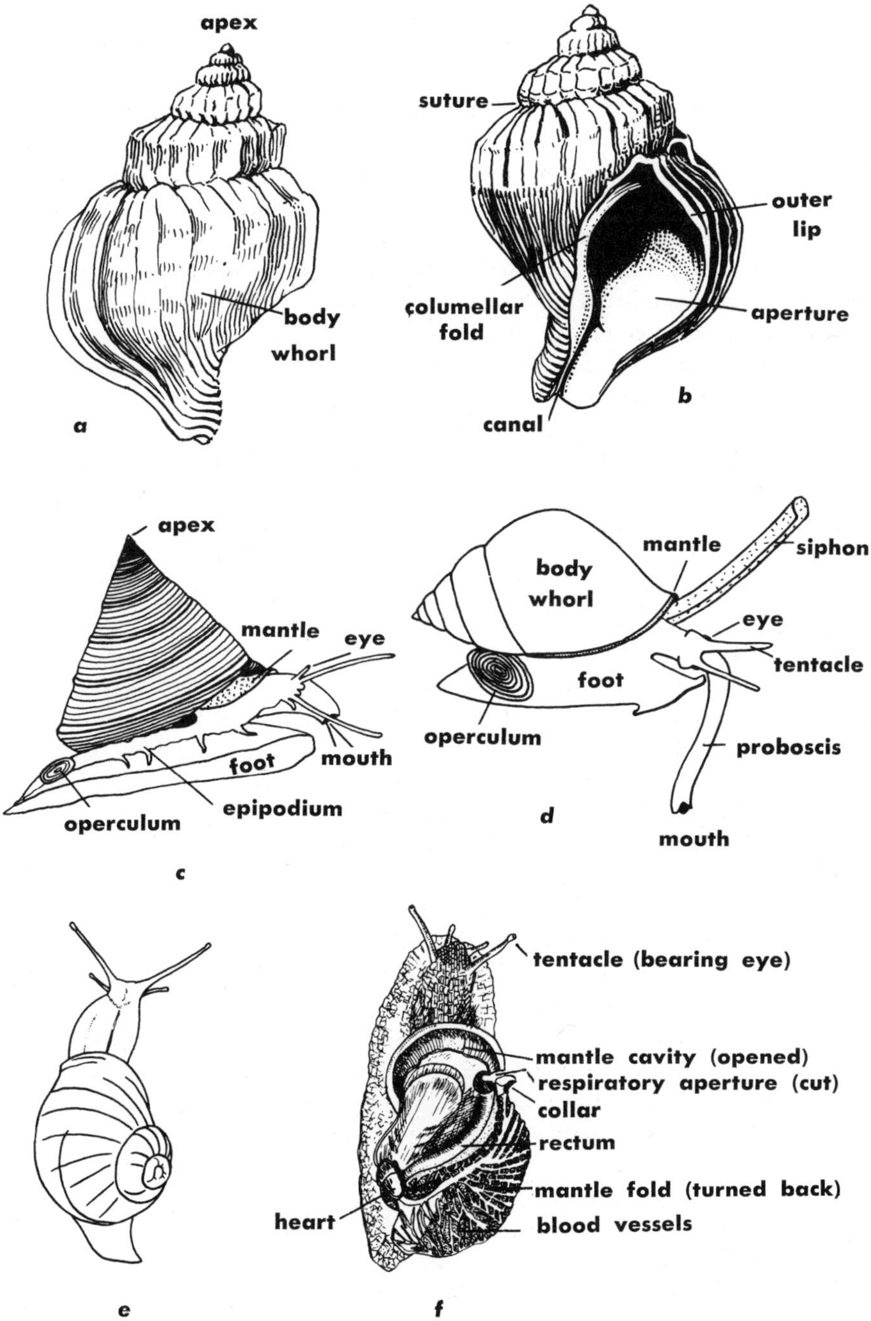

Fig. 104. (Legend on facing page)

other molluscs, is the phenomenon of torsion. This is not the same thing as coiling. Torsion occurs early in development and consists of a 180° counterclockwise rotation of the visceral mass upon the head and foot, with the result that the mantle cavity, ctenidia, and anus, which originally lay to the rear, come to lie back of the head (see Borradaile and Potts, figs. 378–379). In the lower gastropods this process takes place rather abruptly in the veliger stage. Torsion in its fullest expression characterizes the subclass **Prosobranchia**, where the ctenidia lie anteriorly and the nervous system is twisted into a crude figure 8 (the **streptoneurous** condition). The higher groups of gastropods have tended to undo to some extent the extreme effects of torsion. One change has been a straightening out of the nervous system to the **euthyneurous** condition. Euthyneury has been attained in two ways: in the subclass **Opisthobranchia** the body has more or less "unwound" itself in a detorsion; in the subclass **Pulmonata** the body as a whole has retained much of its torsion, but the central nervous system has straightened out by a drawing together in the form of a ring of ganglia about the esophagus, as is seen in the familiar *Helix*.

It is important to remember that coiling is not the same thing as torsion. Coiling of the visceral hump and shell is, indeed, very characteristic of the Gastropoda, although suppressed in a good many. However, it is not unique to gastropods among molluscs, since it was also characteristic of many of the ancient shelled cephalopods and of a few present-day survivors such as *Nautilus*. The loss of coiling is apparent in those gastropods which have assumed the low "limpet form," and in those such as nudibranchs and land slugs which have reduced or lost the shell and flattened the visceral hump.

To understand the basis of classification in gastropods, we must study the ctenidia and the pallial cavity. The most primitive snails, represented by keyhole limpets and abalones, have each one pair of ctenidia, and these are bipectinate, having a double set of flattened filaments along the axis (fig. 105, *a*, *b*). Cilia on the filaments set up the respiratory water flow. In all snails with two ctenidia the shell is slit or pierced with one or more holes, apparently to permit the discharge of waste-laden water at a point not directly over the head. Most snails find one ctenidium more efficient than two, since it permits a respiratory water current to be moved from left to right across the pallial cavity. Animals such as the limpets and turbans (*Acmaea*, *Tegula*) have each a single ctenidium of the primitive bipectinate form (fig. 105, *c*, *d*). Most prosobranchs, however, reduce the ctenidium to the monopectinate form (fig. 105, *e*, *f*), permitting it to be fused with the roof of the mantle cavity, where it is less easily clogged by silt. Often the edges of the mantle form a siphon for the intake of water. In the pulmonates, the ctenidium is lost altogether, and the vascularized lining of the pallial cavity takes its place in respiration (fig. 104, *f*).

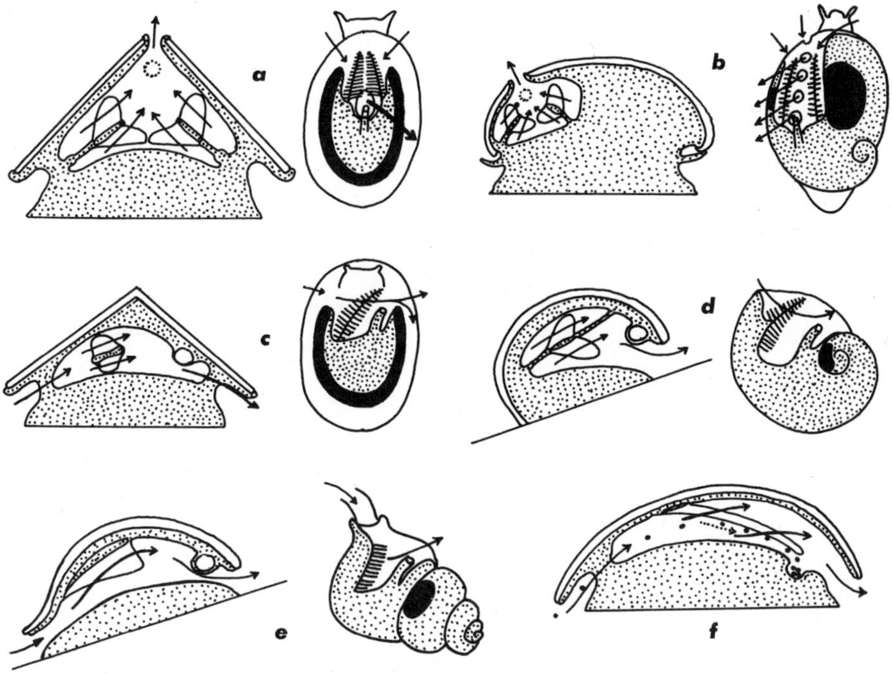

Fig. 105. Ctenidial arrangements in gastropods. Muscle in black; other tissues stippled; shell and ctenidial filaments shown in outline; schematic. *a*, **Zygobranchia** (two-gilled aspidobranchs), e.g., *Diodora*; note the pair of symmetrically placed bipectinate ctenidia, and the accompanying exit hole in the shell; anus marked by (⁘); *b*, **Zygobranchia**, *Haliotis*; notice similarity to *Diodora* except for asymmetry and multiplication of vent holes; c, **Patellacea**, e.g., *Acmaea*; note that single ctenidium permits a crossflow of the respiratory water current, without need for a vent hold in the shell; *d*, **Trochacea**, e.g., *Tegula*; note general similarity of ctenidial arrangement to that in *Acmaea*, although shell is coiled; e, **Pectinibranchia**, e.g., *Nassarius*; note that ctenidium is monopectinate and fused to roof of mantle cavity, thus lessening chance of clogging with silt; siphon aids in clean water intake; *f*, **Pectinibranchia**, showing specialization of the ctenidium for filter feeding, e.g., *Crepidula*; extremely long filaments strain out food and transport it by ciliary action to a ciliated collecting groove along the right side of the body (diagrams mostly modified after Yonge, 1947).

There are several different schemes used in classifying the major gastropod groups. The following outline will indicate the chief divisions, together with terms which are more or less synonymously used to designate them.

Subclass I. PROSOBRANCHIA (= STREPTONEURA): Ctenidia in a forward location; nervous system streptoneurous; sexes separate. The subclass retains the main effects of torsion and includes the great majority of marine snails, with few fresh-water representatives.

Order 1. Aspidobranchia (= Scutibranchia, approximately = Diotocardia): Ctenidia bipectinate, paired or single; heart with two auricles (except in Patellacea). Divisible into several suborders or tribes, of which 3 are especially abundant locally.

Suborder 1. Zygobranchia (= Zeugobranchia): The abalones and keyhole limpets; shell slit or perforated; ctenidia paired (fig. 105, *a*, *b*); two auricles.

Suborder 2. Patellacea (= Docoglossa): The "true" limpets; shell conical; ctenidium single (fig. 105, *c*); one auricle.

Suborder 3. Trochacea: Top shells, turban shells; rather limpetlike internally, but with two auricles; shell coiled; single ctenidium (fig. 105, *d*).

Order 2. Pectinibranchia (=Ctenobranchia, approximately = Monotocardia): Single monopectinate ctenidium; one auricle. A very numerous group, including the majority of marine snails; divisible into about 18 tribes which fall into two suborders.

Suborder 1. Taenioglossa: Marine snails, generally without proboscis or siphon; radula fairly broad; usually herbivorous.

Suborder 2. Stenoglossa: Marine snails with proboscis and siphon (fig. 104, *d*; 105, *e*); radula narrow, often adapted for a carnivorous diet.

Subclass II. OPISTHOBRANCHIA: Ctenidium displaced to right side or rear of body by process of detorsion, or absent; nervous system euthyneurous as result of detorsion; entirely marine; hermaphroditic; tend to reduce or lose shell.

Order 1. Tectibranchia: Usually have a ctenidium; shell often present.

Order 2. Nudibranchia: Ctenidium lost, usually replaced by adaptive gills or dorsal processes; shell absent.

Subclass III. PULMONATA: Mantle cavity used as a lung (fig. 104, *f*); ctenidia lost; nervous system euthyneurous by condensation; mostly fresh-water and terrestrial; hermaphroditic.

(Opisthobranchia and Pulmonata are sometimes placed together as a single subclass, Euthyneura, but since euthyneury seems to have been achieved by different means in the two groups, this combination is unnatural.)

Most generic and specific identification of snails is based upon characteristics of the shell (fig. 104, *a-d*), or by form and color in shell-less types. There follow separate keys to the shelled intertidal gastropods, mostly prosobranchs, (p. 243), and to opisthobranchs (p. 264). Pulmonates as a group are omitted. For those who wish to go further in identification of gastropods, several excellent guides are available. The recently published *Illustrated Key to West North American Gastropod*

Genera, by Keen and Pearson (1952), provides a means of identifying the genera of marine shelled snails, more than two hundred genera being covered. An excellent illustrated glossary is included. For pulmonate land snails and slugs the student should consult the series of papers by Ingram and Lotz (1949–1950), "Land Molluscs of the San Francisco Bay Counties," as well as Mead's (1943) paper on the difficult genus of large land slugs, *Ariolimax*.

Key to Marine Shelled Gastropods

1. Shell with 1 or more perforations; 2 ctenidia . . . . Suborder Zygobranchia 2
1. Shell cap-shaped; apex may be recurved but does not coil . . 9
1. Shell with a coiled spire, (fig. 111) or cowry-shaped (fig. 118, *d*); without perforations 29
1. Shell irregularly twisted, attached; resemble serpulid worm tubes . . . . Family VERMETIDAE 75

2. With single perforation (keyhole limpets) 4
2. With row of several perforations . . . . *Haliotis* 3

3. Exterior of shell black, usually fairly clean and smooth; 5 to 8 holes open (black abalone) . . . . *Haliotis cracherodii*
3. Exterior brick-red, often rough and overgrown; 3 to 5 holes open (red abalone) . . . . *Haliotis rufescens*

4. Perforation on front slope, in front of the recurved apex, wide anteriorly and narrowed posteriorly; shell white, heavily ribbed, with ribs projecting beyond margin . . . . *Puncturella* spp.
4. Perforation at apex 5

5. Shell conical and covering entire animal 7
5. Shell relatively small and flattened, insufficient to cover the animal; shell itself more or less covered by the mantle in life . . . 6

6. Animal large (several inches in length); jet black with yellow sole; mantle usually covers shell; shell brownish-pink to white, 4-5 times as long as its perforation, edges finely crenulated . . . . *Megathura crenulata*
6. Animal smaller (1-2 in.); brownish or yellowish, often with dark markings; mantle covering more or less of shell; shell whitish, about 3 times as long as its perforation, grayish-white with radiating rays of gray or brown . . . . *Megatebennus bimaculatus*

7. Perforation round or oval . . . . *Diodora* 8
7. Perforation elongated, 3 times as long as wide; shell color white, rayed with rose. . . . . *Fissurella volcano*

8. Posterior margin of perforation markedly higher than anterior margin; length of shell may exceed 2 in.; fairly common intertidally . . . . *Diodora aspera*
8. Posterior margin only slightly higher than anterior margin; length of shell not more than 1 in.; generally subtidal and not common . . . . *Diodora murina*

9. With shelf on inner side of shell 10
9. Without such a shelf 12

10. Shelf attached at both sides (fig. 106, *a*, *b*) . . . . *Crepidula* 11
10. Shelf not attached on left side; projecting in tonguelike fashion . . . . *Crepipatella lingulata*

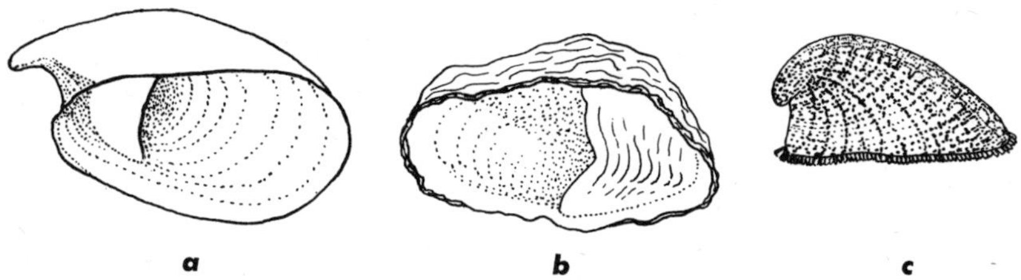

Fig. 106. *a, Crepidula adunca; b, Crepidula nummaria; c, Hipponix tumens.*

11. Shell black or dark brown; apex high and recurved; common on *Tegula* (fig. 106, *a*) . . . . *Crepidula adunca*
11. Shell flat or concave; white but covered with a golden-brown periostracum; internal shelf arching; on rocks and shells (fig. 106, *b*) . . . . *Crepidula nummaria*
11. Shell white with brown markings; low, flat or concave; on rocks or in borer holes, or in apertures of shells occupied by hermit crabs. . . . . *Crepidula perforans* (?)

12. Shell apex marginal or submarginal; shell white, may have gray-brown periostracum; attached by shelly base to rocky substrate . . . . *Hipponix* 13
12. Shell apex not marginal or, if so, shell not white and without periostracum; not attached 14

13. Shell regular to extremely irregular; radial striations obscure or discontinuous; concentric growth lines conspicuous . . . . *Hipponix antiquatus*
13. Shell regular; apex recurved (fig. 106, *c*); radial striations distinct and continuous; concentric striations present but not more conspicuous than radial; periostracum forms bearded fringe around the margin . . . . *Hipponix tumens*

14. Shell white, circular, with low central apex; radial striations intersect less conspicuous concentric lines; a shallow groove extends

from right-hand tip of the U-shaped muscle scar inside the shell to the margin; often considered a pulmonate, but possibly an offshoot of the opisthobranchs (button shell) . . . . *Trimusculus* (= *Gadinia*) *reticulatus*

14. Shells not as above; patelliform (limpetlike) 15

15. Shell thin, conical, with recurved apex; reddish or greenish, very smooth, with lighter radiating lines; has a "siphonal groove" comparable to that of *Trimusculus*, to which it is related; absence of a ctenidium will distinguish it from any local limpets; a scarce, low-tidal form . . . . *Williamia vernalis*

15. Shells generally oval; apex varies in position; a single bipectinate ctenidium in pallial chamber behind head (limpets, figs. 107-110) . . . . Family ACMAEIDAE 16

16. With accessory pallial gills beneath circumference of the mantle, in addition to the usual ctenidium; ribs broad and undulating, branching; margin crenate; shell low, with apex very near anterior end; interior generally dark with lighter, dimly owl-shaped, apical marking; sometimes very large (owl limpet) . . . *Lottia gigantea*

16. Lacking pallial gills; shell various . . . . . *Acmaea* 17

17. Shell with heavy radial ridges running from apex to shell margin, giving margin a distinct scalloped appearance; ridges may be variously spined 28

17. Shell with fine radial ribs or striations running from apex to margin of shell; margin entire, smooth or slightly crinkled, not scalloped; or shell smooth 18

18. Shell distinctly laterally compressed so that sides of aperture are essentially parallel 19

18. Shell with subcircular or oval aperture 21

19. Shell brown, inside and out 20

19. Shell brown outside with fine striations, interior bluish white; shell unstable when placed on flat surface owing to lateral margins being rocker-shaped; on *Laminaria andersoni* . . . *Acmaea instabilis*

19. Shell pure white (frequently overgrown) save for occasional small brown spot at the apex on inner and outer surfaces; about 1/4 in. long; on coralline algae and encrusted *Tegula brunnea* shells; rare . . . . *Acmaea triangularis*

20. Shell horny in appearance, up to 1 in. in length; width about 1/2 length; on *Egregia menziesii* . . . . *Acmaea insessa*

20. Shell minute, fragile; about 1/4 in. long with fine radial striations; width about 1/4 length; on *Phyllospadix* . . . *Acmaea paleacea*

21. Shell of same color inside and out unless altered by erosion or overgrowth of coralline algae; no distinct color pattern . . . 22

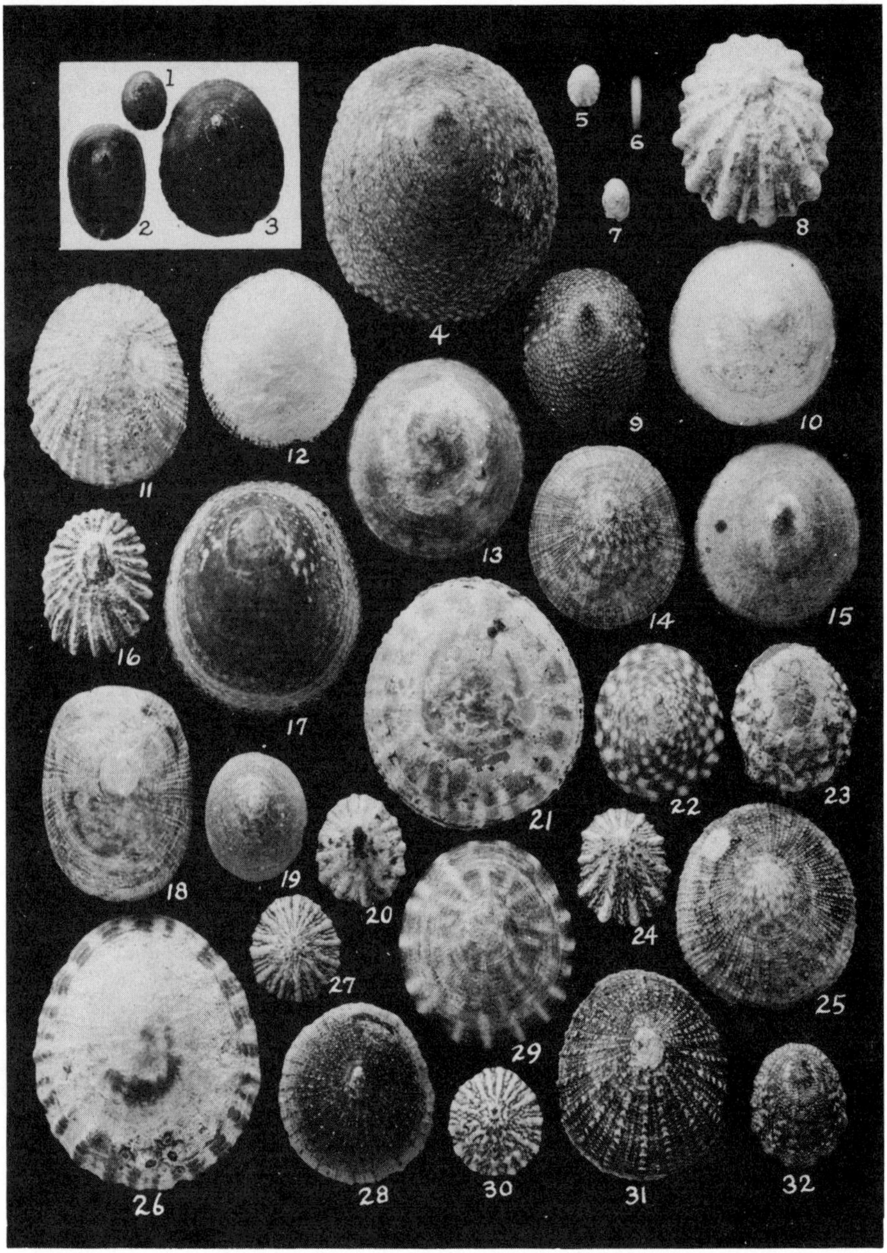

Fig. 107. Shells of various species of *Acmaea* in dorsal view. Listed alphabetically by species: 1, *A. asmi*; 20, 27, 30, *A. conus*; 23, 32, *A. digitalis*; 13, 15, *A. fenestrata cribraria*; 2, *A. insessa*; 18, *A. instabilis*; 11, 12, 25, 28, 31, *A. limatula*; 19, *A. mitra*; 19, *A. ochracea*; 6, *A. paleacea*; 3, 26, *A. pelta*; 4, 9, 17, *A. persona*; 5. *A. rosacea*; 8, 16, 24, *A. scabra*; 14, 21, 22, 29, *A. t. scutum*; 7, *A. triangularis*.

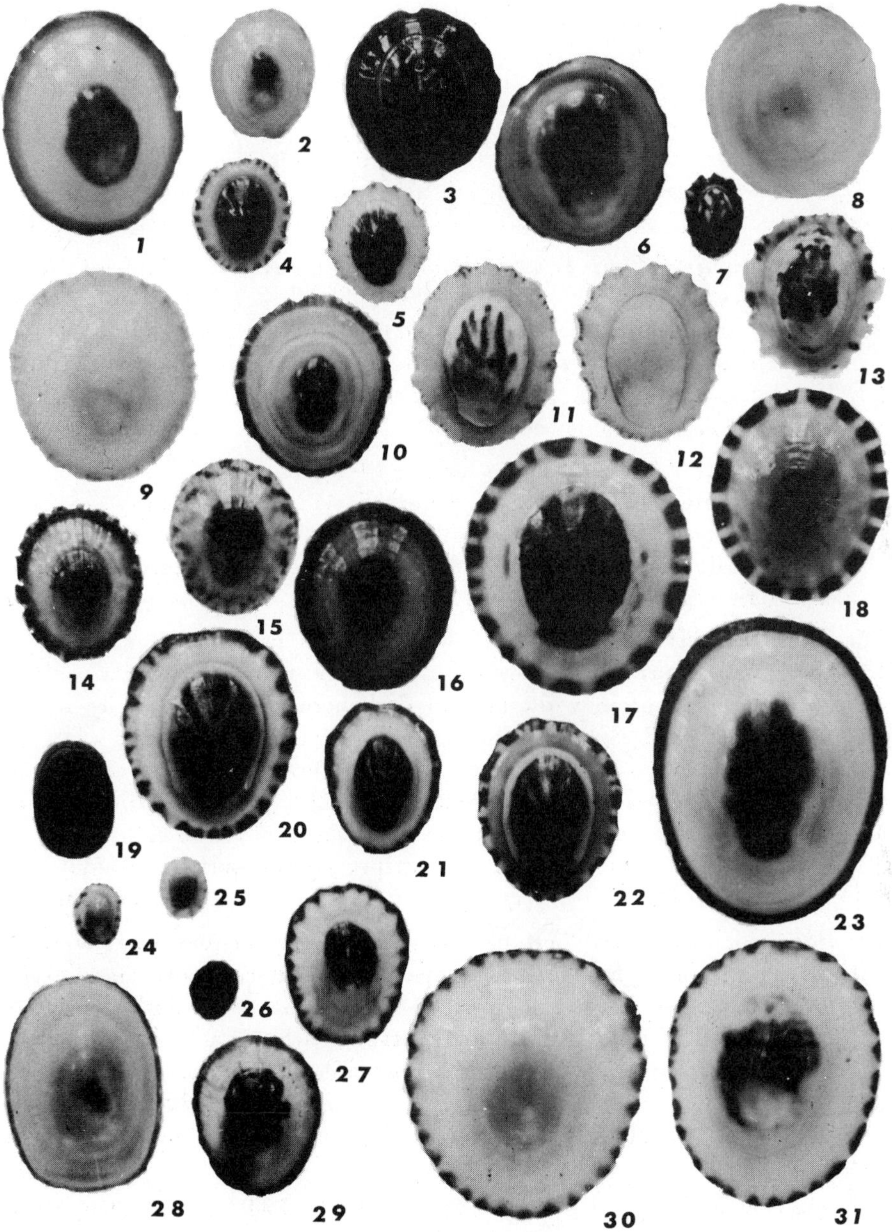

Fig. 108. Inside view of shells of various species of *Acmaea*. Listed alphabetically by species: 26, *A. asmi*; 3, 5, *A. conus*; 21, 22, 24, *A. digitalis*; 4, 6, *A. fenestrata*; 19, *A. insessa*; 28, *A. instabilis*; 9, 10, 14, 15, *A. limatula*; 8, *A. mitra*; 1, 2, *A. ochracea*; 20, 27, 29-31, *A. pelta*; 23, *A. persona*; 25, *A. rosacea*; 7, 11-13, *A. scabra*; 16-18, *A. t. scutum*.

21. Shell color distinctly different inside and out; both surfaces usually with color patterns 23

22. Shell black; small, not greater than 1/2 in. long; usually on *Tegula funebralis* . . . . *Acmaea asmi*
22. Shell white, thick, and heavy; aperture subcircular; no sculpturing; usually overgrown with coralline algae upon which limpet feeds. . . . . *Acmaea mitra*

23. Shell minute, about 1/4 in. to 3/8 in.; rare; white interior, with pink and white external pattern . . . . . . *Acmaea rosacea*
23. Height of shell at least 1/3 of width; usually more (height range from 35 per cent to 60 per cent of shell width) 26
23. Height of shell equal to 1/3 or less of width; upper limit slightly greater than 1/3 (35 per cent or so) 24

24. Width of shell distinctly more than 3/4 length 25
24. Shell fragile; width about 3/4 length with fine, equally spaced, radial riblets; 2 color phases: yellowish, with or without brown apical stain inside; or with dark and white pattern externally, white or bluish within . . . . *Acmaea ochracea*

25. Shell with olive, brown, and white external pattern, often checkerboardlike; interior bluish-white with large oval or owl-like apical brown stain; margin with internal dark border, entire or with alternating dark and light blocks of color; tentacles brownish . . . . *Acmaea testudinalis scutum*
25. Shell with spined riblets alternating with 2–4 finer striations; shell exterior brown or yellowish; interior white or yellowish; with or without apical stain and border stripe; dorsal surface of foot and head gray to black. . . . . *Acmaea limatula*

26. Posterior and lateral surfaces of shell clearly convex when viewed in profile; shell has inflated appearance; exterior with fine olive, brown, and white pattern; interior bluish-white with dark border at margin and oval apical stain; tentacles white; a nocturnal form, found on under-surfaces of high barren boulders and under ledges . . . . *Acmaea persona*
26. Posterior and lateral surfaces of shell not clearly convex unless disturbed by shell irregularities 27

27. Shell subcircular to evenly ovoid; anterior end only slightly more narrow than posterior; width more than 3/4 length; exterior brown with satinlike finish; interior with dark border passing into suffused brown; oval brown stain at apex; on smooth, polished boulders . . . . *Acmaea fenestrata cribraria*

   Intergrades in San Luis Obispo County, California, with southern subspecies, *A. fenestrata fenestrata*.

27. Shell width about 3/4 length; exterior with varied pattern, usually dark and white radiating bands or checkerboardlike; anterior end usually obviously narrower than posterior; interior margin dark or spotted dark and light; inside white or bluish and usually with dark stain behind apex; on rocks, *Postelsia, Egregia* and *Laminaria*; highly variable. . . . . *Acmaea pelta*

28. Shell with anterior surface usually clearly concave; posterior surface convex; apex very far forward, may overhang anterior margin; internal margin black with dark brown oval stain at apex; very common . . . . *Acmaea digitalis*
28. Shell very heavily ribbed, often with well-developed spines on ribs; interior dull, not lustrous, with or without scattered brown markings; dorsum of foot and head with black spots . . . . *Acmaea scabra*
28. Shell with heavy unspined ribs; usually with brown stain inside, posterior to apex; all external surfaces usually plane . . . . form of *Acmaea pelta* (see also 27)

29. Aperture entire, without indication of canal adjacent to columella 30
29. Aperture interrupted by notch or canal or rudiment thereof adjacent to columella 50
29. Aperture with faintly defined, nonconstricted, siphonal area; shell bluntly fusiform, with 2 broad revolving gray bands on white background (a tectibranch; see p. 264, fig. 111, *e*) . . . . *Acteon punctocoelata*

30. Shell **trochoid**, i.e., of general form of a *Tegula* (fig. 111, *a–c*), top-shaped, with blunt conical spire and more or less flat bottom; with pearly luster or white within (this feature should be looked for in very small shells which show the trochoid form less markedly) . . . . Families TROCHIDAE and TURBINIDAE 31
30. Shell **turriform** or awl-like with relatively small aperture (not more than 1/2 total length of shell); spire sharp pointed, usually of more than 5 turns; umbilicus closed 40
30. Shell globular or only slightly elongated (not more elongate than a *Littorina*); if interior of aperture is pearly, refer back to first item under 30 44
30. Shell **bulloid**; thin and earlike, with relatively enormous aperture; spire sunken; in life largely covered by mantle (bubble shell, a tectibranch, see p. 264) . . . . *Haminoea* sp.

31. Lacking tooth on columellar callus; shells sculptured with regular spiral lines, or with pronounced undulated peripheral ridge . . 35
31. Most local species with tooth on columellar callus; shells without spiral lines or sculpture, or with such sculpture weak or irregular . . . . *Tegula* 32

Fig. 109. Shells of various species and one hybrid of the genus *Acmaea* seen in lateral view. Listed alphabetically by species: 1, *A. asmi*; 5, 6, *A. cona*; 3, 10-14, 16-19, *A. digitalis*; 7, 8, *A. mitra*; 28, 29, *A. ochracea*; 9, *A. paleacea*; 23-27, *A. pelta* x *A. digitalis*; 2, *A. rosacea*; 15, 20-22, *A. scabra*; 4, *A. triangularis*.

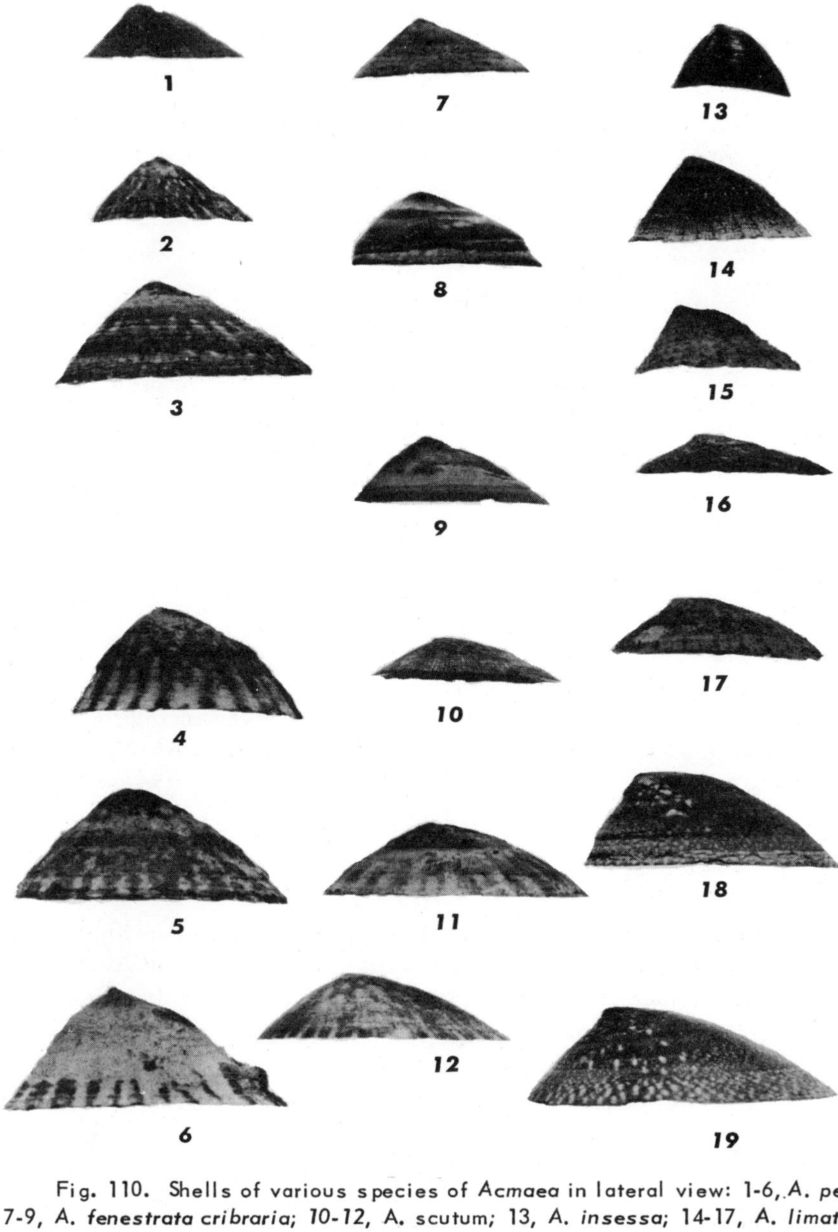

Fig. 110. Shells of various species of *Acmaea* in lateral view: 1-6, *A. pelta*; 7-9, *A. fenestrata cribraria*; 10-12, *A. scutum*; 13, *A. insessa*; 14-17, *A. limatula*; 18-19, *A. persona*.

32. Umbilicus open; base and whorls flat; body whorl angulated; shell pyramidal; usually on floating kelp 33
32. Umbilicus closed; whorls and base rounded (more angular when young); common in rocky intertidal 34

33. With distinct tooth at inner side of aperture (fig. 111, *a*); umbilicus deep, with wall lower on side adjacent to aperture . . . . *Tegula montereyi*
33. Lacking tooth on columellar callus; umbilicus with wall that seems to separate it from aperture . . . . *Tegula pulligo*

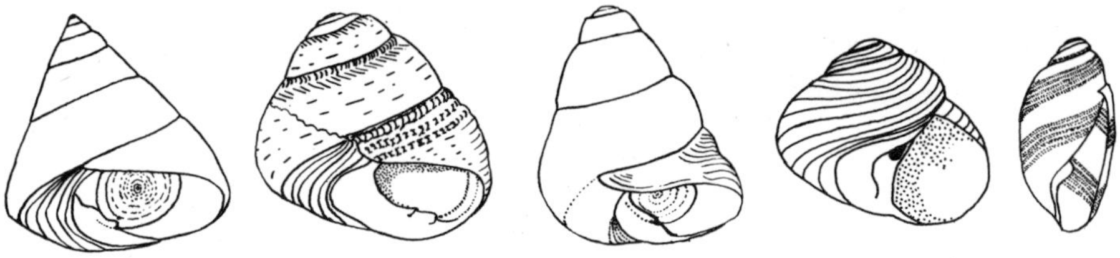

a b c d e

Fig. 111. *a, Tegula montereyi; b, T. funebralis; c, T. brunnea; d, Margarites* sp.; *e, Acteon punctocoelata.*

34. Color black; outer lip overlapping the whorl above (fig. 111, *b*); in upper mid-tidal (black turban) . . . . *Tegula funebralis*
34. Color brownish; outer lip meeting but not overlapping the whorl above (fig. 111, *c*); larger; in lower mid-tidal (brown turban) . . . . *Tegula brunnea*

35. Umbilicus open; sculpture of spiral lines (fig. 111, *d*); small; very pearly within . . . . *Margarites* spp.
35. Umbilicus closed; larger 36

36. Sculpture in form of spiral lines; shell not wider than high; operculum horny . . . . *Calliostoma* 37
36. Sculpture either as spiral lines or as strong marginal undulations; operculum heavily calcified . . . . . Family TURBINIDAE 39

a b c

Fig. 112. *a, b, Calliostoma costatum; c, C. canaliculatum* (after Oldroyd).

37. Whorls rounded, spiral lines smooth; sutures distinct (fig. 112, *a*, *b*); color reddish-brown, often with tinge of blue; intertidal . . . . *Calliostoma costatum*
37. Whorls angulated; sutures less obvious; shell pyramidal; characteristic of floating kelp 38

38. Color ivory with thin revolving dark lines; revolving ridges smooth; base sharply angulated (fig. 112, *c*) . . *Calliostoma canaliculatum*
38. Color golden with purple on lower edges of whorls; revolving ridges beaded; base slightly angulated . . . . *Calliostoma annulatum*

39. Small; aperture and whorls rounded; reddish or dark brown; sculpture of spiral lines; outer surface of operculum pearly . *Homalopoma* spp.
39. Large; reddish; periphery of whorls with an expanded undulated edge that extends outer edge of aperture (not next to columella like an ordinary siphon); operculum heavily calcified with smooth rounded surface. . . . . *Astraea inaequalis*

    This species resembles *Astraea undosa* of southern California; the latter has a more pronounced silky periostracum and 3 heavy ribs on the surface of its operculum.

40. White; whorls beset with numerous varices or lamellae . . . 41
40. Whorls smooth or faintly ribbed 42

41. Varices thin and numerous; extend all the way down around basal whorl (fig. 113); numerous species . . . . . *Epitonium* spp.
41. Varices somewhat fewer and heavier; stop at a basal ridge (at about the line of the suture) leaving basal area of shell free of varices. . *Opalia* spp.

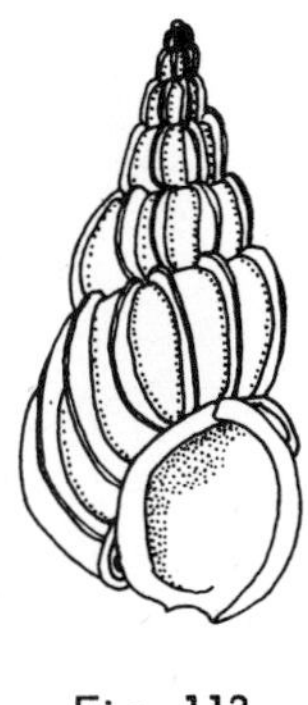

Fig. 113. *Epitonium* sp.

42. Small; white; gleaming and smooth; sutures indistinct; sharp-spired. . . . . . . *Balcis* sp.
42. Brownish; not shining; sutures evident . . 43

43. Minute; aperture rounded and not more than 1/3 length of shell. . . . . *Barleeia* sp.
43. Small, up to 5 mm.; aperture somewhat elongated (nearly 1/2 length of shell); columella with an oblique ridge; found in *Salicornia* marshes (a pulmonate) . . . . *Phytia setifer*

44. Umbilicus open 45
44. Umbilicus closed 47

45. Inner lip with groove extending up into umbilicus (fig. 114, *a*); very small; numerous on eelgrass and kelp. . . . . . *Lacuna* spp.
45. Medium-sized to large; shell globular; light-colored; foot extends to cover much of shell when crawling; predatory upon other molluscs in sand or mud (moon snails) . . . . *Polinices* 46

46. A shoulder on the whorl parallel to the suture; height of shell a little greater than maximum diameter as measured at right angles to the axis . . . . *Polinices lewisii*
46. No shoulder present; spire lower (height of shell less than maximum diameter); brown callus beside umbilicus . . . *Polinices draconis*

47. Medium-sized; externally rather nondescript, sometimes eroded; operculum horny; obvious and abundant in upper intertidal (littorines, periwinkles) . . . . *Littorina* 48
47. Small shells, not fitting above description; numerous forms, only 2 of which are mentioned 49

48. Spire short, of 3 whorls, often brownish and eroded; area of columella polished away, exposing a flat area of underlying shell; diagonal white band inside lower part of aperture (fig. 114, *b*); the commonest snail of the splash zone . . . . *Littorina planaxis*
48. Spire slightly longer, of 4 whorls, generally blackish in life (both littorines show regular mottling if clean, and tend to become lighter in color after drying); no polished area of columella (fig. 114, *c*); common in upper mid-tidal, not as high as *L. planaxis* . . . . *Littorina scutulata*

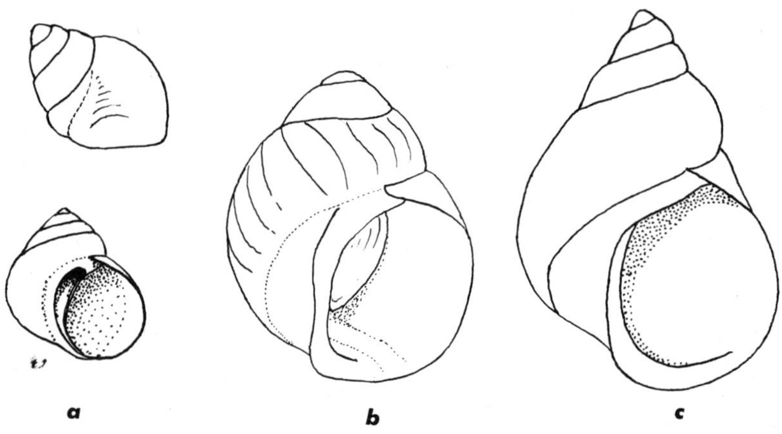

Fig. 114. *a, Lacuna;* above, side view; below, bottom view showing open umbilicus; *b, Littorina planaxis; c, L. scutulata.*

49. Small; externally smooth; sutures not deeply impressed; finely but irregularly streaked or mottled with red; operculum heavily calcified, convex, shining . . . . *Phasianella* spp.
49. Very small; dark; whorls rounded with sutures very deeply impressed; may be abundant on weeds in brackish pools of salt marshes . . . . *Amnicola* (?)

50. Shell obovate (fig. 118, *b*), obconic (fig. 119), cypraeiform (cowrylike, fig. 118, *d*), or bluntly fusiform and shining (fig. 118, *c*) . . . 71

50. Shell not as above; if bluntly fusiform, not shining 51

51. A deep spiral furrow separating body whorl from canal; canal short and ending in a notch (fig. 115) 52
51. Lacking such a furrow. 54

52. Axial ribs very numerous, but not much stronger than spiral lines, forming regular granules or nodes at intersections with each other (fig. 115, *c*); color pale . . . . *Nassarius fossatus*
52. Axial ribs decidedly heavier than spiral lines. 53

53. Spaces between axial ribs not wider than ribs themselves (fig. 115, *b*); color pale . . . . *Nassarius mendicus*
53. Spaces between axial ribs twice as wide as ribs (fig. 115, *a*); variously marked in brown, yellow, and white . . *Nassarius cooperi*

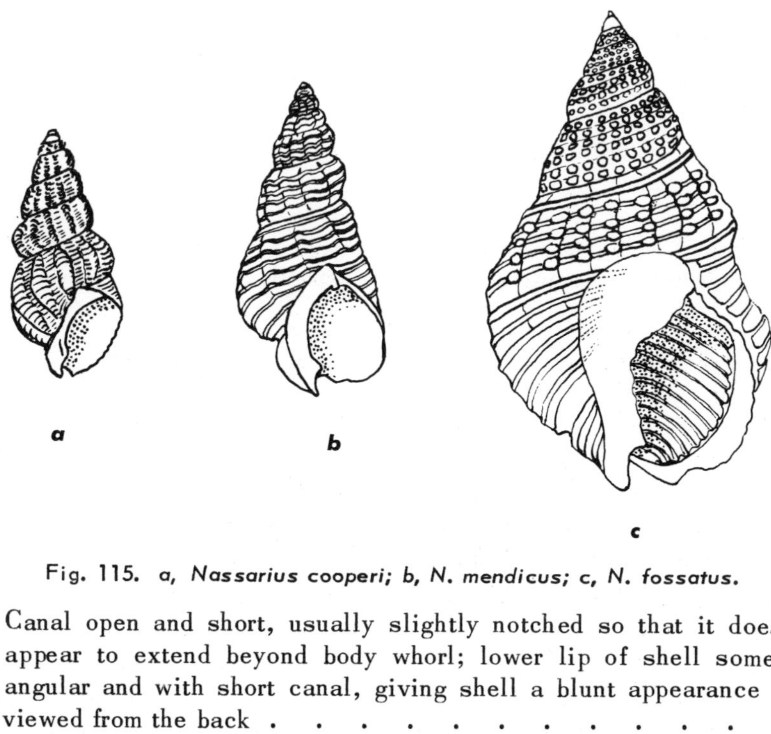

Fig. 115. *a, Nassarius cooperi; b, N. mendicus; c, N. fossatus.*

54. Canal open and short, usually slightly notched so that it does not appear to extend beyond body whorl; lower lip of shell somewhat angular and with short canal, giving shell a blunt appearance when viewed from the back 55
54. Canal open or closed, pronounced or obscure, but projecting enough that shell does not have a distinctly blunt appearance when viewed from the back 58

55. Medium-sized; spire prominent but often blunt or eroded; surface blackish with obscure spiral and longitudinal lines; introduced and extremely abundant in San Francisco Bay (mud snail) . . . . *Nassarius (= Ilyanassa) obsoletus*

55. Small; fusiform; with pointed spire 56

56. Surface smooth; a variety of colors and markings; upper end of aperture somewhat angular. . . . . . . *Mitrella* (= *Columbella*) 57
56. Surface with fine spiral lines . . . . *Mitromorpha* sp.

57. Body whorl with distinct shoulder or keel near the suture, giving aperture a somewhat quadrangular form . . . *Mitrella carinata*
57. As above, but shoulder slightly or not at all developed . . . . *Mitrella* spp.

58. Top of aperture at middle of shell length or higher; spires consequently appear relatively short. Not a sharp division; *Ocenebra* will be considered here, although some species have smaller apertures 59
58. Top of aperture below middle of shell length; spires relatively pronounced 67

59. With tooth on outer lip near lower edge of aperture 60
59. Without such a tooth 61

60. With 3 very large varices or flanges on body and other whorls; canal usually closed and aperture constricted (fig. 116, *a*) . . . . *Purpura foliata*
60. Without flanges; canal open; spiral series of small brown rectangles in interstices of sculpturing; young usually lack spine . . . . *Acanthina spirata*

61. Principal sculpture spiraled; shells with flat or shouldered area on whorls next to suture (figs. 116, *b-d*) . . . . *Thais* 62
61. Principal sculpture in form of axial ribs; whorls rounded, without shoulder near suture 63
61. Sculpture, both spiral and axial, usually pronounced; no shoulders on whorls in those where spiral sculpture predominates; canal may be closed; a difficult group . . . . *Ocenebra* 64

62. Sculpture of numerous, close-set, uniform, smooth spiral ridges separated by narrow furrows; a narrow shoulder makes the suture very distinct; aperture nearly 2/3 height of shell (fig. 116, *b*) . . . . *Thais canaliculata*
62. Sculpture variable, mostly of fairly large, beaded, spiral ridges alternating with smaller ridges; color variable, commonly in spiral pattern; aperture very large, about 3/4 height of shell; spire relatively quite low (fig. 116, *c*); very common in exposed situations . . . . *Thais emarginata*
62. Shell very heavy; sculpture of angular spiral ridges crossed by smaller longitudinal ridges; aperture somewhat more than 1/2 length of shell; spire rather prominent (fig. 116, *d*); most common in semiprotected situations . . . . *Thais lamellosa*

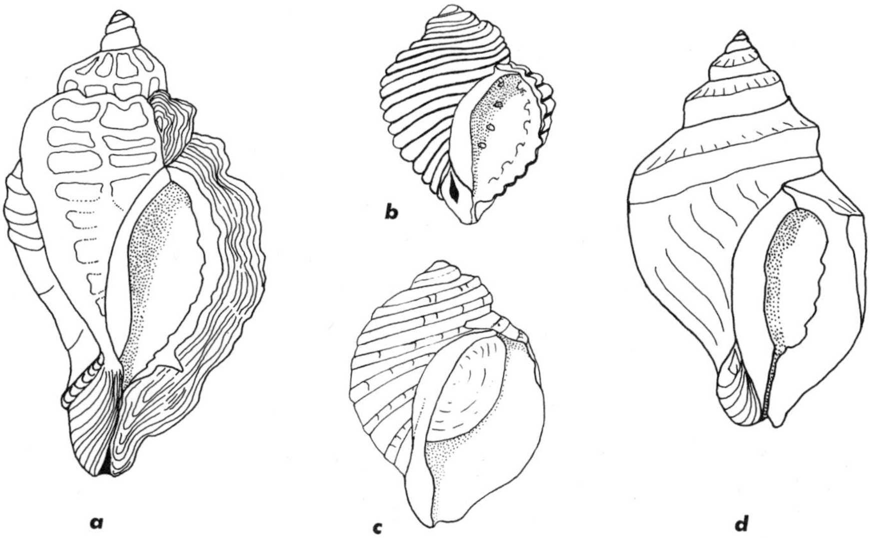

Fig. 116. *a, Purpura foliata; b, Thais canaliculata; c, T. emarginata; d, T. lamellosa.*

63. Small; often varicolored; sculpture in form of numerous discrete slanted axial ridges (on the body whorl these are restricted to upper slope); aperture widest near lower end; outer lips with small teeth; canal distinct but short . . . . *Amphissa* spp.
63. Larger; uniformly colored reddish brown or gray; axial ridges larger, fewer, and crossed by fine spiral ridges; aperture widest at middle, tapering toward canal; outer lip without teeth (oyster drill) . . . . *Urosalpinx cinereus*

64. Height of spire equal to, or greater than, height of body whorl . 65
64. Height of spire distinctly less than that of body whorl 66

65. Sculpturing of spiral ribs crossed by longitudinal (axial) ribs of equal prominence, giving effect of basket weave . *Ocenebra interfossa*
65. Sculpturing of flattened spiral ribs without prominent axial cross ribs . . . . *Ocenebra gracillima*

66. Sculpturing of regular, flattened, spiral ribs not markedly interrupted by longitudinal growth lines; grooves shallow; shell regular in outline . . . . *Ocenebra lurida*
66. Sculpturing of elevated, rounded, spiral ribs, undulated by prominent rounded longitudinal swellings; grooves deep and sharply cut; outline irregular; color ivory with regularly distributed brown markings . . . . *Ocenebra circumtexta*

67. Shells with pronounced spiral sculpture 68

67. Shell of medium size with no spiral sculpture; with a band next to suture resembling a second suture; periphery of whorls set with short axial riblike prominences; color yellowish-brown, lighter on prominences . . . . *Pseudomelatoma torosa*

68. Shell with prominent axial ridges 69
68. Shell without prominent axial ridges 70

69. Medium-sized; canal pronounced and extended; sculpture of spiral lines and prominent axial ribs; ground color dark, with axial ribs yellowish and conspicuous . . . . . . *Fusinus luteopictus*
69. Medium-sized; canal rudimentary; with pronounced longitudinal ribs or varices; characteristically in muddy pools of salt marshes . . . . *Cerithidea californica*

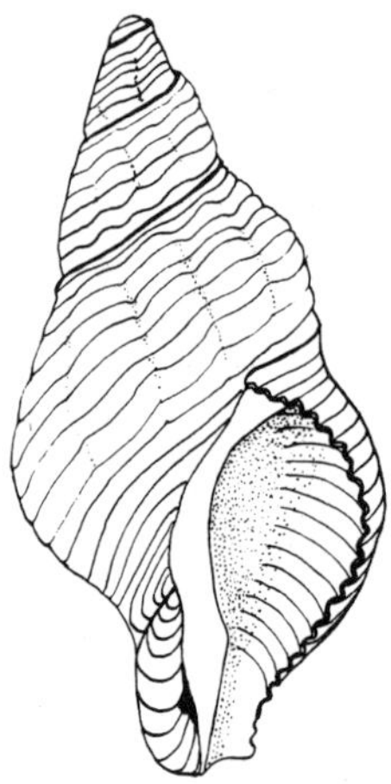

Fig. 117. *Searlesia dira.*

70. Shell fairly large, strong; canal pronounced and extended; sculpture of uniform smooth spiral lines, with low axial ribs; outer lip with corrugations extending far inside, corresponding to external sculpturing (fig. 117); color gray, usually encrusted with coralline algae; not known south of Pigeon Point . . . . *Searlesia dira*
70. Small; canal open and rudimentary; 5–7 flat spiral ridges on each whorl; without axial ribs (fig. 118, *a*) . . . . . *Bittium* spp.

71. Shell cypraeiform (cowry-shaped) or nearly so (fig. 118, *d*); spire concealed 72
71. Shell obconic (fig. 119); spire low but exposed; aperture nearly as long as whole shell . 73
71. Shell obovate or bluntly fusiform (fig. 118, *b*, *c*); aperture at least 1/2 length of shell; shell without spiral sculpture, gray and smooth; a notch at siphonal area of apertural margin; lobes of foot nearly cover shell in life; found burrowing just beneath surface of clean sand . *Olivella* 74

72. Shell tiny; smooth, white, glistening; outer lip thin; common among weeds and gravel of tide pools . . . . *Cypraeolina pyriformis*
72. Shell small; ribbed; outer lip thickened cowry-fashion (fig. 118, *d*); scarce . . . . *Pusula* (= *Trivia*) *californiana*

73. Shell typically obconic (fig. 119, *a*); outer lip thin; aperture without teeth; dull brown or spotted with white . . . *Conus californicus*
73. Shell obconic; outer lip thickened; aperture with teeth on both margins (fig. 119, *b*); surface glossy . . . . *Erato* sp.
73. Shell elongate-obconic, approaching cylindrical; outer lip thin; aperture with teeth on inner margin only (fig. 119, *c*); surface glossy . . . . *Marginella* spp.

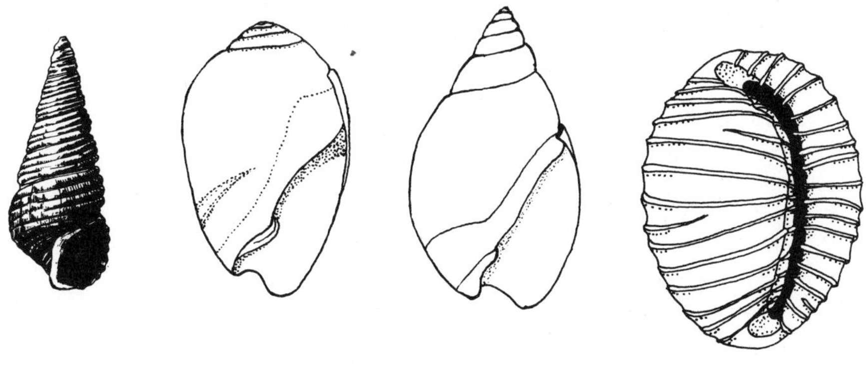

Fig. 118. *a, Bittium; b, Olivella biplicata; c, O. pycna; d, Pusula californiana.*

74. Aperture 2/3 length of shell; spire relatively low; 2 (or 3) columellar folds (fig. 118, *b*); shell without wavy pattern. . *Olivella biplicata*
74. Aperture somewhat more than 1/2 length of shell; spire fairly prominent (fig. 118, *c*); columellar fold usually single; shell with zigzag or wavy pattern . . . . *Olivella pycna*

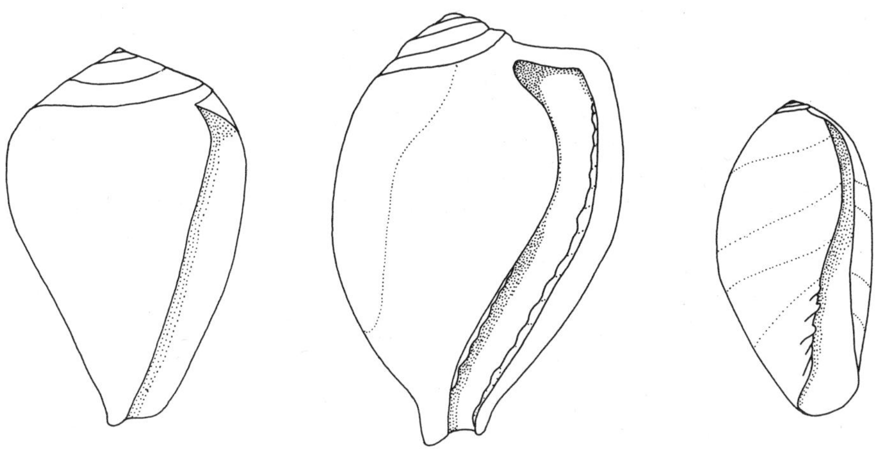

Fig. 119. *a, Conus californicus; b, Erato sp.; c, Marginella sp.*

75. Large (tube up to 1/2 in. in diameter); usually solitary; white, with pronounced longitudinal scaly ribs; operculum lacking; interior of tube smooth . . . . *Aletes squamigerus*
75. Small; clustered; brownish; with transverse growth ridges; operculum well developed, thin and chitinous; interior of tube with a faint sharp ridge along at least part of columellar wall (i.e., on inside of turns) . . . . *Petaloconchus montereyensis*

List of Marine Littoral Shelled Gastropods

Subclass PROSOBRANCHIA
  Order Aspidobranchia
    Suborder Zygobranchia
      FISSURELLIDAE
        Diodora aspera (Eschscholtz, 1833)
        Diodora murina (Dall, 1885)
        Fissurella volcano Reeve, 1849
        Megatebennus bimaculutus (Dall, 1871)
        Megathura crenulata (Sowerby, 1825)
        Puncturella spp.
      HALIOTIDAE
        Haliotis cracherodii Leach, 1817
        **Haliotis wallalensis* Stearns, 1899 (Usually encountered north of Point Reyes)
        Haliotis rufescens Swainson, 1822
    Suborder Patellacea
      ACMAEIDAE
        Acmaea asmi (Middendorf, 1849)
        Acmaea fenestrata cribraria Carpenter, 1866
        Acmaea digitalis Eschscholtz, 1833
        Acmaea insessa (Hinds, 1842)
        Acmaea instabilis (Gould, 1846)
        Acmaea limatula Carpenter, 1864
        **Acmaea limatula mörchii* Dall, 1878 (Occurs in Tomales Bay; possibly only an ecological variant of *A. limatula.*)
        Acmaea mitra Eschscholtz, 1833
        Acmaea ochracea (Dall, 1871)
        Acmaea paleacea Gould, 1851
        Acmaea pelta Eschscholtz, 1833 (= A. cassis)
        Acmaea persona Eschscholtz, 1833
        Acmaea rosacea Carpenter, 1866
        Acmaea scabra (Gould, 1846)
        Acmaea testudinalis scutum Eschscholtz, 1833
        Acmaea triangularis (Carpenter, 1866)
        Lottia gigantea Sowerby, 1843
    Suborder Trochacea
      TROCHIDAE
        Calliostoma annulatum (Martyn, 1784)
        Calliostoma canaliculatum (Martyn, 1784)
        Calliostoma costatum (Martyn, 1784)
        **Calliostoma* spp.

**Margarites lirulatus* (Carpenter, 1864)
**Margarites parcipictus* (Carpenter, 1864)
**Margarites succinctus* (Carpenter, 1864)
Margarites spp.
Tegula brunnea (Philippi, 1848)
Tegula funebralis (Adams, 1854)
Tegula montereyi (Kiener, 1850)
Tegula pulligo (Martyn, 1784)

PHASIANELLIDAE
Phasianella spp.

TURBINIDAE
Astraea inaequalis (Martyn, 1784)
**Homalopoma carpenteri* (Pilsbry, 1888)
Homalopoma spp.

Order Pectinibranchia

Suborder Taenioglossa

EPITONIIDAE
Epitonium (numerous species)
**Opalia chacei* (Strong, 1930)
Opalia spp.

EULIMIDAE
Balcis spp.

ERATOIDAE
Pusula (= *Trivia*) *californiana* (Gray, 1828)
Erato spp.

CERITHIOPSIDAE
**Cerithiopsis* spp.
**Seila montereyensis* Bartsch, 1907

CERITHIIDAE
Bittium (numerous species)
Cerithidea californica Haldeman, 1840

VERMETIDAE
Aletes squamigerus Carpenter, 1856 (Grows singly in this area; in south grows in great intertwined masses; see Plate VIII in Ricketts and Calvin)
Petaloconchus montereyensis Dall, 1919

LITTORINIDAE
Littorina planaxis Philippi, 1847
Littorina scutulata Gould, 1849

LACUNIDAE
Lacuna (several species)

BARLEEIIDAE
Barleeia spp.

HIPPONICIDAE

Hipponix antiquatus (Linnaeus, 1767)

Hipponix tumens Carpenter, 1865

CALYPTRAEIDAE

Crepidula adunca Sowerby, 1825

Crepidula nummaria Gould, 1846

Crepidula perforans Valenciennes, 1846

Crepipatella lingulata (Gould, 1846)

NATICIDAE

Polinices draconis (Dall, 1903)

Polinices lewisii (Gould, 1847)

Suborder **Stenoglossa**

CONIDAE

Conus californicus Hinds, 1844

TURRIDAE

Pseudomelatoma torosa (Carpenter, 1865)

OLIVIDAE

**Olivella boetica* Carpenter, 1863

Olivella biplicata (Sowerby, 1825)

Olivella pycna Berry, 1935

MARGINELLIDAE

Cypraeolina pyriformis (Carpenter, 1865)

Marginella spp.

MITRIDAE

Mitromorpha spp.

FUSINIDAE

Fusinus luteopictus (Dall, 1877)

NEPTUNEIDAE

Searlesia dira (Reeve, 1846)

Urosalpinx cinereus (Say, 1822)

NASSARIIDAE

Nassarius cooperi (Forbes, 1850) [Some workers consider *cooperi* a subspecies of *mendicus*: *N. mendicus cooperi* (Forbes, 1850)]

Nassarius fossatus (Gould, 1849)

Nassarius mendicus (Gould, 1849)

Nassarius obsoletus (Say, 1822) (An introduced species; has usually been called *Ilyanassa obsoleta*, although the consensus is to consider *Ilyanassa* at most a subgenus of *Nassarius*)

**Nassarius perpinguis* (Hinds, 1844)

PYRENIDAE (COLUMBELLIDAE)

**Amphissa versicolor* Dall, 1871

Amphissa (several species)

Mitrella carinata (Hinds, 1844)
Mitrella (several species)
MURICIDAE
Acanthina spirata (Blainville, 1832)
Ocenebra circumtexta Stearns, 1871
Ocenebra gracillima Stearns, 1871
Ocenebra interfossa Carpenter, 1864
Ocenebra lurida (Middendorf, 1849)
**Ocenebra* spp.
Purpura foliata Martyn, 1784
Thais canaliculata (Duclos, 1832)
**Thais canaliculata compressa* Dall, 1915 (This subspecies, which occurs at Carmel, is distinguished by larger size and fewer spiral ridges than in typical *T. canaliculata*)
Thais emarginata (Deshayes, 1839)
Thais lamellosa (Gmelin, 1792)

Subclass OPISTHOBRANCHIA
Order Tectibranchia
ACTEONIDAE
Actaeon punctocoelata (Carpenter, 1862)
AKERIDAE
**Haminoea vesicula* (Gould, 1855) (See p. 264)

Possibly allied to OPISTHOBRANCHIA (see also *Oncidiella*, pp. 264 and 270)
*PYRAMIDELLIDAE, a family containing numerous species, especially of *Odostomia* and *Turbonilla*. Pyramidellids are commensal or parasitic on polychaetes or bivalve molluscs. See Dall and Bartsch (1909) and Fretter and Graham (1949).
TRIMUSCULIDAE (GADINIIDAE)
Trimusculus (= *Gadinia*) *reticulatus* (Sowerby, 1835) (Generally found hanging from underside of ledges. Eggs are laid in jelly masses and hatch to free-swimming veligers)
SIPHONARIIDAE
Williamia vernalis (Dall, 1870)

Subclass PULMONATA
ELLOBIIDAE
Phytia setifer (Cooper, 1872) (Characteristic of *Salicornia* marshes)

SUBCLASS OPISTHOBRANCHIA

In this group **detorsion** has occurred, bringing the anus and the ctenidium (if present) back to the right side or to the rear of the body, and "untwisting" the nervous system to the euthyneurous condition. Opisthobranchs are hermaphroditic, exclusively marine, and show a marked tendency to reduce the shell, which is often completely buried in the mantle in the order **Tectibranchia** and absent in the order **Nudibranchia**, except in larval stages. The opisthobranchs are characterized by a striking variety of form, color, and habits, but are difficult to preserve satisfactorily. Positive identification is often a matter of some difficulty, and only the more common forms are included in the following key.

Key to the More Common Opisthobranchia

by Joan E. Steinberg
University of California, Berkeley

1. Shell present or absent; a series of gills (representing a true ctenidium) present within a mantle cavity on right side or dorsally; a pair of enrolled or tubular rhinophores usually present anteriorly . . . . Order **Tectibranchia** 2
1. Shell absent; true ctenidium absent but body usually provided with dorsal processes and/or secondary gills (branchiae); anterior part of dorsum usually bears a pair of rhinophores; exceedingly diverse in color, form, and size . . . . Order **Nudibranchia** 5
1. Small, flattened, lacking shell; resembling a shell-less limpet; color dark brown with pattern of alternating dark and light patches on margin; sometimes considered a pulmonate, but more likely an offshoot of the Opisthobranchia . . . *Oncidiella* (= *Arctonchis*) sp.

2. Shell present and more or less exposed 3
2. Shell either absent or completely buried in mantle 4

3. Shell oblong with conical spire; spirally grooved; color white with revolving black lines . . . . *Acteon punctocoelata*
3. Shell thin, spire sunken, rather transparent, partly hidden by mantle . . . . *Haminoea* spp.

4. Small, bluish black; no conspicuous external flaps or tentacles; found burrowing in surface of sand flats . . . *Aglaja diomedea*
4. Medium sized (2 in.); green with black longitudinal striping; pointed posteriorly; tentacles and parapodial flaps not pronounced; characteristically found on broad-leaved eelgrass (*Zostera*) . . . . *Phyllaplysia taylori*

4. Large, sometimes reaching 15 in.; brownish to greenish; 2 large parapodial flaps more or less covering back; 2 cephalic tentacles, 2 rhinophores that stand up like ears (sea hare) . . . . *Tethys californica*

5. With a circlet of gill-like branchial plumes about anus on posterior dorsal surface; rhinophores present (fig. 120, *a*); other dorsal processes may be present or absent 6

5. Without a circlet of branchial plumes on posterior dorsal surface; anus often on right side of body; rhinophores usually present (fig. 120, *b*); other dorsal processes may be present or absent. . . . 19

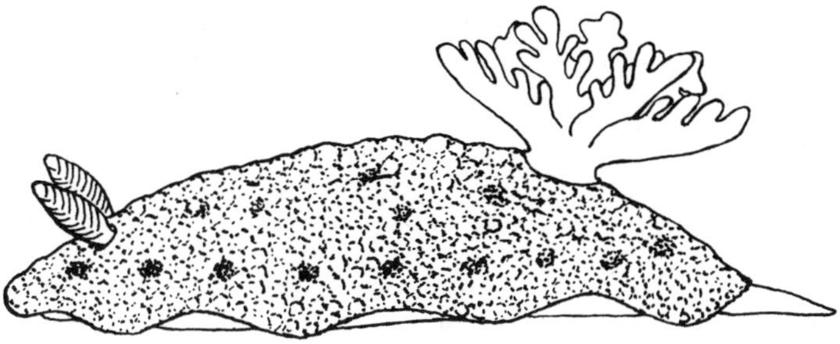

a

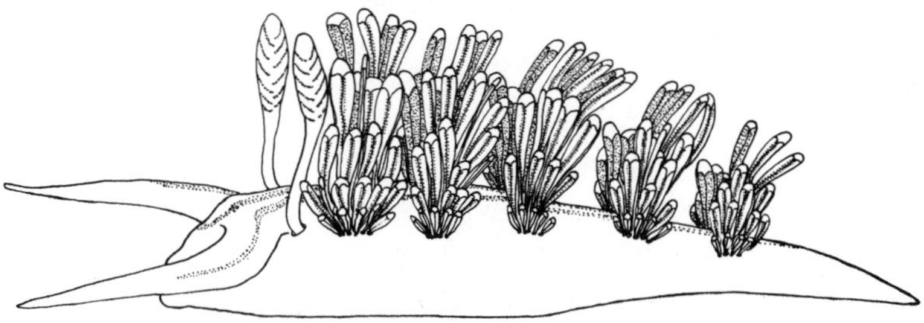

b

Fig. 120. *a*, a typical dorid nudibranch; *b*, *Hermissenda crassicornis*.

6. Branchial plumes fully retractile beneath the body surface; lacking elongate processes on dorsum; body oval and somewhat flattened. 7

6. Branchial plumes contractile but not retractile beneath the body surface; usually with dorsal processes on body 14

7. Mouth opening porelike; body yellow with small contrasting white spots . . . . *Dendrodoris fulva*

7. Mouth opening a vertical slit or inverted T in shape; body yellow or not yellow, but without contrasting white spots 8

8. Color red; mature specimens typically less than 2 cm. in length . 9
8. Color not red; mature specimens more than 2 cm. in length . . 10

9. Two black spots on back, 1 in front of branchiae and 1 behind rhinophores . . . . *Aldisa sanguinea*
9. No conspicuous black spots; often found on red sponges, which it matches closely in color . . . . *Rostanga pulchra*

10. Dorsal surface velvety in appearance, without obvious tubercles 11
10. Dorsal surface set with distinct tubercles 12

11. Dark brown or black rings on dorsum; ground color whitish to dark tan . . . . *Diaulula sandiegensis*
11. Seven to 10 lemon-yellow spots on each side of dorsum; ground color whitish; rhinophores dark . . . . *Cadlina flavomaculata*

12. Color whitish; dorsum with low tubercles, each tipped with lemon-yellow; margin and foot edged with yellow . . *Cadlina marginata*
12. Color yellow to orange, with darkly pigmented areas 13

13. Dark pigment both on dorsum and on tubercles; oral tentacles ear-shaped and grooved externally; branchial plumes dingy yellow . . . . *Archidoris montereyensis*
13. Dark pigment only on dorsum, not on the tubercles; oral tentacles fingerlike; branchial plumes whitish . . . . *Anisodoris nobilis*

14. Body elliptical and flattened; numerous tapering dorsal processes in addition to branchiae; color vivid rose-pink . *Hopkinsia rosacea*
14. Body elongate and sluglike, sometimes humped; usually with a frontal veil bearing short processes 15

15. Body humped; striped longitudinally in black and gray, with yellowish spots; 4 pointed anterior processes . . . . *Polycera atra*
15. Body not striped; usually with numerous dorsal processes . . 16

16. Body white; markings dark brown or orange 17
16. Body orange to brownish, with or without bluish spots . . . 18

17. Dorsal surface thickly set with short, blunt, white tubercles; usually with dark brown spots on body between tubercles. . . . . *Aegires albopunctatus*
17. Middorsal region set with low white tubercles; dorsolateral margins set with club-shaped white processes tipped with orange . . . . *Laila cockerelli*
17. Dorsum and sides set with low orange tubercles; dorsolateral margins set with longer orange processes . . . . . *Triopha carpenteri*

18. Body orange to dark brown with bluish spots that may be inconspicuous in young, making them appear uniformly orange; branchial

plumes orange-red; dorsolateral processes entirely orange-red; usually found in tide pools . . . . *Triopha maculata*

18. Body yellowish-brown with or without bluish spots; branchial plumes whitish; dorsolateral processes tipped with orange-red; usually found in beds of floating kelp . . . . *Triopha grandis*

19. Dorsal processes absent 20
19. Dorsal processes present 21

20. Body oval, elongate, pointed behind; moderately large; mantle dark red or brown with fine longitudinal white lines; branchiae situated ventrally . . . . . . *Armina* (= *Pleurophyllidia*) *californica*
20. Body elliptical, very flat, notched behind; usually about 1 cm. in length; branchiae located posteroventrally on either side of anus; occurs on and closely matches the bryozoan *Membranipora* on floating kelp . . . . *Corambe pacifica*

21. Dorsal processes branched 22
21. Dorsal processes not branched 23

22. Dorsal processes consist of a number of separate, small branchial plumes set on 2 longitudinal dorsolateral ridges; a delicate pattern of chalky white lines upon the more transparent dorsum . . . . *Duvaucelia* (= *Tritonia*) *festiva*
22. Dorsal processes large, stout, extensively branched; animals of large or moderate size and extremely soft; oral veil bears branched processes; rhinophores retractile into sheaths which are branched at apex . . . . *Dendronotus* spp.
22. Dorsal processes palmately branched; head bears 2 palmately branched frontal lobes; animal reddish- to greenish-brown with irregular white spots; coloration closely resembles brown algae such as *Laminaria* on which it is frequently found; average length about 2 cm. . . . . *Hancockia californica*

23. Dorsal processes numerous, tapering and pointed. 25
23. Dorsal processes constricted at base (club-shaped or leaflike), sometimes easily detached 24

24. Dorsal processes few, expanded and leaflike; rhinophores absent; animal possesses curious extensible oral hood used in catching food; typically found in kelp beds . . . . . . *Melibe leonina*
24. Dorsal processes more numerous (but readily detached), inflated, tuberculate on inner surface, largest ones closest to median line; body translucent reddish-brown with white spots; head expanded into conspicuous undulating anterior veil . . . *Dirona picta*
24. Dorsal processes club-shaped, not tuberculate, tipped with white; dorsal surface of body greenish, flecked with brown and white;

very small (less than 1 cm.), on *Obelia* and other hydroids . . . . *Eubranchus* (= *Galvina*) *olivacea*

25. Body purple with rusty-red dorsal processes and rhinophores . . . . *Flabellina iodinea*
25. Body whitish; dorsal processes in 4–5 distinct clusters, translucent gray with brown cores and subterminal orange rings; opalescent blue line runs down median line from oral tentacles, bifurcating 2–3 times to enclose orange areas (fig. 120, *b*) . . . . *Hermissenda crassicornis*
25. Body brownish to pinkish; dorsal processes slightly flattened, grayish brown, tipped with white; not in distinct clusters but cover dorsolateral surfaces like a shaggy coat . . . *Aeolidia papillosa*

List of Opisthobranchia

(Classification of nudibranchs follows O'Donoghue, 1926)

Order Tectibranchia

Acteon punctocoelata (Carpenter, 1864)
Aglaja diomedea (Bergh, 1894)
**Haminoea vesicula* (Gould, 1855)
Haminoea sp.
**Navanax inermis* (Cooper, 1862)
Phyllaplysia taylori Dall, 1900
**Pleurobranchus californicus* Dall, 1900
Tethys californica (Cooper, 1863)

Order Nudibranchia
Section Sacoglossa
Tribe Holohepatica (= Doridacea) (Liver in a single mass)
Superfamily ZONOBRANCHIATAE
**Duvaucelia exsulans* (Bergh, 1894)
Duvaucelia (= *Tritonia*) *festiva* (Stearns, 1873)
**Duvaucelia tetraquetra* (Pallas, 1788)
Superfamily CRYPTOBRANCHIATAE (Branchiae retractile into permanent pockets)
Aldisa sanguinea (Cooper, 1862)
Anisodoris nobilis (MacFarland, 1905)
Archidoris montereyensis (Cooper, 1862) (One of the commonest nudibranchs on this coast; often associated with the sponge *Halichondria*)
Cadlina flavomaculata MacFarland, 1905
Cadlina marginata MacFarland, 1905

Dendrodoris (= *Doriopsis*) *fulva* (MacFarland, 1905)
Diaulula sandiegensis (Cooper, 1862)
**Discodoris heathi* MacFarland, 1905
**Glossodoris* (= *Chromodoris*) *californiensis* (Bergh, 1879)
**Glossodoris* (= *Chromodoris*) *porterae* (Cockerell, 1902)
Rostanga pulchra MacFarland, 1905

Superfamily PHANEROBRANCHIATAE (Branchiae not retractile into permanent pockets)

**Acanthodoris brunnea* MacFarland, 1905
**Acanthodoris columbina* MacFarland, 1926
**Acanthodoris hudsoni* MacFarland, 1905
**Acanthodoris lutea* MacFarland, 1926
Aegires albopunctatus MacFarland, 1905
**Ancula pacifica* MacFarland, 1905
Corambe pacifica MacFarland and O'Donoghue, 1929
Hopkinsia rosacea MacFarland, 1905
Laila cockerelli MacFarland, 1905
Polycera atra MacFarland, 1905
Triopha carpenteri (Stearns, 1873)
**Triopha catalinae* (Cooper, 1863)
Triopha grandis MacFarland, 1905
Triopha maculata MacFarland, 1905

Tribe Cladohepatica (= Aeolidiacea) (Liver divided and in most families ramified)

Aeolidia papillosa (Linnaeus, 1761)
Armina (= *Pleurophyllidia*) *californica* (Bergh, 1862) (Modified for a burrowing life in sand; often obtained by dredging)
**Dendronotus giganteus* O'Donoghue, 1921 (One of the largest nudibranchs on this coast. Like others of this genus it is capable of swimming short distances by an undulating motion of the body.)
Dendronotus spp.
**Dirona albolineata* MacFarland, 1912
Dirona picta MacFarland in Cockerell and Eliot 1905
Eubranchus (= *Galvina*) *olivacea* (O'Donoghue, 1922)
**Fiona pinnata* (Eschscholtz, 1831) (On driftwood bearing gooseneck barnacle colonies; cerata bear branchial membranes)
Flabellina iodinea (Cooper, 1862)
Hancockia californica MacFarland, 1923
Hermissenda crassicornis (Eschscholtz, 1831) (One of the commonest nudibranchs on this coast; occupies widely varying habitats, e.g., rocks, mud, wharf pilings)

**Janolus fuscus* O'Donoghue, 1924. (Resembles *Hermissenda crassicornis* but cerata extend in front of the rhinophores)

Melibe leonina (Gould, 1853)

Section **Ascoglossa**

**Phyllobranchopsis enteromorphae* Cockerell and Eliot, 1905. (Mouth parts modified for feeding on fluids of algae such as *Enteromorpha*, on which it is found)

Of questionable position:

Oncidiella (= *Arctonchis*) sp. (Fairly common on stipes of laminarians, but inconspicuous. For other near-opisthobranchs see p. 263)

Moss Animals

Under the general heading of "Moss Animals" are included two superficially similar but fundamentally unlike phyla: the large assemblage of the true *Bryozoa* and the small group of the *Entoprocta.*

Phylum Entoprocta (Kamptozoa, Calyssozoa)

Members of the small phylum **Entoprocta**, together with those of the large phylum **Bryozoa**, are often called "Moss animals." In the past the entoprocts generally have been included with the Bryozoa and are still so included in much of the taxonomic literature; hence we cover the two groups in a single key in this manual. However, the entoprocts differ fundamentally from the Bryozoa proper in lacking a true coelom, in possessing protonephridia with flame cells, and in having the anus opening within the circlet of tentacles which also surrounds the mouth (hence "ento-proct"). In contrast, the true bryozoans are coelomate animals which lack excretory organs and which have the anus opening outside the circlet of tentacles ("ecto-proct"). In the field, entoprocts are readily recognized by the fact that the individual stalks (fig. 121, *a*, *b*) are separately movable in a characteristic fashion, leading to the common name of "nodding heads." Individuals of the true Bryozoa are not separately movable. When using a microscope, the simplest way to distinguish the two phyla is to notice whether the tentacles are rolled in to the center when withdrawn, as in a hydroid, in which case it is an entoproct, or whether the whole anterior end of the individual body is drawn back into the basal part with the tentacles still pointing forward, in which case it is a bryozoan. There are few local species of entoprocts, inconspicuous because of their small size, but fairly common, forming low mosslike patches beneath boulders or on the bases of hydroids or algae.

Phylum Bryozoa (Polyzoa, Ectoprocta)

The varied and abundant Bryozoa are beautiful objects to study, but not easy, because of their small size, diversity, complexity, and number of species. The key that follows covers only the more easily recognized and commoner forms of the Monterey area. The student is warned not to attempt to force every bryozoan he finds into this key, since almost any collection will include some forms not covered here, or in which identification is a task for a specialist.

The bryozoan colony or **zoöarium** is composed of numerous individuals united in characteristic ways (figs. 121, *c-i*, 122), with the result that the colony may be creeping, erect, flexible, stiff, encrusting, foliaceous, or arborescent. The arborescent forms may be articulated (divided into internodes by nodes or joints) or nonarticulated. The bryozoan individual consists of a zoöecium (plural zoöecia) and a polypide. The zoöecium is the basal part of the individual, consisting of a sack-like body wall over which is secreted an external wall, usually chitinous or calcareous, sometimes membranous or gelatinous. The polypide is the distal extensible part of the individual, including the digestive tract and the lophophore, which bears the ciliated tentacles. The polypide may be extended through an orifice at the upper end of the zoöecium and can be retracted within the zoöecium, in which process the tentacles are drawn back base first in a fashion characteristic only of the true Bryozoa. The orifice may lack a covering (suborder **Cyclostomata**, fig. 121, *c-g*), may be closed by a folded or toothed membrane (suborder **Ctenostomata**, fig. 121, *h, i*), or may be covered by a chitinous operculum (suborder **Cheilostomata**, fig. 122, which includes the majority of bryozoans). The front wall of the zoöecium is often referred to as the aperture, especially when this is membranous or uncalcified; the term "aperture" should not be confused with "orifice." The edge of the orifice may be produced into small laterally placed projections, or there may be a single denticle borne within the lower margin. The lower margin is frequently hollowed out to form a sinus.

A feature important in classification, but difficult to see, is the elevation of the edges of the orifice into a raised collar or peristome, the opening of which is spoken of as a secondary orifice. Close study of a secondary orifice will usually reveal the margin, often bearing a denticle, of the primary orifice within it. In many cases it is advisable to study young individuals at the edges of a colony in order to see the primary orifice and other features which may be hidden or lost in the more heavily calcified older zoöecia. The lower lip of the secondary orifice may be divided by a channel or elevated into a tooth (mucro) or an umbo.

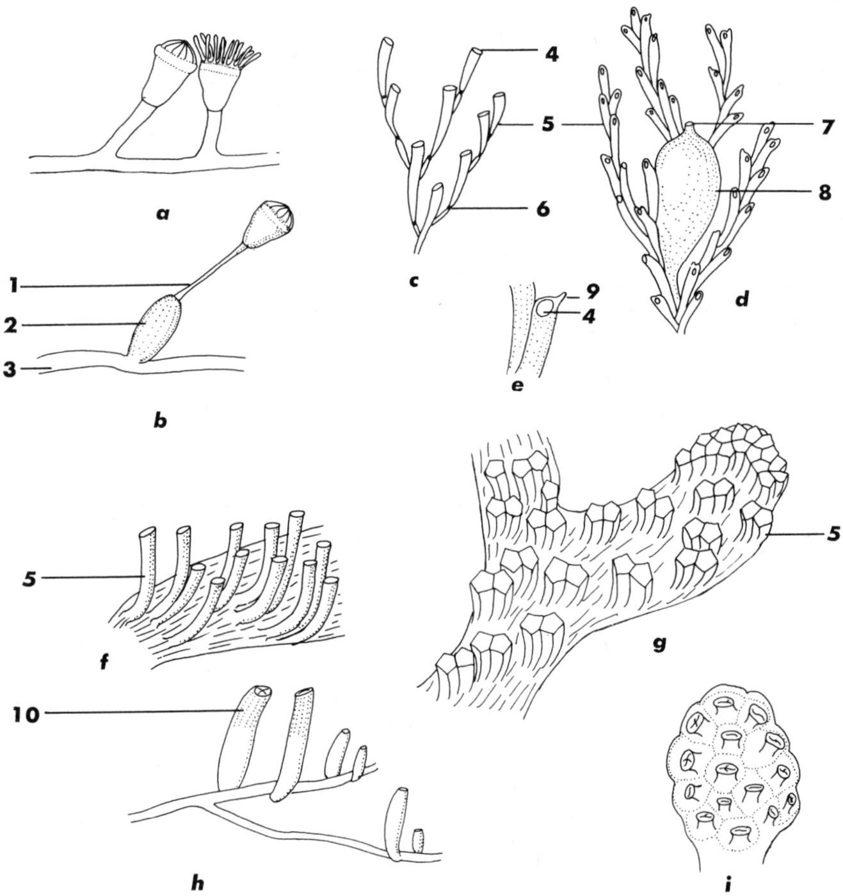

Fig. 121. *a-b*, **Entoprocta**; *a*, *Pedicellina*; *b*, *Barentsia*. *c-g*, **Bryozoa, Cyclostomata**: *c*, *Filicrisia*; *d*, *Crisia*; *e*, *Crisia*, enlargement to show orifice and spine; *f*, *Tubulipora*; *g*, *Idmonea*. *h-i*, **Bryozoa, Ctenostomata**; *h*, *Bowerbankia*; *i*, *Alcyonidium*. 1, Stem; 2, muscular dilation; 3, stolon; 4, orifice; 5, zoöecium; 6, joint; 7, oöeciopore; 8, ovicell; 9, spine; 10, retracted tentacles.

Within the single colony the zoöecia may assume a variety of forms quite different from that of the typical feeding individual. Some are reduced and modified to bird-head-like snapping structures called avicularia (singular, avicularium), which may be stalked or sessile and either frontally or laterally placed upon typical zoöecia. Vibracula are movable whiplike structures, which are also modified zoöecia, as are the prominent spines of such forms as *Flustrella corniculata*. The oöecium or ovicell is an expanded zoöecium in which the embryos develop. Oöecia often take the form of hoodlike chambers above the zoöecia, as in Cheilostomata, or may appear as much-inflated zoöecia, as in the Cyclostomata.

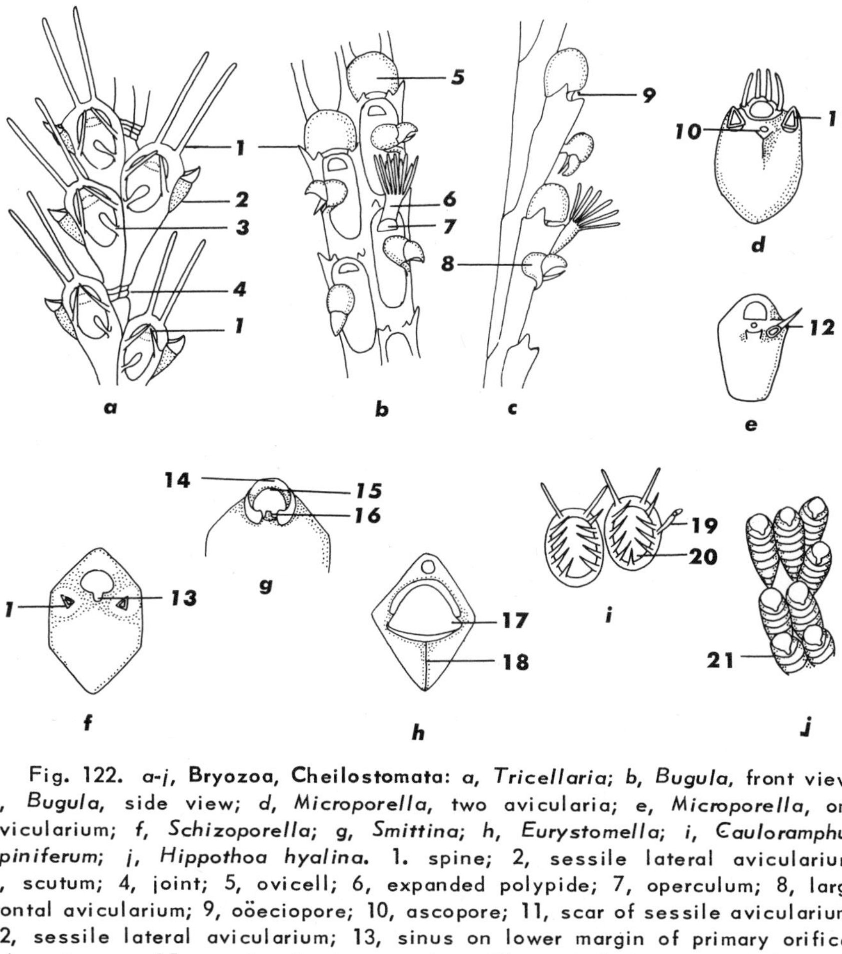

Fig. 122. *a-j*, Bryozoa, Cheilostomata: *a*, *Tricellaria*; *b*, *Bugula*, front view; *c*, *Bugula*, side view; *d*, *Microporella*, two avicularia; *e*, *Microporella*, one avicularium; *f*, *Schizoporella*; *g*, *Smittina*; *h*, *Eurystomella*; *i*, *Cauloramphus spiniferum*; *j*, *Hippothoa hyalina*. 1. spine; 2, sessile lateral avicularium; 3, scutum; 4, joint; 5, ovicell; 6, expanded polypide; 7, operculum; 8, large frontal avicularium; 9, oöeciopore; 10, ascopore; 11, scar of sessile avicularium; 12, sessile lateral avicularium; 13, sinus on lower margin of primary orifice; 14, peristome; 15, margin of primary orifice; 16, tooth (below margin of primary orifice); 17, extended corners of orifice; 18, keel; 19, stalked avicularium; 20, uncalcified front; 21, transverse lines.

A study of figures 121 and 122 should acquaint the student with the general appearance of typical bryozoan structures, but he should remember that these figures do not attempt to show the great variety of form that bryozoans may assume. Most of the species and genera in the following key may be identified with the aid of a 10X hand lens after a little practice, but for serious study it is generally necessary to employ magnifications of 25–35X and good illumination in order to see much detail. Identifications made with the aid of the key should, if possible, be verified by reference to the recent authoritative works of Osburn on West Coast Bryozoa.

Key to the More Easily Recognized Entoprocta and Marine Bryozoa (Ectoprocta)

by Joan C. Rattenbury and R. I. Smith
McGill University, Montreal, and University of California, Berkeley

1. Individuals consisting of bell-shaped heads upon independently movable stalks arising from a creeping stolon. . ENTOPROCTA 2
1. Colony (zoöarium) composed of individuals (zoöecia) not ordinarily divisible into stalk and body regions; individuals not movable . . . . BRYOZOA (ECTOPROCTA) 4

2. Stalk of more or less uniform diameter throughout (fig. 121, *a*) . . . . *Pedicellina cernua*
2. Stalk with 1 or more distinct muscular dilations at which movement takes place (fig. 121, *b*) . . . . *Barentsia* 3

3. Stalks not branched, of fairly stout build; dilations at base. *Barentsia gracilis*
3. Stalks branched with muscular dilations at bases of branches . . . . *Barentsia ramosa*

4. Colony soft or gelatinous; zoöecia lacking calcareous skeletons . . . . CTENOSTOMATA 5
4. Zoöecia wholly or partly encased in stiff skeletons, sometimes horny, and hyaline, usually calcified. 8

5. Colony a low encrustation of close-packed, squat, gelatinous zoöecia, often so closely fused that outlines are indistinct . . . 7
5. Colony of separate upright tubular zoöecia 6

6. Colony forming a dense gray or brown furry coating on mussels, barnacles, and pilings; characteristically in harbors; zoöecia arise from creeping stolons (fig. 121, *h*). . . . *Bowerbankia gracilis*
6. Colony appearing as a delicate whitish fuzz in branchial chambers

and on legs of crabs; individual zoöecia very transparent and set upon slender pedicels arising from creeping stolons . . . . *Triticella elongata*

7. Colony brownish, of close-set zoöecia, between which arise branched, reddish, antlerlike spines up to 1 mm. in height; encrusting algal stipes, corallines, and the like . . . . *Flustrella corniculata*
7. Colony gray, brown, or colorless; lacking spines; sometimes produced into small, stalked lobes; individual zoöecia close-set (fig. 121, *i*); encrusting rocks, legs and carapace of kelp crabs, algal stipes, and so on . . . . *Alcyonidium* sp.

8. Zoöecia in form of distinct whitish tubes with orifice at, or nearly at, the freely projecting upper end of each tube; surface often finely pitted, but large pores and conspicuous sculpturing absent; orifice edges only occasionally produced into spines; opercula and avicularia lacking; colonies may be branching and arborescent, encrusting, or raised into stiff, flattened lobes . . . . Cyclostomata 9
8. Zoöecia almost never resemble simple tubes; of diverse form with orifice at upper end of the "front"; orifice provided with an operculum; spines, avicularia, pores, and sculptured ornamentation usual; colonies arborescent, foliaceous, or encrusting . . . . Cheilostomata 12

9. Colony arborescent, flexible; slightly stiff in appearance; whitish 10
9. Colony encrusting or produced into irregularly branched, flattened, inflexible outgrowths 11

10. Zoöecia uniserial (fig. 121, *c*); joints between each 2 zoöecia may blacken, giving older colonies a characteristic speckled appearance; low, sparse, delicate growths on rocks and among algal holdfasts . . . . *Filicrisia franciscana*
10. Zoöecia biserial (fig. 121, *d*, *e*); joints few, between groups of a dozen or more zoöecia; forms a larger and less delicate colony than *Filicrisia*; whitish, stiff and bushy in appearance . . *Crisia* sp.

11. Zoöecia encrusting; in a fan-shaped or circular mat, with long, free, tubular zoöecia directed upward (fig. 121, *f*) . . . *Tubulipora* sp.
11. Zoöecia closely fused, forming an erect, fanlike or branched, flattened, inflexible growth, from the upper side of which the short, free ends of the oöecial tubes project (fig. 121, *g*) . . . . *Idmonea* and several other genera

12. Colony bushy or arborescent 15
12. Colony in form of single-layered flat growth, either leaflike or perforated; erect or loosely applied to the substratum 13
12. Colony forming an adherent matlike encrustation; if produced into erect lobes these are of 2 layers of zoöecia, back to back . . 19

13. Colony an erect, ruffled, netlike growth; resembles lace, but inflexible and brittle; color orange . . *Phidolopora* (= *Retepora*) *pacifica*
13. Colony in form of greenish to brownish leaflike lobes . . . 14

14. Lobes flexible; zoöecia elongated, slightly widened distally; front membranous; 1 or 2 erect spines on each side of orifice, and 2-5 medially inclined spines on each zoöecial margin; avicularia lacking; in field may be confused with *Tegella* . *Dendrobeania lichenoides*
14. Lobes brittle; zoöecia broadly oval, fronts membranous; margins bearing several spines; large, conspicuous avicularia alternating with zoöecia; often found with, and in field possibly confused with, *Dendrobeania* . . . . *Tegella robertsonae*

15. Colonies small, pale, and delicate; slender stems with elm-tree-like spreading branches; zoöecia flared outward at upper ends, which are fringed with several very long curved spines, contributing to a characteristic iridescence and graceful appearance . *Caulibugula ciliata*
15. Colonies coarse, bushy, often large; brownish, yellowish or purplish 16

16. Colony purplish or deep brown; zoöecia biserial; ovicells globose and relatively conspicuous; avicularia lacking; forms large dense bushy masses on rocks and pilings . . . . . *Bugula neritina*
16. Color some shade of tan or yellowish brown 17

17. Branches flattened, arising in a regular close spiral; color yellow brown; forms large dense colonies; bird-head-like avicularia prominent on fronts of zoöecia (fig. 122, *b*, *c*) . . . *Bugula californica*
17. Branches arising in a generally irregular fashion; colonies brownish, tufted or bushy; small flap (scutum) partly covering aperture; avicularia sessile on upper outer angles of zoöecia 18

18. Colony soft and finely branched, with joints separating groups of three zoöecia; aperture of zoöecium with prominent spines at sides and top margin; only sessile lateral avicularia present (fig. 122, *a*) . . . . *Tricellaria* sp.
18. Colony somewhat stiff and coarsely branched, with joints separating groups of more than 3 zoöecia; small frontal avicularia are present in addition to sessile lateral avicularia; whiplike organs (vibracula) may be present on dorsal side of older parts of colony . . . . *Scrupocellaria* sp.

19. Front of zoöecium membranous, uncalcified 20
19. Front of zoöecium partly or completely calcified 23

20. Zoöecium distinctly oval; walls of adjacent zoöecia not closely appressed 21
20. Zoöecium rectangular, corners angular or rounded; walls of adjacent zoöecia closely appressed 22

21\. Fourteen to 16 spines on margin of zoöecium (fig. 122, *i*); avicularia on short stalks attached among rows of spines; colony brownish yellow in life . . . . *Cauloramphus spiniferum*

21\. Margin of zoöecium granulated; no avicularia; often triangular hollow areas above fronts of zoöecia or on either side . . *Conopeum* spp.

22\. Zoöecium long and rectangular; narrow; margin high, narrow and smooth, bearing a short, blunt spine at each upper angle of zoöecium; colony forming delicate mat on algal fronds and shells . . . . *Membranipora membranacea*

22\. Margin of zoöcium low, bearing small teeth; at each lower angle of zoöecium a large, heavy, blunt spine; the 2 spines at end of a zoöecium may fuse at their bases to form a thick wall; colony forming obvious, white, lacelike encrustation on algal fronds. . . . . *Membranipora tuberculata*

23\. Zoöecia oval to fusiform, front bearing series of transverse lines (fig. 122, *j*); no avicularia; colony hyaline, forming silvery, delicate, circular patches on algal fronds and shells . . *Hippothoa hyalina*

23\. Zoöecia variable in shape; front not showing transverse lines . 24

24\. Zoöecia with primary orifice only; no produced peristome; lower margin of orifice without tooth 25

24\. Zoöecia with produced peristome surrounding primary orifice; a tooth on lower margin of orifice in types considered in this key. . . 29

25\. Orifice semicircular, lower margin complete, corners often contracted, never extended 28

25\. Orifice with sinus on lower margin; avicularia present . . . 27

25\. Orifice wide, semicircular, lower margin slightly concave, corners extended; no spines; generally no avicularia 26

26\. No pores on front of zoöecium; in older specimens a keel passing down front wall from lower border of orifice to lower margin of zoöecium; zoöecia diamond-shaped (fig. 122, *h*); colony rose-colored when alive; operculum large, brown, and obvious . . . . *Eurystomella bilabiata*

26\. Front of zoöecium pierced by several large pores; colony brownish yellow when alive; rarely large avicularia present . . . . *Lyrula hippocrepis*

27\. A sessile avicularium on low mound on either side of orifice (fig. 122, *f*); surface of front marked with fine anastomosing lines; punctured near margins in younger specimens; colony usually salmon pink when alive; often forming large patches on rocks and shells . . . . *Stephanosella biaperta*

(*Schizoporella cornuta* is very similar and perhaps more common, but has usually one avicularium, larger and more frontally placed.)

27. On one side of median line, front rising to a high mound bearing single large avicularium which overhangs the orifice (best seen in young, marginal specimens); in older heavily calcified individuals the orifice may be sunk rather deeply and there are usually large numbers of avicularia scattered over the colony . *Rhynchozoon tumulosum*

28. Zoöecia rhomboidal, white; avicularia either 2, 1 on each side of orifice, or 1 on 1 side of orifice only; a special pore (ascopore) on front just below orifice; ascopore often partly hidden by being on upper slope of an umbo; front wall punctured; variable number of long spines (4-7) on upper margin of orifice. A complex of forms including *Microporella californica* (stouter with 2 avicularia, fig. 122, *d*); and *Microporella ciliata* (more delicate, with 1 avicularium, fig. 122, *e*).

28. Zoöecia without avicularia; a high umbonate process on the front below orifice, with a crescentic pore on its upper surface; surface of front with prominent pores; a small denticle on either side of orifice near lower margin; ovicells globose and radially grooved; colony encrusting and typically rising into yellowish 2-layered foliaceous expansions; common . . . *Hippodiplosia insculpta*

29. Zoöecia with peristome channeled on lower margin; a median hyaline tooth on lower margin of primary orifice (fig. 122, *g*); usually 2 (sometimes 3) protuberances on front below the orifice and a median avicularium placed below these protuberances; usually a row of pores around margin of zoöecium; colony white; encrusting rocks and shells; often heavily calcified . . . . *Parasmittina collifera* and other related spp.

29. Zoöecia with lower border of peristome raised to form a mucro or projection overhanging the orifice; behind mucro is a tooth on lower margin of primary orifice; commonly with large, sessile avicularia, each occupying the position of a complete zoöecium, distributed irregularly throughout the colony . . . . . . *Smittina* sp.

Because of the appalling number and diversity of bryozoans found in the local intertidal, the revisers have felt it best to limit this key to fairly distinctive and readily identifiable genera and species. Any collection, especially of encrusting types, will yield material not covered here. The student wishing to make an intensive study of bryozoans will have to invest in, or borrow, Osburn's 3 volumes.

List of Entoprocta

Barentsia gracilis var. *nodosa* (Lomax, 1886) (Look for the Barentsias under water in tide pools, on the bases of corallines and other algae)
Barentsia ramosa (Robertson, 1900)
**Myosoma spinosa* Robertson, 1900
Pedicellina cernua (Pallas, 1771) (Pedicellina tends to occur on the under surfaces of rocks in lower mid-tidal)

List of Marine Bryozoa

Ctenostomata
Alcyonidium mytili Dalyell, 1847
Alcyonidium polyoum (Hassall, 1841) (Look for this on legs of old, heavily encrusted kelp crabs)
Bowerbankia gracilis var. *aggregata* O'Donoghue, 1926 (Extremely abundant on pilings)
Flustrella corniculata (Smitt) (On stipes of coralline algae)
Triticella elongata (Osburn, 1912) (On legs of various crabs; because of transparency and insensitivity to handling, this is superb class material)

Cyclostomata
Crisia occidentalis Trask, 1857 (The Crisias are found in low intertidal among sponges and hydroids)
Crisia serrata (Gabb and Horn, 1862)
Filicrisia franciscana (Robertson, 1910) (Sparse delicate growths on rocks in mid-tidal)
Idmonea californica d'Orbigny, 1852 (On holdfasts, low intertidal or subtidal)
Tubulipora pacifica Robertson, 1910 (The Tubuliporas are generally found on *Mytilus* shell fragments in lower intertidal)
Tubulipora tuba (Gabb and Horn, 1862)

Cheilostomata
Bugula neritina (Linnaeus, 1758) (Characteristic of harbors and pilings)
Bugula californica Robertson, 1905 (Characteristic of low intertidal, beneath ledges; best demonstration of avicularia)
Callopora horrida (Hincks, 1880)
Caulibugula ciliata (Robertson, 1905) (Look for this under water in lower mid-tidal, often on sponges)
Cauloramphus spiniferum (Johnston, 1832)

**Conopeum reticulum* (Linnaeus, 1767)

**Conopeum* (?) sp. (A form with a single prominent median basal spine, common on breakwater of Berkeley Yacht Harbor)

Dendrobeania laxa (Robertson, 1900)

Dendrobeania lichenoides (Robertson, 1900) (Among *Perophora* in shaded places in low intertidal)

Eurystomella bilabiata (Hincks, 1884) (Unmistakable because of rose color. Keeps well in laboratory. White polypides make a good display.)

Hippodiplosia insculpta (Hincks, 1882) (Common on stipes of red algae and under rocks)

Hippothoa hyalina (Linnaeus, 1758) (Common on fronds of red algae)

Lyrula hippocrepis (Hincks, 1882)

Membranipora membranacea (Linnaeus, 1767) (Characteristic of the floating kelps, *Macrocystis* and *Nereocystis*, just offshore)

Membranipora tuberculata (Bosc, 1802)

Microporella ciliata (Pallas, 1766)

Microporella californica (Busk, 1856)

Parasmittina collifera (Robertson, 1908)

Phidolopora pacifica (Robertson, 1908) (Subtidal, or in deep, sheltered pools)

Rynchozoon tumulosum (Hincks, 1882)

**Schizoporella cornuta* (Gabb and Horn, 1862)

**Scrupocellaria californica* Trask, 1857

Smittina sp.

Stephanosella biaperta (Michelin, 1845)

Tegella robertsonae O'Donoghue, 1926 (Excellent for large avicularia)

**Tricellaria occidentalis* Trask, 1857

Phylum Phoronidea

by Joan C. Rattenbury
McGill University, Montreal

This is a very small phylum of animals wormlike in appearance, each inhabiting a tube of cemented sand grains, shell fragments, and the like. The tube is normally closed at one end. At the exposed end of the animal a double coiled or crescentic organ, the lophophore, supports a double whorl of ciliated tentacles, which when expanded project from the open end of the tube. The body wall is smooth and soft, consisting of a muscular part below the lophophore and a terminal thin-walled region in which lie the gonads. The digestive tract is U-shaped, mouth and anus lying close together at the base of the lophophore. Many species possess a system of blood vessels in which circulate corpuscles containing the red respiratory pigment, haemoglobin. There are two genera in the phylum and both are represented on the coast of central California. *Phoronopsis* is found in the mud flats of Elkhorn Slough, Bolinas Lagoon, Tomales Bay, and Bodega Bay as well as in some more southern locations. *Phoronis* has been found in the Monterey area, on wharf pilings and rocks in Monterey Harbor and on intertidal rocks at China Point (Hopkins Marine Station). What is apparently the same form occurs on pilings outside Berkeley Yacht Harbor. The species has not yet been identified, but shows affinities with *Phoronis vancouverensis* Pixel, 1912, of Vancouver Island, and *Phoronis hippocrepis* Wright, 1856, a widespread European species.

Diagnosis of the Local Forms

1. Occurring in intertidal flats, often in great numbers; sandy tubes straight, vertical, separate; tentacles green; an ectodermal collar at base of lophophore; sexes separate; eggs shed into sea . . . . *Phoronopsis viridis* Hilton, 1930

2. On rocks and pilings or in sand in rock crevices; tubes small, contorted, intertwined to form compact masses; tentacles colorless; no ectodermal collar; hermaphroditic; embryos develop to actinotroch larvae in tentacular crown of parent . . . . . *Phoronis* sp.

Phylum Brachiopoda

A very few small brachiopods may be found at the lowest tidal levels in this region. Most are *Terebratalia transversa* (Sowerby, 1846) a species found in larger numbers in deeper waters off this coast or intertidally farther north.

At first glance the two-valved shell suggests a pelecypod mollusc, such as a young *Hinnites*, but closer inspection reveals a flexible attachment stalk emerging from one valve near the hinge. The valves are actually dorsal and ventral rather than right and left as in a bivalve mollusc, and instead of the molluscan ctenidia or gills, the brachiopod has a pair of spiraled arms (the lophophore), bearing numerous ciliated tentacles by which it feeds. The possession of a lophophore, as well as certain similarities in internal structure and in development, indicates that brachiopods are related more closely to phoronids and ectoproct bryozoans than to any other phyla.

Phylum Echinodermata

The echinoderms are diversified animals having, in the adult, an imperfect radial symmetry and a skeleton of calcareous plates which may be reduced and completely embedded in the body wall. One of the features peculiar to the phylum is the water-vascular system, opening externally by a madreporite and usually associated with tube feet bearing suckers for attachment and locomotion. Frequently the tube feet are represented by tentaclelike structures (in ophiuroids and crinoids) and may be reduced in number.

Of the five present-day classes, the Crinoidea are not represented in the local intertidal area. This group is characterized by the upward direction of the mouth, which is more or less centrally located in a disc bearing a number of branching rays or arms. The rays bear on their oral surface ciliated ambulacral grooves leading to the mouth. The anus is usually on the oral surface of the disc. Of more than six hundred living species of crinoids, some eighty are stalked sessile forms, found only in depths below fifty fathoms; the remainder are free as adults and float or swim about, often in shallow water.

Of the remaining classes, two have rays or arms: the Asteroidea or sea stars and the Ophiuroidea or brittle stars. The Echinoidea or urchins lack arms and have a rigid skeleton and movable spines. The Holothuroidea or sea cucumbers are spineless and soft bodied, the skeletal plates being reduced to minute structures embedded in the body wall.

CLASS ASTEROIDEA

Asteroids are common intertidal forms with rather heavy arms having well-developed ambulacral grooves to which the numerous tube feet are confined. The mouth and ambulacral grooves are directed downward. The anus and madreporite are aboral. Among the characters important in classification are the spines and pincerlike pedicellariae which can be observed externally, and the skeletal plates which must be exposed by removal of the surface tissue. Most asteroids are active predators, chiefly upon barnacles and molluscs, although some, like *Patiria*, are mainly herbivorous.

Key to Asteroidea

by Frances M. Weesner
University of California, Berkeley

1. With 8 to 20 or more arms 2
1. Normally with 5 or 6 arms, rarely with more than 7. As a result of injury and regeneration, individuals with other than the normal number of arms are common 3

2. Usually with 8-13 arms which are firm and rigid . *Solaster dawsoni*
2. Usually with 15-24 arms, limp when exposed, and with softer fleshy covering . . . . *Pycnopodia helianthoides*

3. Aboral surface of disc and arms smooth and leathery, without spines . . . . *Dermasterias imbricata*
3. Aboral surface of disc and arms with spines, not smooth . . . 4

4. With 5 arms 5
4. With 6 arms 10

5. Arms short and broadly joined giving a "webbed" appearance; spines short and arranged in small concentric clusters . . *Patiria miniata*
5. Arms well separated 6

6. Arms elongate and narrow, spines of aboral surface minute and clustered so that each cluster superficially appears to be a single spine . . . . *Henricia leviuscula*
6. Spines of aboral surface individually conspicuous, although they may be arranged in groups or clusters 7

7. Spines of aboral surface in concentric groups along arms, or forming network, or very few and irregularly spaced, sometimes inconspicuous 8
7. Spines of aboral surface rather evenly scattered, not in distinct groups, fairly numerous, conspicuous 9

8. Pale pink in color, body appears weak and flabby, although not actually so; spines short and arranged in groups; marked variation in spines, but color is quite reliable. . . . *Pisaster brevispinus*

8. Color varies (purple to reddish-brown or yellow); spines large, varying in number and disposition; when numerous they form a network; frequently in concentric groups across arms; occasionally few and scattered, irregular in distribution with occasional groups of 2 or 3 . . . . *Pisaster ochraceous*

9. Arms narrow; general color brilliant red; each of large spines with conspicuous collar of pedicellariae; areas between spines generally free of pedicellariae except for an occasional large one, almost the size of the spines . . . . *Evasterias troschelii*

9. Arms somewhat widened at base, each spine with ring of naked skin at base, usually blue in color; beyond this a ring of small pedicillariae surrounding numerous small papillae and more or less continuous between spines . . . . . . *Pisaster giganteus*

10. Seldom more than 1 in. in diameter; splotched or fairly solid olive green or gray on pale background; spines usually longer than wide, not closely packed; pedicellariae relatively few, especially on mid-aboral arm surface, frequently only 1 pedicellaria for every 3 or 4 spines; very few, if any pedicellariae on the disc . . . . *Leptasterias pusilla*

10. Up to 2½ in. in diameter; frequently splotched with pinks and reds; highly variable species, generally with numerous arm spines interspersed with pedicellariae; generally with numerous pedicellariae on disc. . . . . *Leptasterias aequalis*

List of Asteroids

Dermasterias imbricata (Grube, 1857)
Evasterias troschelii (Stimpson, 1862)
Henricia leviuscula (Stimpson, 1857)
Leptasterias aequalis (Stimpson, 1862)
**Leptasterias hexactis* (Stimpson, 1862) (Primarily northern in distribution)
Leptasterias pusilla (Fisher, 1911)
Patiria miniata (Brandt, 1835)
Pisaster brevispinus (Stimpson, 1857)
Pisaster giganteus (Stimpson, 1857)
Pisaster ochraceus (Brandt, 1835)
Pycnopodia helianthoides (Brandt, 1835)
Solaster dawsoni Verill, 1878

CLASS OPHIUROIDEA

Ophiuroids, though less conspicuous in the intertidal than the asteroids, are actually very common. They are relatively small forms occurring in more or less concealed habitats, buried in sand and under rocks or among the fronds and in the holdfasts of algae. They resemble asteroids in their starlike form but may be readily distinguished by the absence of ambulacral grooves. The arms are almost always five in number and are articulated throughout their length so that they are very flexible. Each segment of the arm (fig. 123, *a*, *b*) generally bears a large aboral, "dorsal" and oral, "ventral" arm plate. Laterally each arm segment bears on each side a vertical row of arm spines which vary in number and structure from species to species. On each arm segment a pair of contractile tentacles (actually tube feet) projects through paired tentacle pores lying on either side of the ventral arm plates. The tentacles may be simple or feathery, but lack suckers. When the tentacles are withdrawn, the pores may be closed by one to several tentacle scales.

The mouth is directed downward toward the substrate, and the digestive tract is a simple sac restricted to the disc. The mouth lies in the center of the disc and is surrounded by five jaws (fig. 123, *a*) which may bear on each side one or more oral papillae (fig. 123, *a*, *e*, *f*). The apex of the jaw may be provided with one or more tooth papillae. The inner margin of the jaw usually bears one or several rows of teeth. In each interradius of the oral surface of the disc is a series of five large buccal plates. The genital bursae open by genital slits on either side of the arm bases on the oral surface of the disc. The madreporite is aboral in the larva but moves to an oral position in the adult.

The aboral surface of the disc is entire. It may be granulated, or bear bristles, nodules, or scales. Many of the scaled forms possess five pairs of large radial shields (fig. 123, *b*), lying at the margin of the disc at the base of each arm.

Most of the species autotomize very readily (hence the common name "brittle stars") so that many specimens show arms of varying length and width, resulting from injury and subsequent regeneration. Some caution must be exercised, therefore, when considering arm/disc ratio as a key character. The arm/disc ratio indicates the relative length of the arm compared to the diameter of the disc.

Some mention should be made of the Gorgonocephalidae or "basket stars," which are not encountered intertidally but may be brought up by dredging. In this group the arms are branched dichotomously throughout their length and may be recurved toward the mouth, producing a basketlike effect.

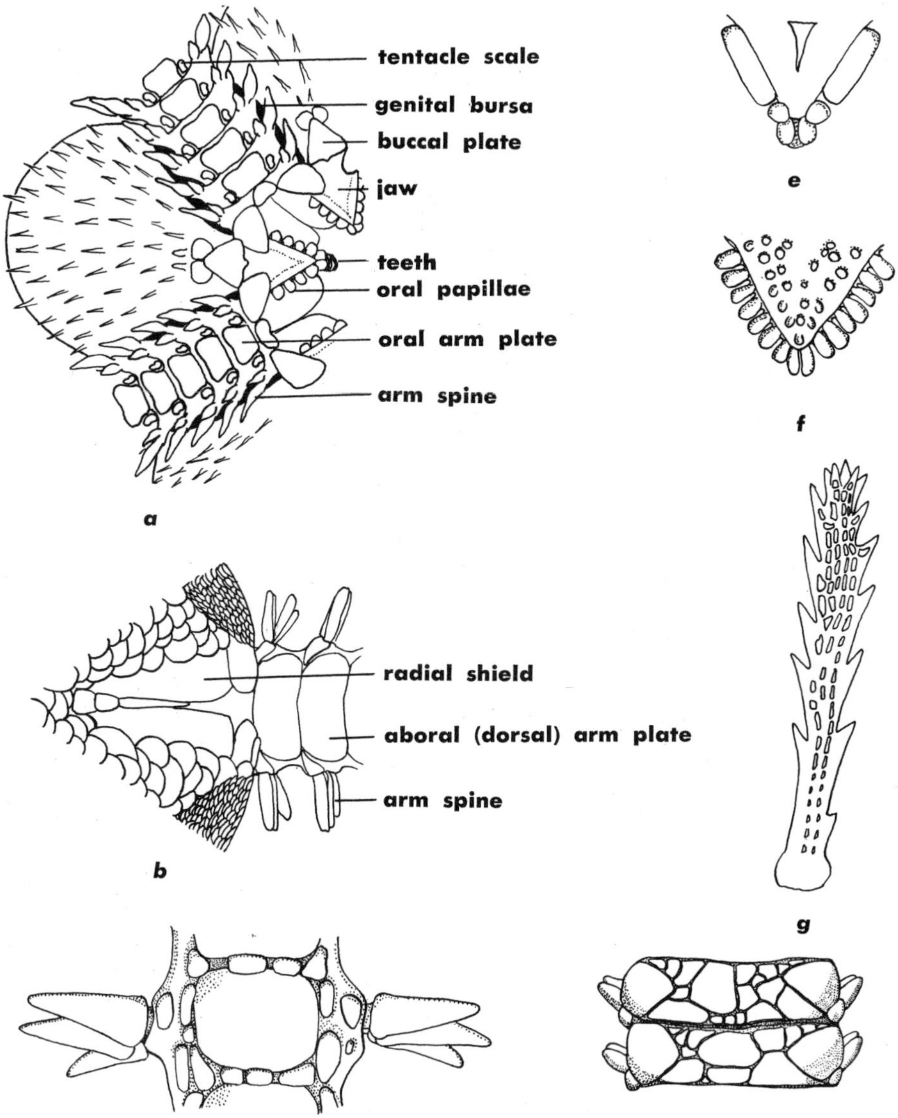

Fig. 123. **Ophiuroidea.** *a.* Oral view of part of the disc of *Ophiopholis aculeata*, illustrating structures referred to in the key; *b*, aboral surface of the disc of *Amphiodia occidentalis*, showing dorsal arm plates and radial shields; *c*, aboral arm plates of *Ophiopholis aculeata*, showing supplementary plates; *d*, aboral arm plates of *Ophioplocus esmarki*; e, jaw of *Amphipholis*; *f*, jaw of *Ophioncus granulosus*; *g*, arm spine of *Ophiothrix spiculata*.

Key to Ophiuroids

by Frances M. Weesner
University of California, Berkeley

1. Arm spine length less than 1/6 width of arm 2
1. Arm spine length at least 1/4 width of arm 3

2. Dorsal arm plates broken up into small irregular plates (fig. 123, *d*); 2 (occasionally 3) arm spines in vertical row on either side of each arm segment; 3-5 (usually 4) oral papillae on either side of each jaw; arm/disc ratio 2–2.5 : 1; disc covered with smooth scales . . . . *Ophioplocus esmarki*
2. Dorsal arm plates entire; 4 arm spines in vertical row on either side of each arm segment; 7-8 oral papillae on either side of each jaw (fig. 123, *f*); arm/disc ratio 1.5 : 1; disc granulated and with nodules . . . . *Ophioncus granulosus*

3. Arm spines with thorny projections along each side (fig. 123, *g*) and almost as long as the arm is wide; 6 arm spines per vertical row on either side of each arm segment; no oral papillae . . . . *Ophiothrix spiculata*
3. Arm spines not thorny, less than 6 per vertical row; with oral papillae 4

4. Disc granulated (not scaled); 5 arm spines per vertical row . . 5
4. Disc covered with scales; 3-4 arm spines per vertical row . . 6

5. Adjacent large dorsal arm plates not in contact, separated by row of small supplementary plates (fig. 123, *c*) . *Ophiopholis aculeatea*
5. Adjacent large dorsal arm plates in contact, not separated by row of small supplementary plates. . . . . *Ophiopteris papillosa*

6. Adjacent dorsal arm plates not in contact, at least in distal half of the arm, so that lateral plates may be seen between; outermost (or peripheral) oral papillae twice as broad as high (fig. 123, *e*) . . . . *Amphipholis* 7
6. Adjacent dorsal arm plates in contact throughout length of arm; outermost oral papillae not as above 8

7. Arm/disc ratio 4 : 1; ovoviviparous . . . *Amphipholis squamata*
7. Arm/disc ratio 7 : 1; oviparous . . . . *Amphipholis pugetana*

8. Dorsal arm plates about as broad as long; 4 paddlelike arm spines per vertical row, extending well up into the aboral arm surface; without tentacle scales . . . . (Unidentified)
8. Dorsal arm plates distinctly broader than long; 3 tapering arm spines per vertical row; with 1-2 tentacle scales 9

9. Without conspicuous radial shields; arm/disc ratio 3 : 1 or 4 : 1 . . . . *Ophionereis eurybrachyplax*

9. Radial shields conspicuous; arm/disc ratio more than 5 : 1 . . . 10
10. Arm/disc ratio 7 : 1 or 8 : 1; 1 tentacle scale; ovoviviparous. . . . . *Amphiodia* sp.
10. Arm/disc ratio 10 : 1 or more; 2 tentacle scales; oviparous . . . . , ... *Amphiodia occidentalis*

List of Ophiuroids

Amphiodia occidentalis (Lyman, 1860)
Amphiodia sp.
Amphipholis pugetana (Lyman, 1860)
Amphipholis squamata (Delle Chiaje, 1828)
Ophionereis eurybrachyplax Clark, 1911
Ophioncus granulosus Ives, 1889
Ophioplocus esmarki Lyman, 1874
Ophiopholis aculeata forma *kennerlyi* (Lyman, 1860)
Ophiopteris papillosa (Lyman, 1875)
Ophiothrix spiculata LeConte, 1851

CLASS ECHINOIDEA

Echinoids have a rigid test which may be globular or flattened and which is armed with numerous articulated spines and pedicellariae. There are five paired ambulacral areas of tube feet extending from the apical pole on the aboral surface to the mouth on the oral surface. Only five intertidal representatives of this class are encountered along the Pacific Coast between Baja California and Vancouver. These include a single Clypeastroida (sand dollar) and three members of the Centrechinoida (sea urchins). In the south, an occasional heart urchin, *Lovenia cordiformis* Agassiz (order Spatangoida), may be encountered in sandy flats at low tide. This has a secondary bilateral symmetry, the test being rounded but somewhat heart-shaped owing to a sinking of one of the ambulacral areas below the level of the others. The mouth lies at the base of this furrow. The anus is excentric in position, lying on the side opposite the ambulacral depression. The spines are long and sharp.

In the order Clypeastroida the test is very flat, and the spines are short and dense. The madreporite is aboral, and the anus is almost always on the oral surface of the disc. In *Dendraster excentricus* (Eschscholtz) the madreporite is off-center, and two of the ambulacral zones are shorter than the other three. The anus opens near the margin of the test on the oral surface. This form occurs in large subtidal beds in sandy areas.

In the order Centrechinoida the test is globular and somewhat flattened on the oral surface. The madreporite and anus open on the mid-aboral surface. The spines are large.

Diagnosis of Common West Coast Sea Urchins

1. *Strongylocentrotus pupuratus* (Stimpson) occurs in tremendous numbers in surf-swept, low-tide areas. It is rarely more than three inches in diameter and has a deep purple color. In areas where sandstone or shale forms the substrate it is often found imprisoned in individual pot-holes where it lies with its oral surface toward the substrate. In areas of granitic rock it may be found in small depressions and crevices, or more or less free on the surface.
2. *Stronglyocentrotus franciscanus* (Agassiz) is a large form, usually four to five inches in diameter. The spines are relatively long and not as densely packed as in *S. purpuratus*. The color varies from a brick red to purple. This species is usually encountered in deep tidal channels and pools.
3. *Stronglyocentrotus dröbachiensis* (Müller) occurs from Washington northward. It may be distinguished from *S. purpuratus*, whose range it overlaps, by its green hue and more numerous, shorter spines.

CLASS HOLOTHUROIDEA

Holothuroids or sea cucumbers are soft bodied, rather sacklike in appearance, with the body elongated in the mouth-anus axis. The skeletal plates are reduced to minute structures embedded in the body wall. Some of the species show a clear pentamerous symmetry with five bands of tube feet; others assume a secondary bilateral symmetry and there is a distinct "sole" bordered by tube feet. In still other forms the tube feet may be indiscriminately scattered over the body surface and in some (e. g., *Leptosynapta*) no tube feet are present. The mouth is surrounded by branching, contractile tentacles, which represent modified tube feet. The madreporite is internal, opening into the general coelom.

Key to Holothuroidea

by Frank P. Filice and James C. Cannan
University of San Francisco and University of California, Berkeley

(The following key includes certain forms common and conspicuous in the Friday Harbor area, Washington, but not found in central California.)

1. Tube feet absent 2
1. Tube feet present 3

2. Wormlike, delicate, white or semitransparent . . *Leptosynapta* spp.
2. Not wormlike; stout body tapering to a "tail"; color in life purplish to dark brown . . . . *Caudina* spp.

3. True tube feet with suckers found only on "ventral" surface; venter easily distinguished from dorsum by form and color 4
3. Tube feet not confined to venter; venter not sharply different from rest of surface 6

4. Body cylindrical; large (up to 16 in.), with prominent papillae on dorsal surface; color in life mottled reddish-brown; usually subtidal . . . . *Stichopus californicus*
4. Venter a distinct flattened "sole"; smaller (less than 5 in.) . . 5

5. Dorsal surface with firm, granulate, easily visible plates; ground color orange; 1-5 in . . . . *Psolus chitonoides*
5. Dorsum without visible plates; color usually bright red; less than 1 in. in length . . . . *Thyonepsolus nutriens*

6. Body color black or very dark, at least dorsally 7
6. Body color white or cream; may be mottled with brown spots . . 8
6. Body color other (brown, salmon, orange, etc.) 9

7. Seldom more than 1 in. long; 8 large tentacles and 2 smaller ones; found in exposed mussel-bed areas, under corallines, among byssal threads of mussels, and so on; often quite numerous ("tar spot") . . . . *Cucumaria curata*
7. Length usually 1-4 in.; 10 large tentacles (specimens in Friday Harbor area are larger and lighter). . . . . *Cucumaria lubrica*

8. Basic body color white or cream with small brown spots, which may be widespread or restricted to tentacular crown . . . . *Cucumaria piperata*
8. Color in life white; tube feet rigid and unretractile, giving animal a bristly appearance; tube feet with spicules and restricted to ambulacra; long, branched yellow tentacles . *Eupentacta quinquesemita*

9. Body color salmon, pink, to dark brown; coloration uniform over dorsal and ventral surfaces; tube feet restricted to ambulacra . 10

9. Anterior and posterior tips reddish; usually white ventrally with dorsal red or orange-red stripe; tube feet evenly scattered over surface . . . . *Pachythyone rubra*
10. Body pale yellow to pink; tube feet numerous, nonrigid, and retractile . . . . *Cucumaria* sp.
10. Body color reddish to dark brown or mottled; large (up to 10 in.); tentacles orange-red to purple . . . . *Cucumaria miniata*

List of Holothuroidea

Order Aspidochirotida
Stichopus californicus (Stimpson, 1857)
Order Molpadida
**Caudina chilensis* (J. Müller, 1850) (A sand dweller; rare in this area)
Order Paractinopoda
**Leptosynapta albicans* (Selenka, 1867) (Several other species may also occur in the Friday Harbor region; see Heding, 1928)
Order Dendrochirotida
Cucumaria curata Cowles, 1907 (Occurs in great numbers under corallines and under mussels on exposed ledges)
**Cucumaria fisheri* Wells, 1924
**Cucumaria lissoplaca* Clark, 1924
Cucumaria lubrica Clark, 1901
Cucumaria miniata Brandt, 1835 (Abundant and conspicuous in the Friday Harbor area. We have collected it a few miles south of Point Arena, but not in the area chiefly covered by this manual)
Cucumaria piperata (Stimpson, 1864) (Rare in central California; abundant at Friday Harbor)
**Cucumaria populifera* (Stimpson, 1864) (= *C. tenuicoriata* Wells)
**Cucumaria trachyplaca* Clark, 1924
Eupentacta quinquesemita (Selenka, 1867) (*Cucumaria chronhjelmi* Théel, 1886, is a synonym)
Pachythone rubra (Clark, 1901) (Formerly *Thyone rubra*)
Psolus chitonoides Clark, 1902 (Common at Friday Harbor; not found in central California)
Thyonepsolus nutriens Clark, 1901 (Bears young upon its back; often found in kelp holdfasts)

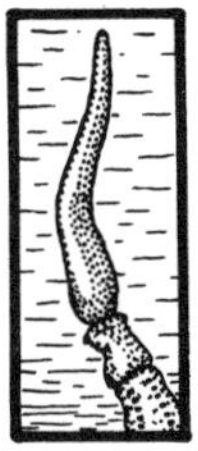

Phylum Hemichordata

by Theodore H. Bullock
University of California, Los Angeles

Among the several groups showing "chordate" characteristics is an inconspicuous series of forms which stand somewhat apart and are generally designated a phylum, **Hemichordata**. The free-swimming larva of some hemichordates, known as a tornaria, is strongly reminiscent of an echinoderm larva. The adults usually possess pharyngeal clefts, comparable to those of an amphioxus, but the remaining two features which characterize Chordata, notochord and dorsal tubular central nervous system, are poorly developed, and there is some doubt about their homology with notochord and central nerve tube of the chordates. The phylum is divided into two classes, one of which, **Pterobranchia**, consists of minute, deep-sea, sessile organisms. The other, **Enteropneusta**, although rare in this area, is well-known, and descriptions may be found in numerous texts. Bullock (1945) gives a good diagrammatic figure of the internal anatomy. Enteropneusts may be recognized in the field by the presence of an anterior, bulbous to elongated, nonretractile proboscis, and back of this a collar. The remainder of the body is soft and wormlike, but unsegmented.

A number of species, mostly still in manuscript, are known from the West Coast. Several species of *Saccoglossus*, readily recognized by their long proboscis (more than six times longer than wide), occur in mud in protected situations, but none is common except *S. pusillus* in southern California and a new species in Willapa Harbor, Washington. A species of the manuscript genus *Mesoglossus* has been repeatedly taken in small numbers in central California under rocks lying in coarse sand in protected tidal channels or lagoons on the outer coast. At least one species each of *Balanoglossus*, *Glossobalanus*, *Schizocardium*, and *Stereobalanus* occurs in scattered localities but is rare. Specimens found may be valuable and should be carefully preserved. The West Coast fauna is under study and is unusually rich in variety, with representatives of half the genera and all three of the known families. The group has been monographed recently in convenient form, both taxonomically and by organ systems (van der Horst in Bronn, *Klassen und Ordnungen des Tierreichs*). See also Dawydoff in Grassé, *Traité de Zoologie*, volume 11.

Phylum Chordata

Although the most familiar chordates are vertebrates (subphylum Vertebrata) a great many of the group are actually invertebrates (subphyla Urochordata and Cephalochordata). These latter groups are often lumped with the phylum Hemichordata and spoken of as "protochordates," but it is generally held at present that the hemichordates are sufficiently distinct to be separated from proper chordates.

The Chordata are characterized by the combination, at least in larval stages, of three distinctive features: (1) the presence of a dorsal rod, the notochord, (2) possession of pharyngeal clefts ("gill slits"), and (3) a tubular and dorsally placed nerve cord. This combination of features is most easily seen in the well-known "amphioxus" (subphylum Cephalochordata). This group is not represented in the area covered by this manual, and we shall not discuss it because accounts of it can be found in numerous texts.

The best known subphylum of chordates is the Vertebrata, which might properly be considered beyond the scope of this manual. However, intertidal pools so frequently contain fish that a key to the more common of these is made available (p. 310). The student of invertebrate life is also likely to be rewarded by the sight of such air-breathing vertebrates as sea lions, seals, whales, and occasionally sea otters, as well as a host of bird species which frequent our shores. Some knowledge of these groups is both intellectually and esthetically satisfying, regardless of one's special interests; hence we include a few useful references to them in the bibliography.

The third chordate subphylum, Urochordata, is abundantly represented in our intertidal fauna, and receives detailed treatment in the following pages.

SUBPHYLUM UROCHORDATA (=TUNICATA)

This subphylum is divided into the three classes **Larvacea**, **Thaliacea**, and **Ascidiacea**. The first two classes are pelagic and not ordinarily encountered intertidally. The **Ascidiacea** are sessile animals, abundant on the sides and undersurfaces of rocks low in the intertidal zone where they occur in company with such forms as hydroids, sponges, and bryozoans.

Simple ascidians occur singly and are ovoid or somewhat irregular in shape. The body lies encased in a protective outer tunic or test, provided with a pair of apertures which are often borne on tubular extensions of the body and tunic called siphons. When the animal is undisturbed a current of water enters the body through the oral aperture and leaves by the atrial aperture. This current brings food to the ascidian and carries off waste products. Simple ascidians reproduce only by sexual means, each egg developing into a swimming tadpole larva which settles, metamorphoses, and develops into a solitary adult.

Colonial ascidians, though basically similar to simple ascidians in structure and sexual reproduction, also produce offspring by budding. As a result, several to many individuals are usually found connected together in a colony. The form of the colony varies greatly. In **social ascidians** the individuals tend to be largely separated from one another, but in the **compound ascidians** the small individuals or zoöids are embedded in a continuous mass of common tunic which frequently has a characteristic growth form. The separation of ascidians into simple, social, and compound forms is a matter of convenience and has no taxonomic significance. Some colonial species form colonies which are almost intermediate in form between the social and the compound types, and some simple ascidians are rather closely related to colonial species.

In some compound ascidians the zoöids are scattered more or less at random throughout the common test, and both apertures of each zoöid open independently to the outside. In other compound species, however, the zoöids are arranged in several recognizable clusters or systems within each colony. The atrial apertures of all zoöids in a single system empty into a common cloacal cavity or pit in the test, which in turn opens to the outside through a pore.

The artificial key presented makes use of external features as far as this is possible, but ascidians are so variable in size, shape, and color that it is often necessary to refer to the internal anatomy in identifying specimens. Internal features are best seen in relaxed, preserved individuals. Ascidians should be brought back from the field in jars of sea water and placed in finger bowls or pans of sea water to which is then added Epsom salts (about one heaping teaspoon per liter). After twelve to twenty-four hours narcotization, the animals should be fixed for at least an hour in 10 per cent formalin before study. Such treatment permits easy observation of features which may be nearly impossible to make out in delicate, contractile living forms or in contracted preserved specimens.

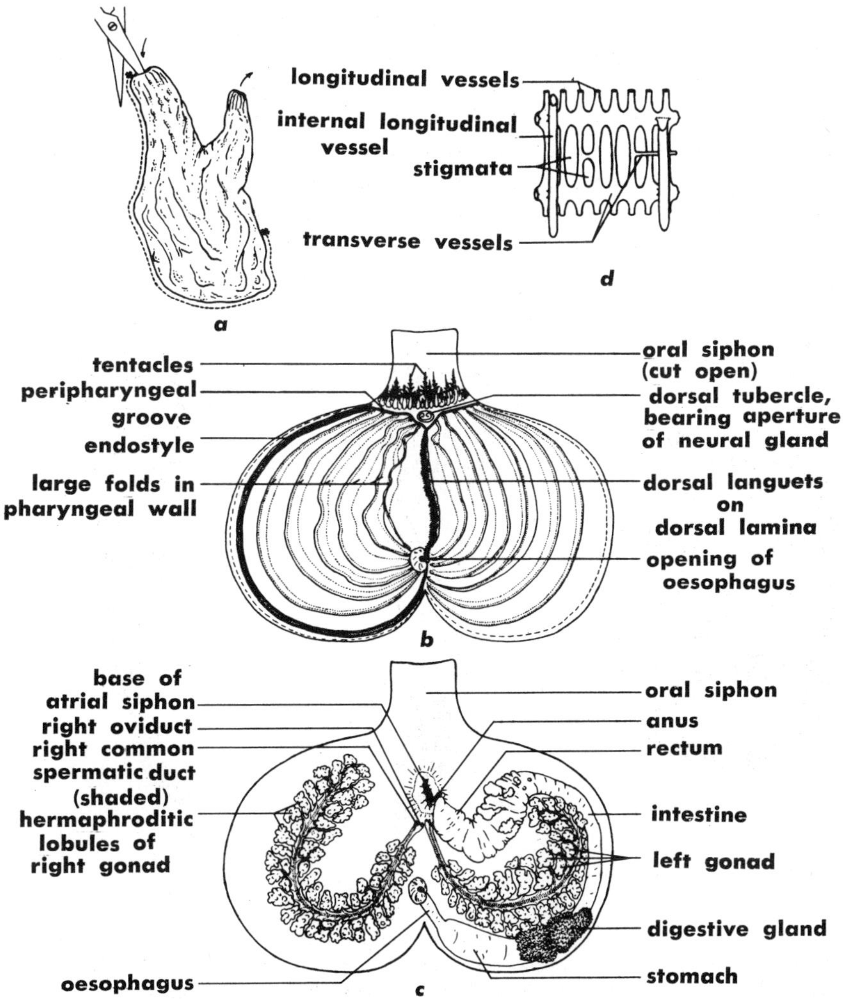

Fig. 124. Method of dissection and anatomy of taxonomic importance in a representative simple ascidian (*Pyura haustor*). *a*, Method of dissection: cut along dotted line; *b*, specimen cut open along midventral line and spread open to show the inner surfaces of oral siphon and pharynx; *c*, same view as in *b* above, but with pharynx removed, tentacles and dorsal tubercle not shown; *d*, a small area (one "mesh") of the pharyngeal wall, much enlarged (diagrammatic).

Where it is necessary to refer to internal anatomy, simple ascidians are conveniently dissected in the following manner (fig. 124, *a*). Insert one point of a pair of scissors into the oral siphon and cut downward along, or very slightly to the right of, the median line (for simple ascidians the median sagittal plane is defined roughly by the positions of the apertures: a plane passing downward through the centers of the two apertures will divide the body bilaterally). Continue the cut around the base of the body, cutting through the tissues of the tunic, mantle, and wall of the pharynx, until the two attached halves of the body may be spread apart like an opened book. The opened animal should be pinned down in a pan and covered with water. The inner surface of the pharynx thus exposed shows many features of taxonomic importance (fig. 124, *b*, *d*). The gut usually lies on the left side of the body and is exposed by separating the pharyngeal wall from the mantle to which it is loosely

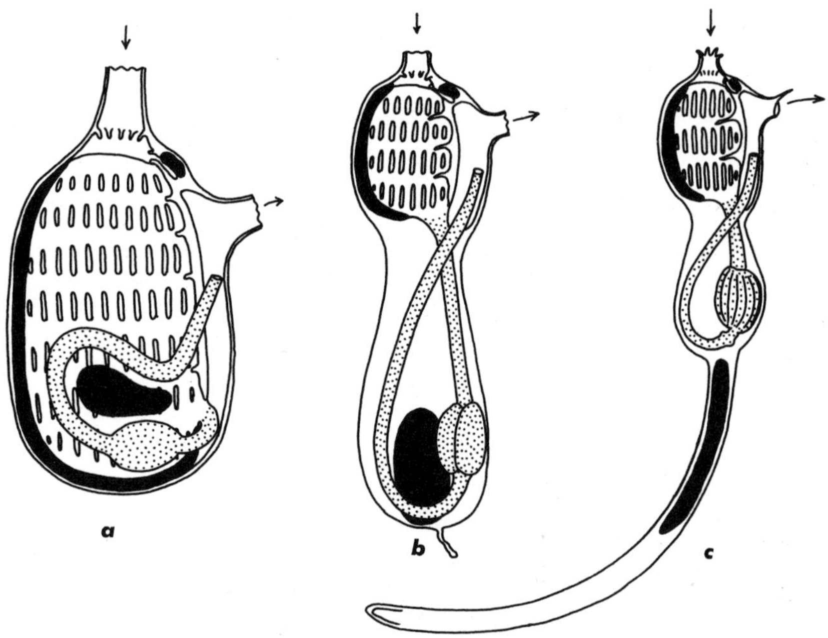

Fig. 125. Diagrams showing common variations in body form and in arrangement of gut and gonads in ascidians: gut stippled; brain, endostyle, and gonads in solid black. *a*, Intestinal loop lying beside the pharynx on the left, body not divided into regions (holosomatous), gonads on one or both sides, attached to gut or to mantle. This condition is common in simple ascidians and is found in numerous colonial species. *b-c*, zoöids of colonial ascidians showing the subdivision into more or less distinct regions (merosomatous condition): *b*, a generalized polycitorid zoöid (e.g., *Clavelina, Eudistoma*) showing division of body into two regions, thorax and abdomen; c, a generalized synoicid zoöid (e.g., *Amaroucium, Synoicum*) showing division of body into three regions, thorax, abdomen, and postabdomen.

attached (fig. 124, *c*). Gonads will be found attached to the gut or to the mantle on one or both sides of the body, and are similarly exposed.

For the anatomical study of social and compound ascidians it is only necessary to remove the relatively small and transparent individual zoöids from the surrounding tunic. For compound ascidians make a slice through the colony parallel to the long axes of the zoöids. Observe how the zoöids lie, then remove several for study in a dish of water under a dissecting microscope or good hand lens. In most cases zoöids can be removed intact by grasping them with forceps by their anterior ends and gently pulling them out of the tunic. Body organization and anatomy are shown in figures 125 and 126.

A few species have been included more than once in the key. In some of these (*Metandrocarpa dura, Perophora annectens*), parts of some colonies might be considered either social or compound in arrangement; in others (*Cystodytes* spp.) spicules may be present and conspicuous in some colonies and absent in others; and at least one species (*Eudistoma molle*) undergoes a rather conspicuous seasonal change in superficial appearance. Where possible, only large and sexually mature specimens should be selected for study. Juvenile individuals and colonies, and old degenerating colonies, will always present difficulties.

Key to the Littoral Ascidians of the Central California Coast

by Donald P. Abbott
Hopkins Marine Station of Stanford University, Pacific Grove

1. Individuals occur singly; each has a separate complete test or tunic, and is not organically joined to other individuals of same species: **simple or solitary ascidians** 2
1. Individuals more or less separate but occurring in groups or clusters, with adjacent individuals connected basally by stolons or by a sheet of tunic: **social ascidians** 12
1. Individuals small, numerous, and completely embedded in a common test mass which may vary in shape from a thin, flat sheet to a tall, stalked lobe: **compound ascidians** 17

2. Tunic attached by a distinct, narrow stalk 3
2. Tunic without a stalk, attached directly by the side or base . . 4

3. Body above stalk elongate; test with longitudinal ridges and grooves but without hairs or spiny projections; intertidal species common on protected rocks and on pilings in bays and harbors . . . . *Styela montereyensis*
3. Body above stalk rounded or oval; test without ridges or grooves but with distinct hairs or spines which may be branched; usually a subtidal species uncommon in intertidal region . *Boltenia villosa*

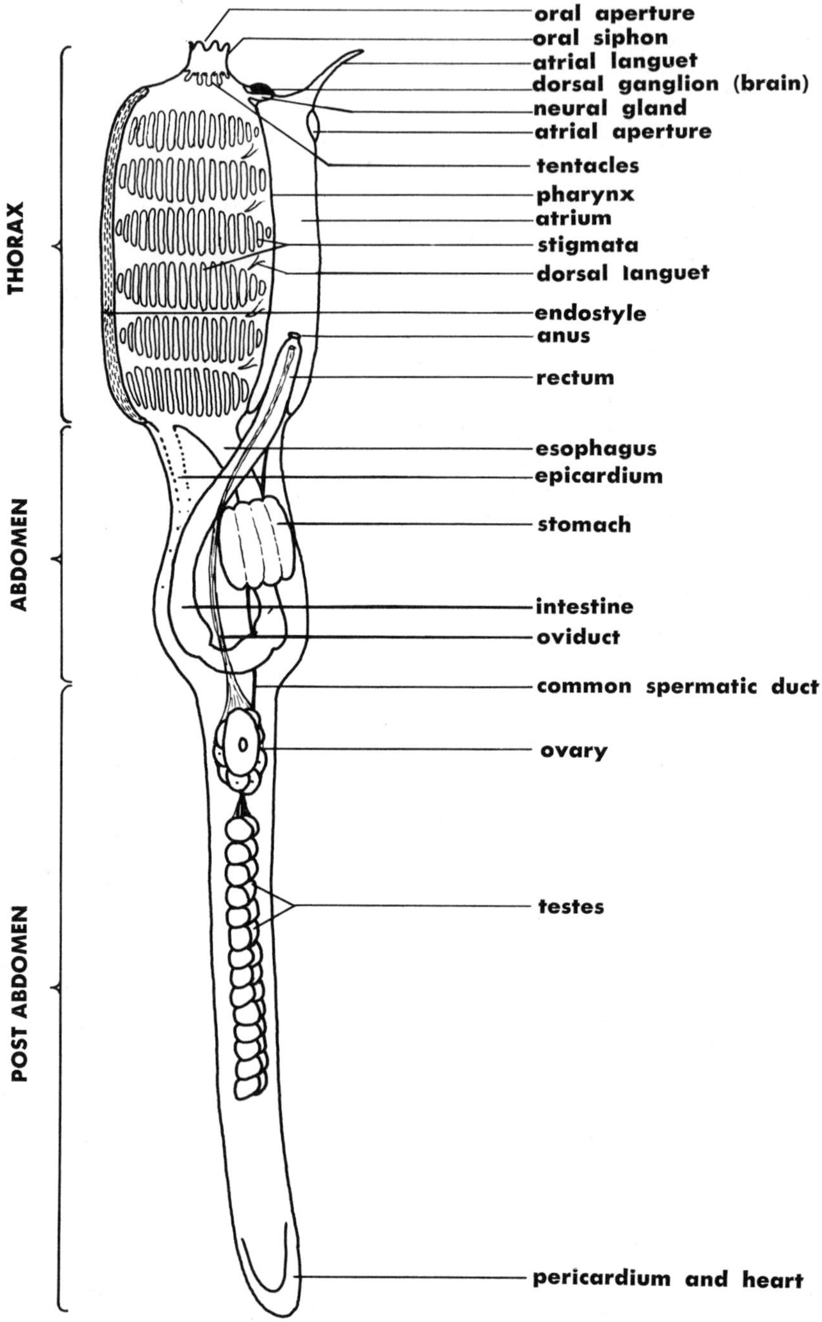

Fig. 126. Diagram of generalized compound ascidian zoöid with a postabdomen.

4. Each aperture surrounded by a ring of 6 distinct, triangular plates in the test; uncommon . . . . *Chelyosoma productum*
4. Apertures not surrounded by such plates 5

5. Individuals elongate; apertures borne at opposite extremes of the rather tubular body and pointing in opposite directions; test attached by one side, whitish, opaque; rare . . . . *Pyura mirabilis*
5. Individual shaped variously; apertures close together or well separated but not borne at opposite extremes of body 6

6. Test pale green to olivaceous, transparent to translucent, its consistency varying from that of soft gelatin to that of soft cartilage; inner surface of pharynx without large, longitudinal, inwardly projecting folds 7
6. Test tough and leathery, thin and tough, or membranous, opaque due to pigment and/or external encrusting matter; inner pharyngeal wall with 4 or more conspicuous, longitudinal, inwardly projecting folds on each side 8

7. Tunic soft and gelatinous, even after preservation; shape elongate, usually attached by one end; apertures both at free end of body; intestinal loop and gonads posterior to pharynx; on rocks and pilings in bays and harbors; common in some areas . . *Ciona intestinalis*
7. Tunic with consistency of stiff gelatin or soft cartilage, 0.5–2 mm. in thickness; body ovoid, often laterally compressed, and attached by one side; oral aperture anterior, atrial siphon arising from around middle of body; intestinal loop and gonads lateral to pharynx; found in bays and along protected rocky coast; common to abundant . . . . *Ascidia ceratodes*

8. Body oval, greatest dimension 10 mm. or less; test thin, membranous, colorless, and completely and evenly encrusted with sand; pharynx with 7 folds on each side; some stigmata arranged in whorls, especially anteriorly and on the folds; uncommon . . *Molgula verrucifera*
8. Size variable, usually larger; tunic tough and leathery or thin and tough, usually pigmented to some extent with red and/or brown; encrustation on tunic, if present, not as above; pharynx with 4-6 folds on each side; stigmata all arranged in straight, parallel rows 9

9. Entire test relatively smooth, bright pink to pearly white in color; body oval, somewhat depressed, and attached by a broad area basally; inner wall of mantle on side adjacent to intestine (left side) bearing at least 5 tubular gonads; uncommon . . . . *Cnemidocarpa finmarkiensis*
9. Test often wrinkled, brown to gray-white in color; shape and attachment variable; not more than 3 gonads on left side of body . . 10

10. Test very tough and hard, deeply and irregularly wrinkled, and often partly encrusted with debris; in expanded specimens the apertures are bright red and borne on prominent tubes (fig. 124, *a*); tentacles conspicuously branched; pharynx with 6 folds on each side; common to uncommon . . . . *Pyura haustor*
10. Test leathery or thin and tough, usually with some wrinkles or irregular protuberances; apertures usually rusty brown or purplish, borne close to the body surface or on stubby tubes; tentacles unbranched; pharynx with 4 folds on each side 11

11. Test brown to pale gray, thin and tough; 2 tubular ovaries on each side of body; common on wharf pilings and rocks in bays . . . . *Styela truncata*
11. Test rusty brown, leathery but flexible; ovaries tubular, 2-3 on left, 4-7 on right side of body; uncommon, so far collected only at Point Richmond, San Francisco Bay . . . . *Styela plicata*

12. Individuals oval, globular, or hemispherical in shape; tunic red, orange, gray-brown, or greenish, and not sand encrusted; intestinal tract lies beside, and close to, the pharynx 13
12. Individuals elongate, tubular, sometimes laterally compressed; tunic colorless or tinged with olive or brown, and usually transparent at least at the distal end; visible parts of the zoöids may be bright pink, orange, or pale gray; intestinal tract lies in a long extension of the body posterior to the pharynx 15

13. Test pale green, transparent; individuals 2-3 mm. in diameter, resembling tiny green grapes; many individuals occur together, growing from a tangle of branching stolons in a loose sheet or in compact clusters; common on rocks and algae . . . *Perophora annectens*
13. Test gray-brown, sometimes with red apertures; adjacent individuals united by thin sheet or strand of test or seemingly separate; body shape hemispherical and depressed; individuals up to 12 mm. in diameter and 3-4 mm. high; on undersides of rocks; uncommon . . . . *Alloeocarpa* sp.
13. Test red to orange, opaque; individuals up to 6 mm. high, ovoid, growing in colonies on rocks or seaweed 14

14. Adjacent individuals seemingly separate but obscurely joined basally by fine stolons or a thin sheet of test; colonies growing on sides and undersurfaces of intertidal rocks; common *Metandrocarpa taylori*
14. Adjacent individuals laterally fused, giving somewhat the appearance of a compound ascidian; usually growing on algae; generally subtidal but occasionally washed up on beaches *Metandrocarpa dura*

15. Test not sand encrusted or only slightly so; part of thorax (especially the endostyle) bright pink in life; ratio of greatest diameter

to length (as measured on test) 1:2.5 to 1:5.5 (average 1:4) for adult specimens; 16-20 rows of stigmata; abundant . . . . *Clavelina huntsmani*

15. Test generally heavily encrusted with sand except near distal tip, occasionally free of sand entirely; thorax of zoöids orange or pale gray; ratio of greatest diameter to length (as measured on test) 1:7 to 1:17 (average 1:10) for adult specimens; 7-13 rows of stigmata 16

16. Pharynx bright orange in life; free parts of zoöids in sandy tubes up to 20 mm. long and 1.5 mm. in diameter; in life the colony appears as a series of well-spaced; bright orange specks on a sandy background; stomach smooth walled; 7 rows of stigmata; common on sandy rocks and among roots of eelgrass . . . *Pycnoclavella stanleyi*

16. Pharynx pale gray or colorless in life; zoöids in tubes (usually sand encrusted) up to 50 mm. long and 2-4 mm. in diameter; colony appears in life as a mass of pale sandy and partly transparent "worm" tubes; stomach wall with distinct longitudinal ridges; usually 12-13 rows of stigmata; common . . . . . *Euherdmania claviformis*

17. Small calcareous spicules embedded in test (examine slice made through colony parallel to long axes of zoöids; use hand lens or scope) 18

17. Calcareous spicules not present, though test is sometimes impregnated with sand 21

18. Spicules consist of stellate or spiny globes, embedded in enormous numbers throughout the test and giving the whole colony a chalky appearance; colony thin, encrusting; zoöids 1-2 mm. long. . . 19

18. Spicules disc-shaped (often with a short spine in the center of one surface) or consisting of irregular crystals; spicules confined to the deeper layers of the test, frequently grouped about the abdomens of the zoöids, and often clearly seen only after slicing through the colony; zoöids up to 4 mm. long 20

19. Colony in life grayish or white tinged with lavender; atrial siphon usually with margin flared out like a funnel and projecting from near the base of the thorax; 3 rows of stigmata; common. . . . . *Trididemnum opacum*

19. Colony white, sometimes tinged with pink in life; atrial siphon without a flaring margin, small, often reduced to a simple perforation in the mantle; 4 rows of stigmata; abundant . *Didemnum carnulentum*

20. Spicules usually numerous, consisting of small discs with a short spine on one side, and arranged as a layer around the lower parts of each zoöid; colony flat, encrusting, up to 10 mm. thick; tunic translucent and pinkish to colorless; common. . . *Cystodytes* sp.

20. Spicules generally few, sometimes as above but often consisting of (or in addition to) irregular white crystals scattered at random deep in the test; larger colonies massive and encrusting, the free surface of the test often sculptured into low, smooth, irregularly arranged ridges; colony up to 30 mm. thick; test translucent, pinkish to gray in life; common . . . . *Cystodytes lobatus*

21. Body of zoöid oval or subspherical, not divided into thorax and abdomen (fig. 125, *a*) 22
21. Body of zoöid divided into 2 regions (fig. 125, *b*): a thorax containing the pharynx and distal tip of the intestine, and an abdomen of variable length containing the esophagus, stomach, intestinal loop, heart, and gonads; no postabdomen present, although the gonads may extend slightly below the bottom of the intestinal loop during the breeding season (usually summer). 23
21. Body of zoöid divided into 3 regions (fig. 125, *c*, 126): thorax, abdomen, and a postabdomen, consisting of a sizable extension of the body posterior to the bottom of the intestinal loop and containing the heart and gonads; postabdomen sometimes detached and lying free in the deeper parts of the test 31

22. Test red and leathery, opaque; adjacent zoöids never arranged in systems; pharynx with about 12 rows of stigmata; generally subtidal but occasionally washed up on beaches attached to algae . . . . *Metandrocarpa dura*
22. Test pale green, transparent; zoöids never arranged in systems; colony resembling a tightly fused mass of tiny green grapes; pharynx with 4 rows of stigmata; common on intertidal rocks in the Monterey Peninsula area. . . . . *Perophora annectens*
22. Test dark purple to blackish, with the exposed anterior tips of the zoöids golden in life; zoöids always arranged in systems; zoöids dark purple flecked anteriorly with gold in life, fading to deep uniform purple after preservation; pharynx with 4 rows of stigmata; uncommon in most areas, and found only in the lowest part of the intertidal zone. . . . . *Botryllus tuberatus*

23. Atrial aperture borne at the tip of a distinct tube or siphon whose length is usually at least as great as its diameter; margin of atrial siphon usually lobed or toothed, but never with an atrial languet 25
23. No distinct tubular atrial siphon (except in some very small and immature zoöids); atrial aperture a simple hole or gap in the mantle; a tonguelike process (the atrial languet) is always present, extending outward from the anterior or anterolateral lip of the atrial aperture 24
23. Neither tubular atrial siphon nor atrial languet present; atrial aperture consisting of a large gap in the mantle, exposing a large area

of the pharynx dorsally and laterally; colony delicate, thin, gelatinous, encrusting; test transparent colorless to olive green, speckled with small zoöids 0.5-2 mm. in length; common under rocks on open coast and on floats in bays and harbors . . . *Diplosoma pizoni*

24. Smaller colonies mushroom-shaped and pedunculate, larger colonies prostrate; zoöids usually in recognizable rounded or stellate systems; atrial aperture and atrial languet symmetrically placed; color of colony exceedingly variable and often bright; abundant . . . . *Distaplia occidentalis*

24. Colonies in the form of clusters of shovel-shaped or club-shaped lobes, each lobe consisting of a thick, somewhat compressed head or blade attached by an elongate stalk; zoöids borne on only one surface of the head, and here more or less regularly arranged in parallel rows; larger zoöids always with a wide, grossly asymmetrically placed atrial languet; colony pale cream to light orange-brown; uncommon in most areas . . . . . . *Distaplia* sp.

25. Test dark brown, purplish, tan, or gray, very tough and hard, and impregnated at all levels with sand, though often not encrusted with sand externally; colony flat, prostrate, generally 10-20 mm. thick; zoöids arranged in circular systems about common cloacal pits in the test; abundant . . . . *Eudistoma psammion*

25. Colony white, gray, yellowish or pinkish; consistency soft to firm but never hard, tough, and leathery; embedded sand grains, when present, restricted to deeper parts of test; zoöids not arranged in systems 26

26. Colony consisting of large, low, rounded mounds; test rather soft, opaque, and milky white, bearing conspicuous bright red zoöids; common in summer and autumn . . . . . . *Eudistoma molle*

26. Colony, test, and zoöids not as above 27

27. Zoöids with 4 rows of stigmata; test stiff owing to the presence of many turgid, bladderlike cells in tunic (seen under 10-15x); commensal amphipods almost never present in surface layers of test . 28

27. Zoöids with 3 rows of stigmata; test softer, no bladder cells present; commensal amphipods often present in superficial parts of test in larger colonies. (Species of the genus *Eudistoma*. These are not all well understood or clearly characterized and are often difficult to identify. *E. ritteri* is the commonest species in central California) 29

28. Colony flat, encrusting, up to 10 mm. thick; test translucent, pinkish to colorless; zoöids with part of the intestine brownish in freshly preserved or living specimens; common . . . . *Cystodytes* sp.

28. Colony massive, encrusting, with the free surface produced into

irregular, low, smooth ridges; test translucent, pinkish to gray; intestine of zoöids without pigmentation; common . . . . *Cystodytes lobatus*

29. Larger colonies occur in great encrusting sheets produced at intervals into projecting lobes of variable size and shape; small colonies in the mid-tide zone may consist of groups of small clear knobs on stubby, sandy stalks; test pale yellow to gray, with surface free of sand except as mentioned above; sand usually embedded in deeper parts of the test; very variable, very common; larger colonies likely to be confused in the field with *Amaroucium californicum* . . . . *Eudistoma ritteri*
29. Colony flat or massive and rounded, but never with projecting lobes as above; test free of embedded sand 30

30. Colony flat, encrusting, firm, up to 10 mm. thick; zoöids average about 3 mm. in length; common . . . . *Eudistoma diaphanes*
30. Colony consisting of large, low, rounded mounds, rather soft and up to 25 mm. thick; zoöids average more than 10 mm. in length; common . . . . *Eudistoma molle*

31. No atrial languet; atrial aperture borne at the tip of a distinct, tubular atrial siphon whose distal margin is lobulated or toothed 32
31. Atrial languet always present, arising from, or near, the anterior lip of the atrial aperture; tubular atrial siphon present or absent 34

32. Colony formed of a cluster of scarlet or crimson globular lobes, borne erect on short peduncles; stomach wall with a single longitudinal groove down one side, otherwise smooth or minutely tuberculate; esophagus curves and enters stomach from the side; uncommon in most areas . . . . *Synoicum* sp.
32. Colony formed of a cluster of orange or brownish, erect lobes; stomach wall with several conspicuous longitudinal ridges separated by grooves, never tuberculate; esophagus enters the anterior end of the stomach (includes 2 species of *Sigillinaria* which may intergrade) 33

33. Larger lobes up to 15 mm. wide across the top and reaching 30 mm. in height; upper surface of lobes relatively free of sand; bright orange zoöids clearly visible in the transparent tunic . . . . *Sigillinaria pulchra*
33. Larger lobes seldom more than 5-7 mm. in width across the top, often less; lobes slender, tapering toward the base, usually occurring in close-packed clusters which form low mounds on the rocks; lobes completely encrusted with sand, obscuring both tunic and zoöids . . . . *Sigillinaria aequali-siphonis*

34. Stomach wall bearing a single longitudinal groove down one side,

otherwise smooth and lacking obvious ridges and tubercles . . 35

34. Stomach wall with several large, obvious longitudinal ridges separated by grooves, regularly arranged or broken and irregular . . 36

35. Typical larger colony consisting of a single, thick, flattened lobe, up to several inches across, attached at one margin by a short, stout stalk; smaller colonies form globular, pedunculate heads; tunic not sand encrusted except in smallest colonies, olive brown to orange in color; zoöids seldom more than 5 mm. long; common . . . . *Polyclinum planum*

35. Typical colony consisting of a cluster of orange, club-shaped lobes up to 60 mm. long, tapering toward the attached bases, and usually sand encrusted; zoöids 10-40 mm. long; common . . . . *Synoicum par-fustis*

36. Larger colonies massive, in the form of irregular encrusting sheets or thick, flat-topped slabs attached by a large area basally; free surface of test usually without any encrusting sand; atrial languet not cleft into teeth at the distal end 37

36. Colonies always consisting of clusters of elongate, stalked, sand-encrusted lobes; atrial languet often distally cleft to form 3 teeth 38

37. Colony generally yellowish but may be any color from white to reddish-brown; exceedingly variable in form, usually encrusting over, and taking more or less the shape of the surface on which it grows; often forming great, irregular sheets on rocks, here and there raised into ridges or lumps but rarely forming conspicuous projecting lobes; 8-12 rows of stigmata; this is the commonest ascidian on the central California coast and is often confused with *Eudistoma ritteri* . . . . *Amaroucium californicum*

37. Colony appearing reddish to pink, largely due to pigmentation of the zoöids; colonies in the form of broad, thick, usually flat-topped slabs or cakes, not in encrusting sheets; 13-15 rows of stigmata; common. . . . . *Amaroucium solidum*

38. Sand grains encrusting colony surface but not deeply embedded in test matrix; 17-21 rows of stigmata; common . . . . *Amaroucium propinquum*

38. Test heavily encrusted and impregnated with sand; 5 rows of stigmata . . . . *Amaroucium arenatum*

List of Ascidians

Order Aplousobranchia

Amaroucium arenatum Van Name, 1917
Amaroucium californicum Ritter and Forsyth, 1917
Amaroucium propinquum Van Name, 1945
Amaroucium solidum Ritter and Forsyth, 1917
Clavelina huntsmani Van Name, 1931
Cystodytes lobatus (Ritter, 1900)
Cystodytes sp.
Didemnum carnulentum Ritter and Forsyth, 1917
Diplosoma pizoni Ritter and Forsyth, 1917
Distaplia occidentalis Bancroft, 1899
Distaplia sp.
Eudistoma diaphanes Ritter and Forsyth, 1917
Eudistoma molle (Ritter, 1900)
Eudistoma psammion Ritter and Forsyth, 1917
Eudistoma ritteri Van Name, 1945
Euherdmania claviformis (Ritter, 1903)
Polyclinum planum (Ritter and Forsyth, 1917)
Pycnoclavella stanleyi Berrill and Abbott, 1949
Sigillinaria aequali-siphonis (Ritter and Forsyth, 1917)
Sigillinaria pulchra (Ritter, 1901)
Synoicum par-fustis (Ritter and Forsyth, 1917)
Synoicum sp.
Trididemnum opacum (Ritter, 1907)

Order Phlebobranchia

Ascidia ceratodes (Huntsman, 1912)
Chelyosoma productum Stimpson, 1864
Ciona intestinalis (Linnaeus, 1767)
Perophora annectens Ritter, 1893

Order Stolidobranchia

Alloeocarpa sp.
Boltenia villosa (Stimpson, 1864)
Botryllus tuberatus Ritter and Forsyth, 1917
Cnemidocarpa finmarkiensis (Kiaer, 1893)
Metandrocarpa dura (Ritter, 1896)
Metandrocarpa taylori Huntsman, 1912 (Hitherto usually called *M. michaelseni* Ritter and Forsyth, 1917)
Molgula verrucifera Ritter and Forsyth, 1917
Pyura haustor (Stimpson, 1864)
Pyura mirabilis (von Drasche, 1884)
Styela montereyensis (Dall, 1872)
Styela plicata (Lesueur, 1823)
Styela truncata Ritter, 1901

INTERTIDAL FISHES

Some larger fishes, which are never present at low water, may enter the intertidal at high tide. Other species are characteristic of the shores and are seldom found elsewhere. Still others, although characteristic of offshore waters (or even of the open sea, like the sunfish *Mola mola*) are seen from time to time in the intertidal or in harbors. The key that follows covers the fishes that may be taken along the shores with dip nets or by guddling, and includes a few rare forms from tide pools, but not the common fishes that are caught only by angling or with seines.

Terms used in the key which need explanation are defined below and/or illustrated in figure 127. Most of the definitions used are taken or modified from Walford (1931).

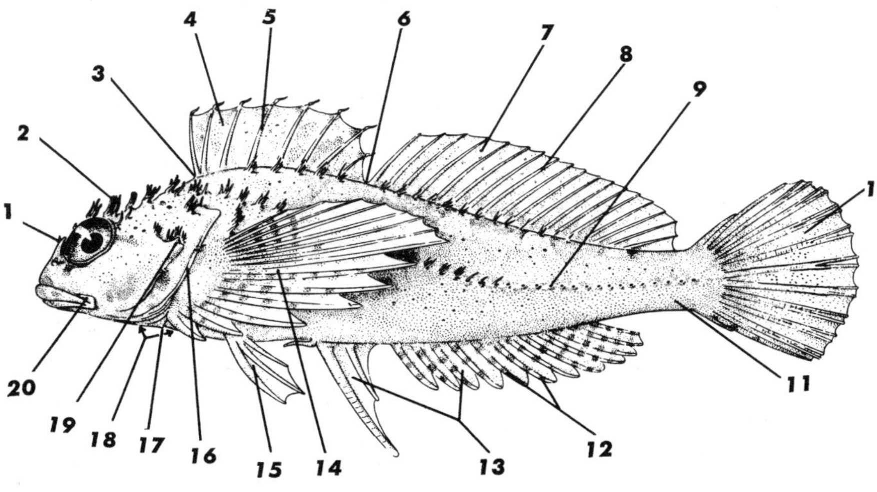

Fig. 127. *Oligocottus snyderi* Greeley, a common tide-pool fish (from Bolin, 1944), labeled to show structures of systematic importance. 1, nasal spine; 2, cirri; 3, origin of first dorsal fin (dorsal origin); 4, first dorsal, or spinous dorsal fin; 5, spine; 6, junction of first and second dorsal fins; 7, second dorsal, or soft dorsal fin; 8, fin ray; 9, lateral line; 10, caudal fin; 11, caudal peduncle; 12, rays of anal fin; 13, first and second anal fins, often contiguous to form a single anal fin; 14, pectoral fin; 15, pelvic or ventral fin; 16, operculum or gill cover; 17, gill membrane; 18, isthmus—the region, not a specific structure; 19, preopercular spine; 20, maxillary.

Glossary of Anatomical Terms

Adipose fin. A soft, fleshy, usually small second dorsal fin that has no supporting rays running through it.

Anal fin. The unpaired fin on the mid-line of the under side of the body, just behind the vent.

Anal origin. The point where the anterior margin of the anal fin arises from the ventral surface of the body.

Anal spine. The spine or spines at the front of the anal fin.

Caudal fin. The tail fin.

Caudal peduncle. That part of the body behind the anal fin, which holds the tail fin.

Ctenoid scale. A bony fish scale with the posterior margin and/or surface bearing teeth or spines.

Depth. The greatest vertical diameter of a fish.

Dorsal fin or "dorsal." The unpaired fin or fins on the mid-line of the back.

Dorsal scale band. In some fishes (e.g., some cottids), a band of scale rows extending from the nape backward along the body on either side of the dorsal fins.

First dorsal fin or "first dorsal." The most anterior of the dorsal fins, where more than one is present.

Gill membrane. The fleshy extension of the lower part of the gill covers or opercula. These membranes are supported by hard parts, and may be connected to each other across the isthmus ventrally, or may be separate. They may be connected to the isthmus or be free from it.

Isthmus. The region between the gill openings ventrally on the chest and throat.

Lateral line. A longitudinal line of modified scales bearing pores, running horizontally along the side of a fish, and overlying a sensory canal; not to be confused with colored stripes.

Maxillary. One of a pair of bones which form the upper part of the upper jaw; the rear ends move downward when the mouth is opened.

Opercula. The somewhat movable structures composed of several bones, which cover the gill openings on either side.

Orbit. The eye socket.

Pectoral fins or "pectorals." The more dorsal, and also usually the more anterior of the paired fins.

Pelvic fins or ventral fins. Paired fins placed below and usually posterior to the pectoral fins.

Preopercular margin. The posterior margin of the most anterior of the bones forming the operculum. The margin is often visible externally as a ridge on the opercular surface, running more or less parallel to the opercular margin but anterior to it, and often bearing one or more spines.

Preopercular spine. A spine on the preopercular margin.

Ray. See soft rays.

Second dorsal fin or "second dorsal." The more posterior of the dorsal fins, when two are present.

Soft dorsal fin or "soft dorsal." A dorsal fin supported by soft rays; usually a second dorsal fin.

Soft rays or rays. The biramous supporting rods of a fin, which are usually branched distally and are composed of many small parts placed end to end. (Compare with **spines.**)

Snout. That part of the head in front of the eyes.

Spines. The stiff, usually sharply pointed rods, not composed of separate parts placed end to end, which support part or all of some fins. Also, any sharp projecting point.

Standard length. The length of the body, measured from the most anterior part of the head to the tip of the hypural. "The latter point . . . may be determined with fair accuracy in larger individuals by bending the caudal fin from side to side and noting the position of the abrupt wrinkle formed at the base of the rays." (Bolin, 1944)

Suborbital stay. A projection from one of the bones forming the lower margin of the orbit. (See key, no. 15.)

Ventral fins or "ventrals." See **pelvic fins.**

Counts of the spines and rays supporting particular fins are often given in abbreviated notations; examples are explained below. "Dorsals VIII(VII-IX)—19(17-20)" means: there are two dorsal fins, which are completely separated from one another. The heavy dash, indicating complete separation of the two fins, separates the counts for the first and second dorsals (if the dorsals are connected by a membrane, the dash is replaced by a comma). The first dorsal is supported entirely by spines (in roman numerals), the number of spines averaging eight (VIII), but showing a variation of from seven (VII) to nine (IX). The second dorsal is supported entirely by soft rays (always indicated in arabic numerals), the number present averaging nineteen, but varying from seventeen to twenty in the series of specimens examined. "Anal III,6" means: the anal fin is supported anteriorly by a series of three spines, followed by a row of six soft rays.

Key to Intertidal Fishes

by Rolf L. Bolin
Hopkins Marine Station of Stanford University, Pacific Grove

1. Eyes unsymmetrical, both on same side of head 2
1. Eyes symmetrical, one on each side of head 4

2. Dorsal and anal fins marked by conspicuous blackish bars; dorsal with fewer than 63 rays, anal with fewer than 47 . . . . *Platichthys stellatus*
2. No conspicuous blackish cross bars on fins; dorsal with more than 73 rays, anal with more than 53 3

3. Eyes on right side of head (except in rare reversed individuals); anterior dorsal rays elongated and connected by membrane only at their bases . . . . *Psettichthys melanostictus*
3. Eyes on left side of head (except in rare reversed individuals); anterior dorsal rays not elongated, connected by membrane to their tips . . . . *Citharichthys stigmaeus*

4. Adipose fin present behind first rayed dorsal 5
4. No adipose fin 6

5. Mouth large, maxillary reaching vertical of hind margin of orbit . . . . *Spirinchus starksi*
5. Mouth small, maxillary scarcely reaching vertical of anterior margin of pupil . . . . *Hypomesus pretiosus*

6. A sucking disc on belly 7
6. No sucking disc on belly 12

7. Pelvic fins alone forming sucking disc 8
7. Basal parts of pectoral fins together with pelvics forming the sucking disc 11

8. Sucking disc closely applied to belly, the pelvic rays appearing as short, heavy, quadrangular pads . . . . . . *Liparis florae*
8. Sucking disc in form of a conical cup attached to belly only at its base; pelvic rays normal 9

9. Eyes obsolescent, under skin . . . . . . *Lethops connectens*
9. Eyes well developed, fully functional 10

10. Scales rather large, about 25 along lateral line . . . . *Coryphopterus nicholsi*
10. Scales too small to count . . . . *Clevelandia ios*

11. Body broad, tadpole shaped; dorsal rays about 13 . . . . *Sicyogaster meandrica*
11. Body slender, elongate; dorsal rays 4–6 . . *Rimicola eigenmanni*

12. Body encased in a series of bony rings; mouth at end of long tubular snout . . . . *Syngnathus californiensis*
12. Body not encased in bony rings; mouth not as above 13

13. Pelvic fins present 14
13. Pelvic fins absent 50

14. First dorsal composed of very short spines entirely unconnected by membrane . . . . *Aulorhynchus flavidus*
14. No separate first dorsal composed of isolated spines 15

15. Suborbital stay present (this bony bridge may be felt by running a needle down across cheek between eye and preopercular margin). 16
15. No suborbital stay developed 42

16. Sides of body clearly marked by several longitudinal lateral lines 17
16. None or only a single longitudinal lateral line (transverse branches present in one species) 18

17. Scales covering suborbital stay; 2 pairs of cirri on head, the 2d pair on nape far behind the conspicuous supraorbital flaps, minute and difficult to see. . . . . *Hexagrammos decagrammus*
17. Area over suborbital stay naked; only supraorbital pair of cirri present . . . . *Hexagrammos superciliosus*

18. Pelvics with 4 or 5 soft rays 19
18. Pelvics with 3 soft rays 28

19. Anal with 3 spines at its anterior end. 20
19. Anal composed entirely of soft rays 24

20. Dorsal XVI,15; anal III,13; mouth small, maxillary reaching about to vertical from anterior edge or orbit . . . *Oxylebius pictus*
20. Dorsal XIII,13-16;anal III,6-9; mouth large, maxillary reaching to or beyond vertical from hind margin of pupil 21

21. Anal III,6; color black and yellow, tending toward vertical bars . . . . *Sebastodes chrysomelas*
21 Anal III,8-9; color uniform, no bars developed 22

22. Lower jaw much projecting, its tip continuing dorsal profile of head when mouth is closed; maxillary extending beyond posterior margin of orbit . . . . *Sebastodes paucispinus*
22. Lower jaw not markedly projecting; maxillary not reaching beyond vertical of posterior margin of orbit 23

23. Peritoneum white . . . . *Sebastodes melanops*
23. Peritoneum black . . . . *Sebastodes mystinus*

24. Body with strongly ctenoid scales 25
24. Body entirely naked 27

25. First dorsal XVII-XVIII, separate from second dorsal; anal 22-24 . . . . *Jordania zonope*
25. First dorsal X-XI, strongly joined to second dorsal; anal 18-20 . 26

26. Dorsal scale band of 7-8 rows . . . *Hemilepidotus spinosus*
26. Dorsal scale band of 4-5 rows . . *Hemilepidotus hemilepidotus*

27. Gill membranes free from isthmus; pelvics I,5 . . . . *Scorpaenichthys marmoratus*
27. Gill membranes joined to isthmus; pelvics I,4 *Leptocottus armatus*

28. Area between dorsal fins and lateral line with well-developed scales in oblique or longitudinal bands or covering entire area . . . 29

28\. Area between dorsal fins and lateral line naked or with minute prickly scales not arranged in definite bands 34

29\. Anus much nearer pelvic base than anal origin *Orthonopias triacis*
29\. Anus not notably advanced in position 30

30\. Conspicuous ctenoid scales on top of head 31
30\. No scales on top of head 33

31\. No cirrus on upper anterior margin of orbit; dorsal scale band originating about under base of third dorsal spine, separated from scales of head by a naked area or by scales so minute and scattered that they do not obscure the definite origin of the band . . . . *Artedius fenestralis*

31\. A well-developed cirrus on upper anterior margin of orbit; dorsal scale band more or less merging with squamation of head. . . . 32

32\. Second dorsal 12-14; anal 10; scales extending under entire orbit and present even on snout. . . . . *Artedius creaseri*
32\. Second dorsal 16-18; anal 12-14; scales extending only under posterior part of orbit, if at all; no scales on snout *Artedius harringtoni*

33\. Dorsal scale band with 24-29 oblique scale rows, 6-11 scales in longest row; no scales behind opercular flap . . *Artedius lateralis*
33\. Dorsal scale band with 39-49 oblique scale rows, 10-18 scales in longest row; a few small scales just behind opercular flap between pectoral base and lateral line . . . . . . *Artedius corallinus*

34\. Pectoral fins united ventrally, of 21-24 rays . . . *Synchirus gilli*
34\. Pectoral fins entirely separate, of 12-17 rays. 35

35\. Anus immediately in advance of anal origin 36
35\. Anus in middle 1/3 of the distance between pelvic base and anal origin 39

36\. Body covered with minute prickly scales; preopercular spine simple . . . . *Oligocottus rimensis*
36\. Body without visible scales; preopercular spine bifid to quadrifid except in very young 37

37\. No cirri on nasal spines and none on body above lateral line. . . . . *Oligocottus maculosus*
37\. A well-developed cirrus on nasal spine and tufts of cirri along base of dorsal fin 38

Fig. 128. Typical intertidal fishes (from Bolin, 1944). *a, Clinocottus analis*, one of the commonest tide-pool species, occurring even in high pools; *b, C. recalvus*, another common tide-pool species; *c, Oligocottus rimensis*, less common in tide pools; *d, Leptocottus armatus*, common in bays and brackish waters.

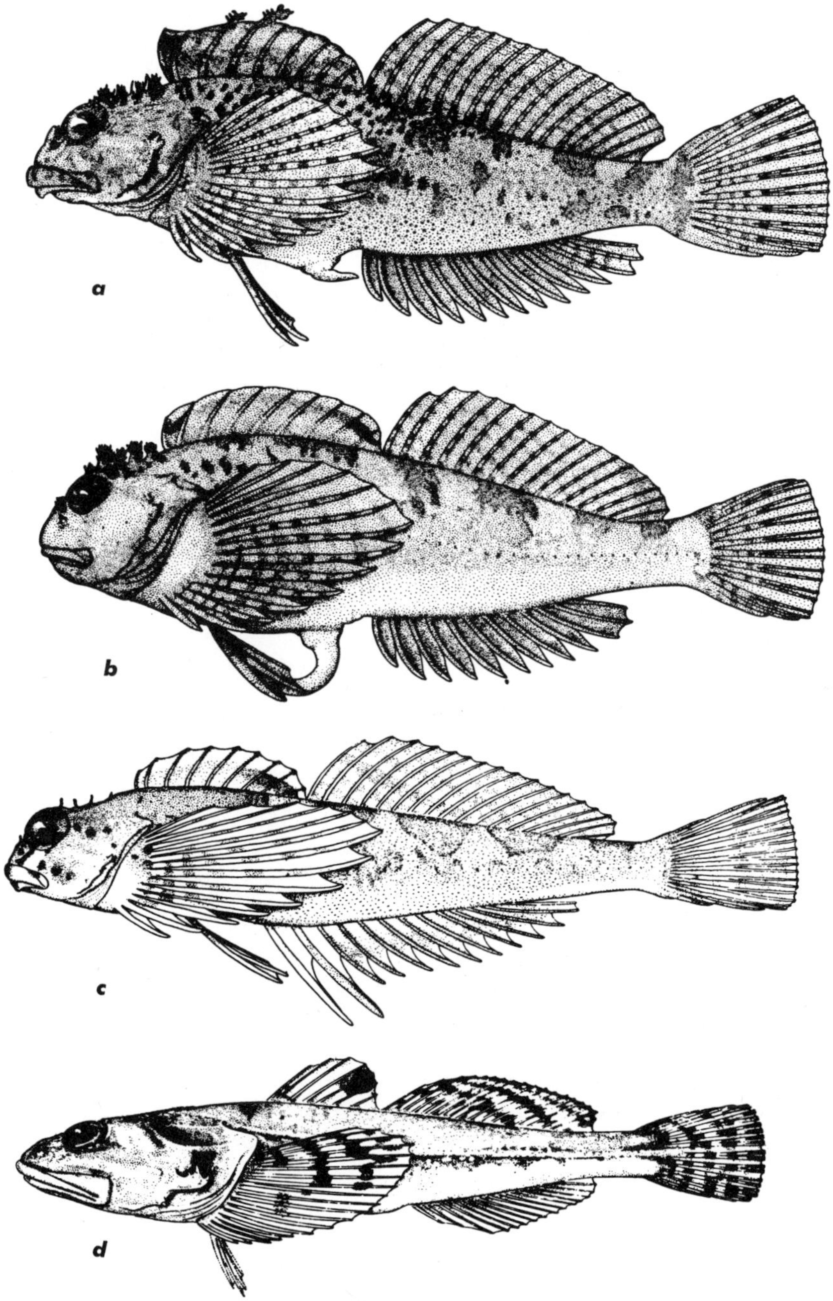

Fig. 128. Typical intertidal fishes. (Legend on facing page)

38. No cirri on maxillary and none on suborbital stay; preopercular spine usually bifid in adults . . . . *Oligocottus snyderi*
38. One to 4 cirri on end of maxillary and a small tuft of cirri on suborbital stay; preopercular spine usually trifid in adults . . . . *Oligocottus rubellio*

39. Cirri and minute prickly scales present between dorsal fins and lateral line; preopercular spines bifid or trifid . *Clinocottus analis*
39. Neither cirri nor scales between dorsal fins and lateral line; preopercular spines simple 40

40. Head moderately pointed and angular, not hemispherical; upper lip strictly terminal; a small fleshy tubercle in median line of groove which limits upper lip dorsally; no cirri behind opercular flap between pectoral base and lateral line . . *Clinocottus embryum*
40. Head very bluntly rounded, hemispherical; upper lip inferior except in juveniles; no fleshy tubercle in groove bordering upper lip; a patch of cirri behind opercular flap between pectoral base and lateral line 41

41. No cirri in anterior half of interorbital space . *Clinocottus recalvus*
41. Cirri in anterior half of interorbital space in specimens more than 35 mm. in standard length (juveniles impossible to differentiate from *C. recalvus* by means of a key) . . . *Clinocottus globiceps*

42. Pelvics with 5 soft rays 43
42. Pelvics with fewer than 5 soft rays 45

43. Dorsal and anal fins evenly rounded, without any stiff and clearly differentiated spines anteriorly . . . . . *Icichthys lockingtoni*
43. Dorsal clearly differentiated into 2 parts; both dorsal and anal with stiff spines anteriorly 44

44. Dorsal VIII-IX, 12-16; distance from dorsal origin to pelvic base 42-50 per cent of standard length; distance from upper end of pectoral base to dorsal origin 7.2-9.1 per cent of length of dorsal base . . . . *Micrometrus minimus*
44. Dorsal VII-IX, 16-19; distance from dorsal origin to pelvic base 36-41 per cent of standard length; distance from upper end of pectoral base to dorsal origin 5.5-7.1 per cent of length of dorsal base . . . . *Micrometrus aurora*

45. Head and body with several rows of conspicuous photophores . . . . *Porichthys notatus*
45. No photophores developed 46

46. Pectoral girdle with a small upturned hook on anterior margin (readily seen by lifting operculum) 47
46. No upturned hook on pectoral girdle 49

47. Caudal fin forked; anal II, 31-35 . . . *Heterostichus rostratus*
47. Caudal truncate or rounded; anal II, 24-28 48

48. Soft dorsal with rounded profile, with 7-10 rays all evenly spaced . . . . *Gibbonsia metzi*
48. Soft dorsal with angular profile, with 5-8 rays of which 2 or 3 in posterior half of the fin are separated by markedly enlarged interspaces. . . . . *Gibbonsia montereyensis*

49. Well-defined transverse rows of pores at right angles to main longitudinal lateral line; no dense hairlike growth of cirri on head . . . . *Plagiogrammus hopkinsi*
49. No transverse lines of pores as above; a dense hairlike growth of cirri covering entire top of head . . . . *Chirolophis nugator*

50. Body disc-shaped; dorsal and anal fins higher than long . *Mola mola*
50. Body long and slender; dorsal and anal much longer than high . 51

51. Gill membranes free, gill openings extending forward ventrally; a prominant longitudinal fold of skin extending just below pectoral base and slightly above base of anal . . . *Ammodytes personatus*
51. Gill membranes joined to each other or to isthmus; no longitudinal fold of skin as above 52

52. Gill membranes attached to isthmus, gill openings restricted to sides of head; body in front of anal origin naked, behind anal origin scaled . . . . *Anoplarchus purpurescens*
52. Gill membranes connected to each other, forming a fold across isthmus; body either completely naked or scaled 53

53. A well-developed naked spine at anterior end of anal fin (this spine, although fitting into a sheath formed in anterior end of fin, is clearly visible without dissection by forcing the membrane back with a needle and pulling spine forward). 54
53. No naked spine at anterior end of anal, all skeletal elements completely covered by membrane 55

54. Anteror surface of anal spine channeled . . *Apodichthys flavidus*
54. Anterior surface of anal spine convex . . . *Xererpes fucorum*

55. Posterior part of dorsal fin composed of needlelike spines which may be detected by drawing a finger forward across their tips (fairly hard pressure may be required in some cases) 56
55. Posterior part of dorsal composed of soft rays 58

56. Dorsal origin over pectoral fin; pectoral fin slightly larger than eye . . . . *Phytichthys chirus*
56. Dorsal origin behind pectoral fin; pectoral fin slightly smaller than eye 57

57. Dorsal origin about 1 orbital diameter or less behind tip of pectoral and about twice as close to tip of snout as to anal origin . . . . *Xiphister mucosus*
57. Dorsal origin 2 or more orbital diameters behind tip of pectoral and about midway between tip of snout and anal origin . . . . *Epigeichthys atro-purpureus*

58. Body covered by readily visible scales . *Cebidichthys violaceus*
58. Body naked . . . . *Scytalina cerdale*

List of Fishes

OSMERIDAE
Spirinchus starksi (Fisk)
Hypomesus pretiosus (Girard) (Surf smelt)
BOTHIDAE
Citharichthys stigmaeus Jordan and Gilbert
PLEURONECTIDAE
Psettichthys melanostictus Girard (Fringe sole)
Platichthys stellatus (Pallas) (Starry flounder)
CENTROLOPHIDAE
Icichthys lockingtoni Jordan and Gilbert
EMBIOTOCIDAE
Micrometrus aurora (Jordan and Gilbert)
Micrometrus minimus (Gibbons)
SCORPAENIDAE (Rockfishes)
Sebastodes paucispinus (Ayres) (Bocaccio rockfish)
Sebastodes mystinus (Jordan and Gilbert) (Black rockfish)
Sebastodes melanops (Girard)
Sebastodes chrysomelas (Jordan and Gilbert) (Black and yellow rockfish)
HEXAGRAMMIDAE (California sea trout)
Hexagrammos decagrammus (Pallas)
Hexagrammos superciliosus (Pallas)
OXYLEBIIDAE
Oxylebius pictus Gill
COTTIDAE
Scorpaenichthys marmoratus Girard (Cabezone)
Jordania zonope Starks
Hemilepidotus spinosus (Ayres)
Hemilepidotus hemilepidotus (Tilesius)
Artedius creaseri (Hubbs)
Artedius harringtoni Starks

Artedius fenestralis (Jordan and Gilbert)
Artedius corallinus (Hubbs)
Artedius lateralis (Girard)
Orthonopias triacis Starks and Mann
Oligocottus rimensis (Greeley)
Oligocottus maculosus Girard
Oligocottus snyderi Greeley
Oligocottus rubellio (Greeley)
Clinocottus analis (Girard)
Clinocottus embryum (Jordan and Starks)
Clinocottus recalvus (Greeley)
Clinocottus globiceps (Girard)
Leptocottus armatus Girard
Synchirus gilli Bean

LIPRIDIDAE

Liparis florae (Jordan and Starks)

AULORHYNCHIDAE

Aulorhynchus flavidus Gill

SYNGNATHIDAE

Syngnathus californiensis Storer

GOBIIDAE

Coryphopterus nicholsii (Bean)
Clevelandia ios (Jordan and Gilbert)
Lethops connectens Hubbs

BATRACHOIDIDAE

Porichthys notatus Girard (Midshipman fish; guards masses of pea-sized yellow eggs beneath rocks)

GOBIESOCIDAE

Rimicola eigenmanni (Gilbert)
Sicyogaster meandrica (Girard)

CLINIDAE

Gibbonsia montereyensis Hubbs
Gibbonsia metzi Hubbs
Heterostichus rostratus Girard

CEBIDICHTHYIDAE

Cebidichthys violaceus (Girard)

STICHAEIDAE

Chirolophis nugator (Jordan and Williams)
Anoplarchus purpurescens Gill
Plagiogrammus hopkinsii Bean
Phytichthys chirus (Jordan and Gilbert)
Xiphister mucosus (Girard)
Epigeichthys atro-purpureus (Kittlitz)

PHOLIDAE

Apodichthys flavidus Girard

Xererpes fucorum (Jordan and Gilbert)

SCYTALINIDAE

Scytalina cerdale Jordan and Gilbert

AMMODYTIDAE

Ammodytes personatus Girard

MOLIDAE

Mola mola (Linnaeus) (Giant sunfish)

Marine Algae and Flowering Plants

No zoölogist can study the intertidal profitably without a speaking acquaintance with the common and conspicuous plants that so plainly mark off the zones of the shore line. Most of these are algae—red, green, or brown—but a few flowering plants also occur in the marine habitat. Chief among these are members of the pond-weed family (Potomogetonaceae). The broad-leaved eelgrass *Zostera* is found in bays and estuaries. Two species of *Phyllospadix*, the surf grass or narrow-leaved eelgrass, occur in the rocky intertidal, and when growing on rocks (not in pools) form a useful indicator of the "zero" tidal level. A key to the local species is given in the revised edition (1952) of Ricketts and Calvin (p. 406).

The intertidal algae generally occur in bands or zones which furnish extremely useful indicators of intertidal zonation, and a knowledge of the commonest larger forms is indispensable to ecological field studies. An excellent discussion of intertidal zonation is found in chapter 6 (by J. W. Hedgpeth) in the 1952 edition of Ricketts and Calvin.

The algal flora of the Pacific Coast of North America is one of the richest in the world, rivaled only by those along the coasts of Japan, Australia, New Zealand, and South Africa. For the Monterey Peninsula alone, Smith (1944) lists 385 species, a number representing an estimated 80 per cent of the species known for the coast. Of this bewildering variety, only those forms most abundant, commonly noticed, strikingly different, or ecologically important have been selected for the following key. The key is based on external morphological characters and attempts to use only obvious features and to avoid specialized terms. This leaves much to be desired for critical determinations (the names of certain genera likely to be confused with those in the key are occasionally included in parentheses). As a check on identifications, reference may be made to the excellent figures and complete keys in G. M. Smith,

Marine Algae of the Monterey Peninsula (1944). Our figures 129-133 are reproduced from Dr. Smith's monograph by permission. These do not duplicate those reprinted in Ricketts and Calvin (1952). Certain identification of many red algae requires a detailed microscopic examination of sections by a specialist.

Key to the Genera of Common Conspicuous Algae

by Isabella A. Abbott
Hopkins Marine Station of Stanford University, Pacific Grove

1. Plants bright grass green or dark spinach green 2
1. Plants olive-green, tan, brown, red, purple, or pink 6

2. Forming clumps of free, smooth, expanded sheets of varying width and height, with ruffled margins (fig. 129, *a*, *b*); occurring in upper mid-tidal zone; "sea lettuce". . . . . *Ulva*
2. Form not as above 3

3. Plants forming groups of convoluted or straight soft tubes (fig. 129, *c*, *d*); especially common in estuarine and brackish-water habitats . . . . *Enteromorpha*
3. Plants growing in tufts or appressed to rock surface; not estuarine 4

4. In low, compact, grass-green, pincushionlike tufts, occurring well up in intertidal zone . . . . *Cladophora*
4. Not as above; surface of thallus feltlike; very dark green (*Codium* spp.) 5

5. Thallus prostrate, adhering closely to rock surfaces (fig. 129, *f*); frequently so dark green as to appear almost black . . . . *Codium setchellii*
5. Growing in erect tufts; branches slender, cylindrical, and repeatedly branched dichotomously (fig. 129, *e*) . . . . . . *Codium fragile*

6. Thallus prostrate, encrusting the substrate 7
6. Not prostrate or encrusting 11

7. Plants conspicuously and heavily calcified to form a brittle crust; usually in pastel shades of pink or purplish (encrusting coralline algae) . . . . *Lithothamnion* (see also *Melobesia, Fosliella,* and *Lithophyllum*)
7. Plants not so calcified; color darker 8

8. Exposed surface bearing concentric rings or ridges; yellow-brown to brownish-black . . . . *Ralfsia*

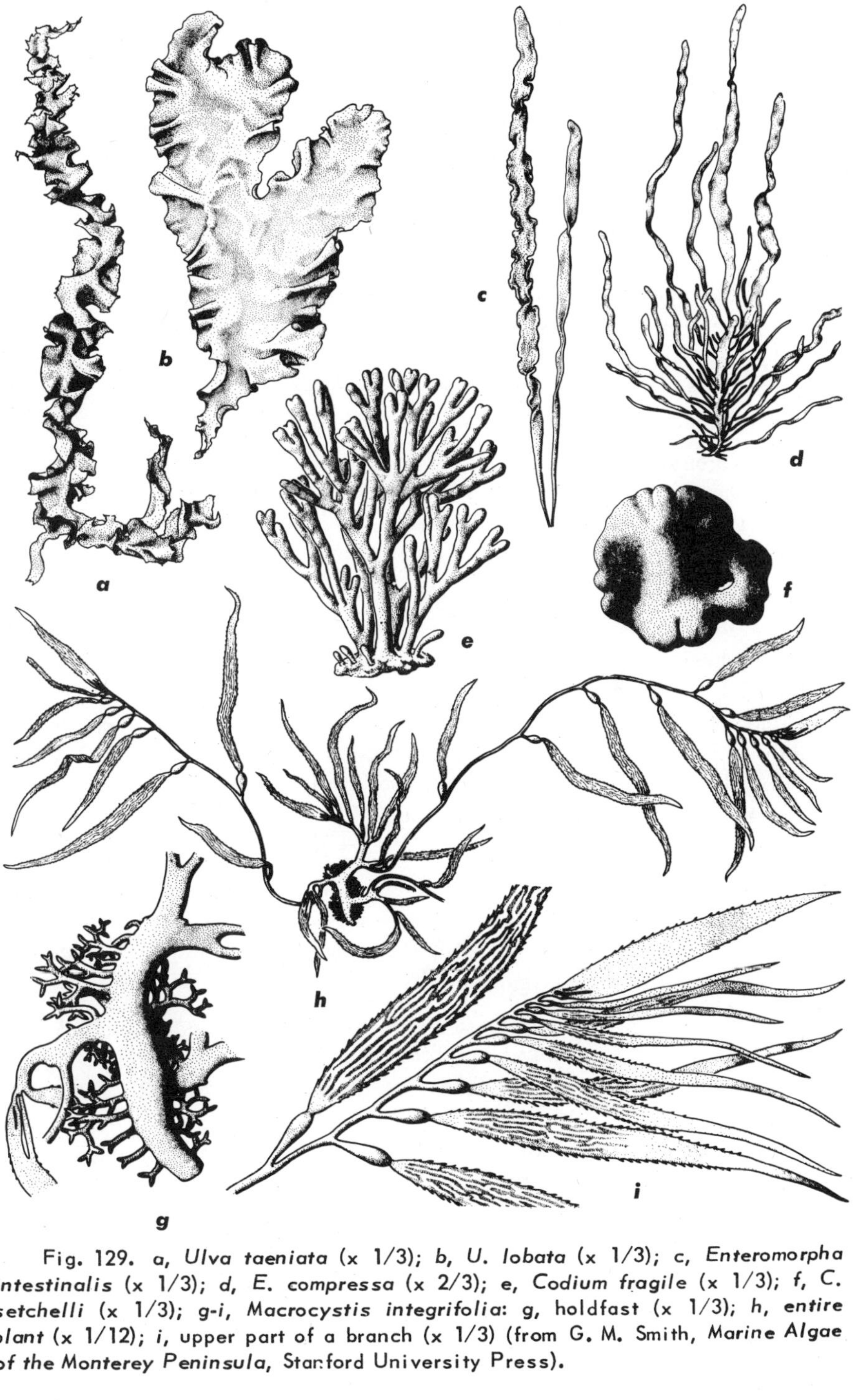

Fig. 129. *a*, *Ulva taeniata* (x 1/3); *b*, *U. lobata* (x 1/3); *c*, *Enteromorpha intestinalis* (x 1/3); *d*, *E. compressa* (x 2/3); *e*, *Codium fragile* (x 1/3); *f*, *C. setchelli* (x 1/3); *g-i*, *Macrocystis integrifolia*: *g*, holdfast (x 1/3); *h*, entire plant (x 1/12); *i*, upper part of a branch (x 1/3) (from G. M. Smith, *Marine Algae of the Monterey Peninsula*, Stanford University Press).

8. Exposed surface without concentric rings or ridges; surface smooth or convoluted 9

9. Growing only on shells of gastropods of the genus *Tegula*; dark purplish-red . . . . *Peyssonnelia*
9. Always growing on rocks 10

10. Very dark green to greenish-black; surface feltlike in texture, never convoluted unless growing on a convoluted substrate (fig. 129, *f*) . . . . *Codium setchellii*
10. Very dark red to reddish-black; surface leathery in texture, never convoluted unless growing on a convoluted surface . . . . *Petrocelis* or *Hildenbrandtia*
10. Light to dark brown (not red); surface smooth in texture, usually convoluted in irregular ridges and folds (fig. 131, *e*) . *Petrospongium*

11. Plants conspicuously calcified, branched and stiffly jointed; usually reddish-pink, pink, lavender, or chalky white; plants generally less than 1 ft. long (nonencrusting coralline algae) 12
11. Plants not calcified; color and size variable 14

12. Plants slender, stiff, with segments of most smaller branches more or less cylindrical (fig. 132, *c*); conceptacles (appearing as small hemispherical bumps) only on terminal segment of the branchlets in fertile plants, and only 1 conceptacle to a segment . . *Corallina*
12. Plants thicker, heavier, more robust, with segments of most smaller branches not cylindrical but conspicuously flattened and flared laterally into small "wings"; conceptacles not confined to tips of branchlets, and frequently more than 1 to a segment (fig. 132, *a, b, d*) 13

13. Conceptacles occurring only on the flattened faces of segments (fig. 132, *a, b*). . . . . *Bossea*
13. Conceptacles occurring on both the margins of lateral wings and on the flattened faces of segments (fig. 132, *d*) . . . . *Calliarthron*

14. Plants tan, yellow-brown, or brownish, sometimes slightly tinged with green (as in *Fucus*) but predominant color is always brown or tan; size variable, up to about 130 ft. long (large kelps and other brown algae) 15
14. Plants olive green (with purplish bases), or red, pinkish, or purplish (including very dark purple); plants always less than 1 yd. long 26

Fig. 130. *a-b, Nereocystis luetkiana: a,* float and bases of blades (x 1/40); *b,* entire plant (x 1/60); *c, Costaria costata* (x 1/6); *d, Dictyoneurum californicum* (x 1/6); *e, Lessoniopsis littoralis,* plant with most of blades removed (x 1/8); *f, Alaria marginata* (x 1/4) (from G. M. Smith, *Marine Algae of the Monterey Peninsula,* Stanford University Press).

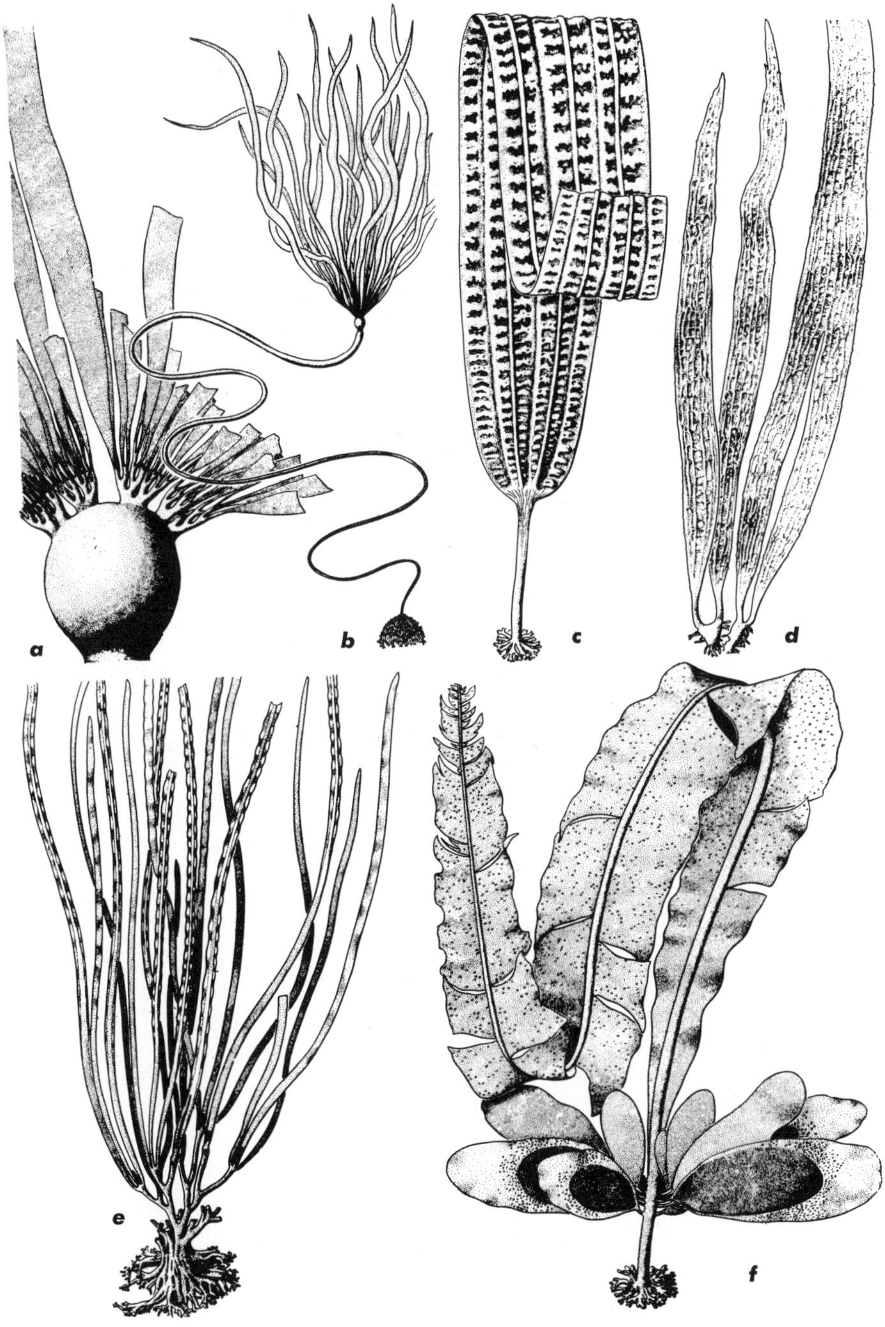

Fig. 130. (Legend on facing page)

15. Plant with an unbranched, whiplike stipe up to about 130 ft. long, terminating in hollow bulbous float bearing elongate blades distally (fig. 130, *a*, *b*); commonly in offshore kelp beds or washed ashore; occasional intertidal individuals are much smaller . *Nereocystis*
15. Plant otherwise; habitat variable 16

16. Plant appearing like small palm tree; with erect, trunklike stipe bearing terminal cluster of blades; on exposed, surf-beaten rocks; the sea palm . . . . *Postelsia*
16. Plant not as above; habitat variable 17

17. Plant consisting of numerous stout straps up to several yards long anchored by common holdfast; hundreds of small blades borne along sides of each strap, some blades modified as small floats; very common in rocky intertidal . . . . *Egregia*
17. Plants otherwise 18

18. Whole plant regularly dichotomously branched; less than 2 ft. long; on rocks 2–6 ft. above zero tide level 19
18. Whole plant not dichotomously branched; size variable but up to 80 ft. or more in length; restricted to lower intertidal regions (+1.0 ft. and below) and kelp beds 21

19. Branches bladelike, conspicuously thin and flattened, and with distinct midrib . . . . *Fucus*
19. Branches cylindrical or somewhat flattened but fleshy and thick; no midrib 20

20. Plants usually 4-8 cm. tall (occasionally up to 15 cm.); lower branches distinctly flattened but rather thick and fleshy . . . . *Pelvetiopsis*
20. Plants 15-40 cm. tall; lower branches slender and cylindrical or slightly flattened . . . . *Pelvetia*

21. With a single, distinct, elongate, swollen float at base of each blade; blades large, corrugated, and with toothed margins (fig. 129, *g-i*); plants found in low intertidal and kelp beds; up to 65 ft. long . . . . *Macrocystis*
21. Floats otherwise or absent; blades lacking regularly toothed margins; habitat variable 22

22. Terminal branches of thallus slender, many bearing linear series of rounded, close-set floats (each series resembling row of peas in pod; lost in winter); plants up to 25 ft. long; in low intertidal and kelp beds . . . . *Cystoseira*
22. No floats of any sort; plants not in kelp beds 23

23. Axis (arising from a holdfast) bearing lateral blades or branches 24

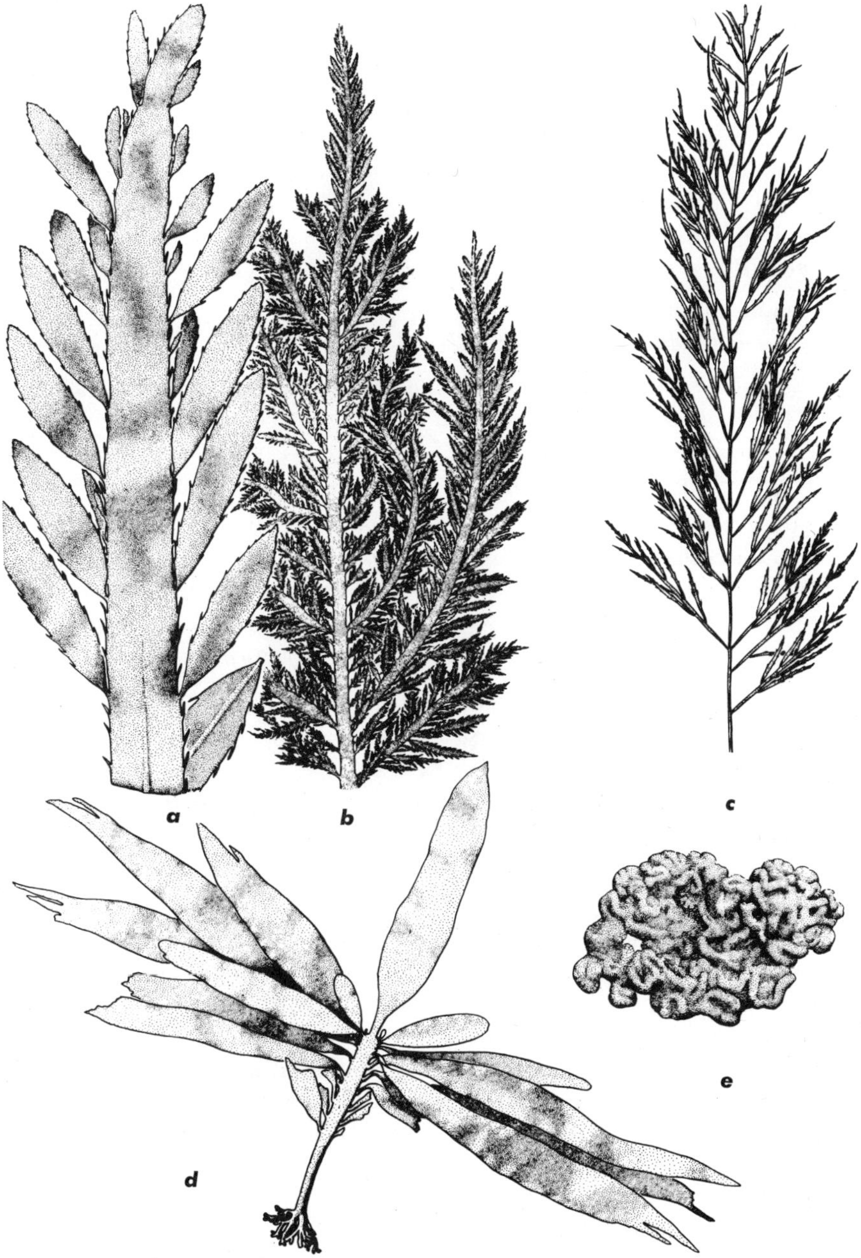

Fig. 131. *a, Desmarestia munda* (x 2/9); *b, D. herbacea* (x 1/3); *c, D. latifrons* (x 2/9); *d, Pterygophora californica* (x 1/2); *e, Petrospongium rugosum* (x 1/9) (from G. M. Smith, *Marine Algae of the Monterey Peninsula*, Stanford University Press).

23. Axis or axes (arising from a holdfast or prostrate rhizome) never bearing lateral branches or blades . . . . *Laminaria* (see also *Costaria* and *Dictyoneurum*, fig. 130, *c*)

24. Axis terminating in, or bearing distally, a single very prominent blade, larger than any of lateral blades 25

24. Axis not terminating in single blade conspicuously larger than lateral blades (fig. 131, *a-c*); with sharp, distinctive odor; this plant liberates quantities of strong acid when detached—do not place it in a container with other organisms . . . *Desmarestia*

25. Terminal blade with midrib (fig. 130, *f*) . . . . *Alaria*

25. Terminal blade lacking midrib (fig. 131, *d*) . . . *Pterygophora*

26. Plants consisting of elongate, unbranched hollow sacs or bladders, up to 10 in. long; olivaceous to purplish in color; in mid-tide zone . . . . *Halosaccion*

26. Plants not so formed 27

27. Plants with a short, stout, erect stipe centrally perforating one or more disc-shaped or circular concave blades, the whole resembling a red mushroom or shallow red goblet (fig. 132, *e*); low intertidal . . . . *Constantinea*

27. Plants not so shaped, generally bladelike or bushy 28

28. Plant form variable, but blades (if any) or branches of the thallus always bearing many elongate papillae or conical, pointed spines 29

28. Plant form variable, but never bearing either elongate papillae or conical, pointed spines; fertile female plants in some may form small, low, less-than-hemispherical bumps on blades or branches . . 30

29. Plants 4–6 cm. high; erect, profusely branched; all branches cylindrical and bearing conical, pointed spines (fig. 132, *f*); dark red to nearly blackish; occurring on rocks 3–6 ft. above zero tide level . . . . *Endocladia*

29. Plants larger, up to 3 ft. long; thallus of broad blades or narrow branches, but always bearing elongate papillae—in some species so thickly grouped as to suggest a turkish towel (fig. 133, *a-e*); usually found below the +3.5 ft. tidal level; color variable, usually some shade of red or purple, tinged with green or brown; several species . . . . *Gigartina*

Fig. 132. *a, Bossea orbigniana* (x 1); *b, B. corymbifera* (x 1); *c, Corallina chilensis* (x 2/3); *d, Calliarthron setchelli* (x 2/3); *e, Constantinea simplex* (x 1/2); *f, Endocladia muricata* (x 1-1/3); *g-h, Botryoglossum farlowianum:* g, entire plant (x 1/3); *h,* segment of blade (x 2/3) (from G. M. Smith, *Marine Algae of the Monterey Peninsula,* Stanford University Press).

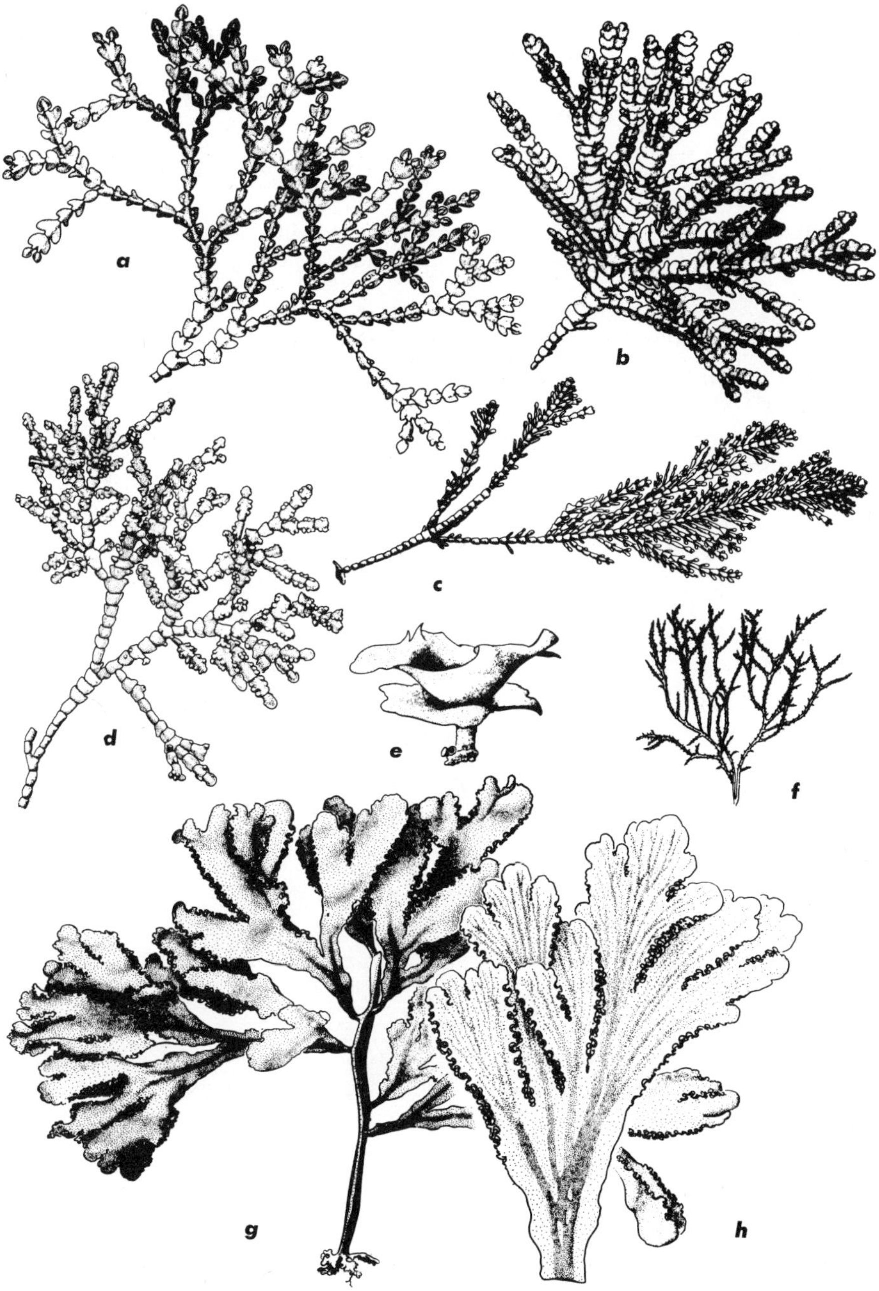

Fig. 132. (Legend on facing page)

30. Thallus "leafy," consisting of broad, simple or partly dissected or subdivided blades; blades always bearing pattern of conspicuous ridges or "veins" 31
30. Thallus variable; blades or branches never bearing ridges or veins 32
31. Blades simple, without frilled margins, but with anastomosing network of veins or ridges (fig. 133, *g*) . . . . . . *Polyneura*
31. Blades distally dissected and with margins narrowly and tightly frilled; veins not anastomosing, but branching and fanning out distally (fig. 132, *g*) . . . . *Botryoglossum*
32. Plants forming clumps of paper-thin, smooth, broad or elongate sheets with ruffled margins (fig. 133, *f*); in form (but not color) very much like *Ulva*; species occur on rocks up to 4 ft. above zero tide level, or as epiphytes on larger plants . . . . . . *Porphyra*
32. Plants otherwise, not *Ulva*-like 33
33. Plants unbranched, forming clumps of tough, leathery, paddlelike or lanceolate blades, each arising from a stubby stipe (fig. 133, *h*); olive green to purplish and often iridescent; blades of fertile plants bearing numerous small bumps (cystocarps); very common in middle and low intertidal zones . . . . *Iridaea* (=*Iridophycus*)
33. Plants much branched; blades either absent or narrow and lanceolate, borne laterally on the main axis or on branches; margins of branches or blades frequently bearing tiny bladelets; thallus surface smooth . . . . *Prionitis*

List of Common Intertidal Algae

Chlorophyta (= Chlorophycophyta, Green algae)
- *Cladophora hemisphaerica* Gardner
- *Cladophora trichotoma* (C. A. Agardh) Kützing
- *Codium fragile* (Suringar) Hariot
- *Codium setchellii* Gardner
- *Enteromorpha compressa* (Linnaeus) Greville
- *Enteromorpha intestinalis* (Linnaeus) Link
- *Ulva lactuca* Linnaeus
- *Ulva taeniata* (Setchell) Setchell and Gardner

Fig. 133. *a*, *Gigartina californica* (x 1/3); *b*, *G. agardhi* (x 1/3); *c*, *G. cristata* (x 1/3); *d*, *G. canaliculata* (x 1/2); *e*, *G. boryi* (x 1/4); *f*, *Porphyra perforata* (x 1/3); *g*, *Polyneura latissima* (x 1/2); *h*, *Iridaea flaccidum* (x 1/4) (from G. M. Smith, *Marine Algae of the Monterey Peninsula*, Stanford University Press).

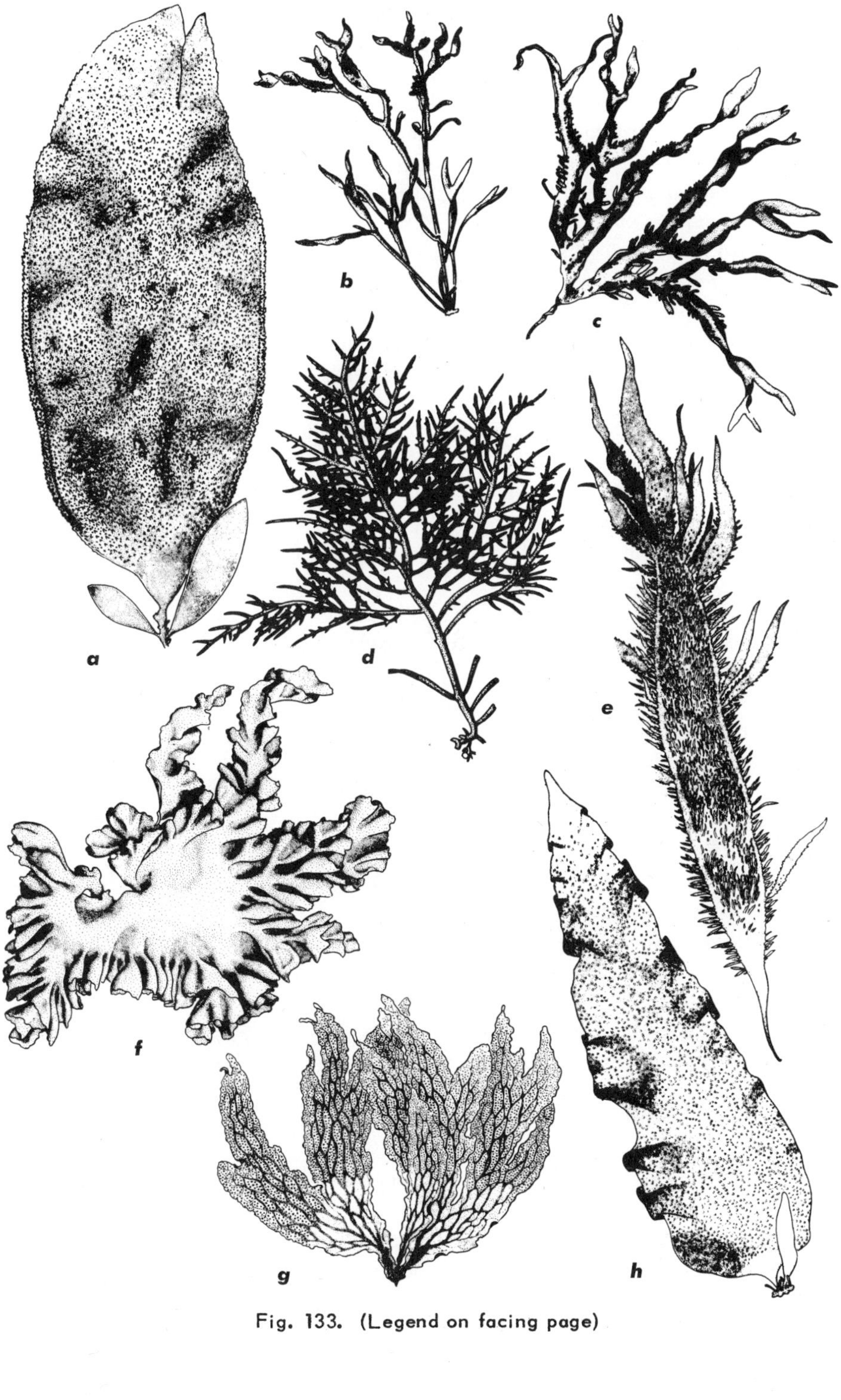

Fig. 133. (Legend on facing page)

Phaeophyta (= Phaeophycophyta, Brown algae)

Alaria marginata Postels and Ruprecht
Costaria costata (Turner) Saunders
Cystoseira osmundacea (Menzies) C. A. Agardh
Desmarestia herbacea (Turner) Lamouroux
Desmarestia latifrons (Ruprecht) Kützing
Desmarestia munda Setchell and Gardner
Dictyoneurum californicum Ruprecht
Egregia menziesii (Turner) Areschoug
Fucus furcatus C. A. Agardh
Laminaria andersonii Eaton
Laminaria farlowii Setchell
Laminaria sinclairii (Harvey) Farlow
Macrocystis integrifolia Bory
Macrocystis pyrifera (Linnaeus) C. A. Agardh
Nereocystis luetkeana (Mertens) Postels and Ruprecht
Pelvetia fastigata (J. G. Agardh) De Toni
Pelvetiopsis limitata (Setchell) Gardner
Petrospongium rugosum (Okam.) Setchell and Gardner
Postelsia palmaeformis Ruprecht
Ralfsia pacifica Hollenberg

Rhodophyta (= Rhodophycophyta, Red algae)

Bossea corymbifera Manza
Bossea dichotoma Manza
Bossea gardneri Manza
Bossea orbigniana (Decaisne) Manza
Botryoglossum farlowianum (J. G. Agardh) De Toni
Calliarthron cheilosporioides Manza
Calliarthron setchelliae Manza
Constantinea simplex Setchell (Common at Dillon Beach; rare south to Monterey area)
Corallina chilensis Decaisne
Endocladia muricata (Postels and Ruprecht) J. G. Agardh
Gigartina agardhii Setchell and Gardner
Gigartina boryi Setchell and Gardner
Gigartina californica J. G. Agardh
Gigartina canaliculata Harvey
Gigartina corymbifera (Kützing) J. G. Agardh
Gigartina harveyana (Kützing) Setchell and Gardner
Halosaccion glandiforme (Gmelin) Ruprecht
Hildenbrandtia occidentalis Setchell
Iridaea (=*Iridophycus*) *coriaceum* Setchell and Gardner
Iridaea (=*Iridophycus*) *flaccidum* Setchell and Gardner

Iridaea (=*Iridophycus*) *heterocarpum* (Postels and Ruprecht) Setchell and Gardner
Lithothamnion californicum Foslie
Lithothamnion pacificum Foslie
Petrocelis franciscana Setchell and Gardner
Peyssonnelia pacifica Kylin
Polyneura latissima (Harvey) Kylin
Porphyra lanceolata (Setchell and Hus) G. M. Smith
Porphyra naiadum C. L. Anderson
Porphyra occidentalis Setchell and Hus
Porphyra perforata J. G. Agardh
Prionitis andersonii Eaton
Prionitis lanceolata Harvey
Prionitis lyalii Harvey

Field Studies

No picture of organisms that ignores their physical and organic environment can be even approximately complete. Studies of dead animals or their parts or even of living animals in the laboratory, indispensable as they are, give but a partial picture, and for fuller understanding we must seek firsthand knowledge of living organisms in their natural settings. Field trips are of prime importance in gaining such knowledge and understanding. They make possible a study of the environment itself, of the distribution of organisms within specific habitats, of the behavior and interrelationships of species, and of the influence of physical and biotic factors on the distribution of organisms. Only through information gained in field studies is it possible to establish correlations between the structure and behavior of an organism on the one hand, and its habitat and ecological niche on the other. One purpose of this manual is to make practical the inclusion of extensive field studies of the marine intertidal within a course in the biology or natural history of invertebrates for advanced undergraduates or for graduate students.

The approach and spirit of these studies is essentially ecological, although limitations in time, training, and equipment seldom permit us to utilize the quantitative and experimental methods characteristic of modern ecology. Rather, the procedure is one in which ecological surveys or reconnaissances are made, on the basis of which we try to set up preliminary hypotheses as to the role and importance of different factors in determining the distribution of animals. These hypotheses are based only on readily observable and relatively obvious facts relating to the structural and behavioristic adaptations of organisms, the presence and nature of other biota, and the physical nature of the environment. Such a factor as degree of exposure at a given spot, for example, may be estimated on a relative basis by determining the vertical level in the intertidal, the nature and arrangement of the local substrate, and the position with regard to the action of waves, wind, rain, and sun. The importance of the effects of exposure may then, in a preliminary way, be evaluated through a study of the animals present (and of those conspicuously absent), and an appraisal of structural and behavioristic adaptations that make successful life possible under the observed conditions of exposure. Such studies encourage student and teacher alike to think of organisms in ecological terms; and they provide the basis for a degree of insight into the interrelationships between organisms, the variables which limit the distribution of organisms, and the factors which control the formation of characteristic biotic communities. They supplement, but do not replace, the more conventional laboratory studies of invertebrate organisms, and they make these laboratory studies far more meaningful.

GENERAL DIRECTIONS FOR FIELD WORK

The following directions apply primarily to the studies of marine intertidal benthos, and should be suitably modified for exploration of other environments. For fresh-water studies the equipment and exercises outlined in Needham and Needham (1951) are useful.

Equipment Recommended

Students will need a geologist's pick, or a hammer and cold chisel; putty knife or similar scraper; forceps (most conveniently worn on a string around the neck); jars and vials of various sizes which may be firmly capped (preferably screw-topped); hard-backed notebook and automatic pencil; hand lens worn on a cord about the neck (6–10 power is ample; higher powers do not permit ready inspection of objects within vials); and some sort of bag, fishing creel, or basket for carrying equipment, bottles, and specimens (the Army surplus canvas buckets are excellent). Hip-length rubber boots and warm clothing are highly desirable. On trips to soft-bottomed areas, shovels and sieves, about one of each per five students, are essential. Small vials may be more safely and conveniently carried if packed in tins or waxed-paper milk cartons.

Suggestions for Collecting

Plan your program beforehand, especially if for the seashore, as every minute of the low tide must be made to count. Go out at once to the water's edge. Follow the tide out and work back as the tide drives you in. On rocky coasts take care not to be trapped on offshore rocks by the incoming tide.

On your first trip to the intertidal the impact of quantity and variety of animal life will usually lead to more or less excited but aimless collecting of the most obvious and conspicuous species. Get down among the rocks and examine them closely. Extend your search by turning over rocks, lifting algal masses, examining the under surfaces of overhanging ledges, chipping away cracked or soft rock, and so on.

Destroy as little as possible either of animal life or of the natural conditions of the environment. *Replace overturned rocks in their original position!* The fauna living on the lower surface of the rock or beneath the rock cannot survive when exposed to air and wave action; neither can organisms originally on the upper surface survive if turned underneath. Carelessness in this by students (and abalone hunters) can markedly alter a good collecting ground within the space of a few years.

On subsequent trips select specific sites that you examine systematically, either to gather data on the aggregation of species occurring together in a given habitat, or to see what representatives of a given systematic group occur there, or both. The most rewarding places for collecting are: (1) deep, shaded tide pools near low-tide level, (2) on and in seaweed, particularly in the holdfasts of larger brown algae and eelgrass, (3) in the crevices and cavities of rock and within rock, particularly if the rock be soft, porous, or covered with coralline algae, (4) under rocks, and (5) in sand and mud if they are not subject to constant movement by tides and currents.

On all field trips, collections should be made of animals not already known, forms not in laboratory demonstration sets of specimens (where these are used), and species needed for special studies. It is well to work on very common forms first, leaving rarities for later identification and study. Do not collect unless you plan to use the specimens for study or identification or for your class or personal collection. Avoid collecting great numbers of the same species.

Small rocks, especially favorable pieces from larger rocks, and holdfasts of kelp from the outer low-tide level may be brought in and the animals removed and cared for at leisure. This requires but one set of notes in the field for many specimens. Also, numerous specimens from the same small habitat (e.g., a rock crevice or a low-level tide pool) may be placed in a single container with proper notes, to be sorted later. Care must be taken, however, to segregate certain types of organisms. Most worms, for example, secrete mucus and tend to form a tangled, often inextricable, mass with other organisms. This is particularly true of nemertean worms, which should be kept separately. Nudibranchs often liberate quantities of mucus, which entangles small or delicate organisms; and some sponges, when cut or broken, liberate materials deleterious to other animals crowded in the same jar. Certain animals, particularly crabs, sometimes tear others. Voracious animals such as large isopods often devour smaller forms, and the spines of echinoderms form a place of escape and entanglement for many smaller, mobile animals. Fragile forms such as hydroids and aborescent bryozoans and sensitive forms such as colonial ascidians, which may be rendered unsuitable for study by rough handling, should be kept only with other inactive forms or by themselves. No jar should ever be crowded or packed with specimens. Intelligent care is necessary if satisfactory collections are to be obtained by putting numerous specimens in the same container in the field.

Especially in hot weather, problems are raised if animals must be transported any great distance to the laboratory. Packing jars loosely in cold, wet seaweeds in a basket or wet canvas bucket will maintain the necessary coolness for some time, especially if the carrier can be kept shaded but exposed to moving air. In a great many cases intertidal animals will survive much longer in jars very loosely packed with damp seaweed than in jars filled with water. Exposure to damp air is not harmful to an intertidal animal, and a jar of air contains twenty times as much oxygen as the same jar full of water.

Field Notes

If specimens are taken merely for immediate study or practice in identification, field notes are not necessary. Any specimen to go into a permanent collection, however, *must* be accompanied by certain information to make it of value to any future student. Experience has shown that most collections ultimately are used for study whether that was the original intention of the collector or not, and many troublesome errors and difficulties have arisen owing to lack of information on specimens in such collections. The difficulty of keeping proper field notes and correlating these with a large collection of specimens on a field trip is considerable, especially under the conditions of wet hands, lack of a place to sit, the presence of rain or spray, lack of time, and lack of sufficient containers to segregate specimens. Careful planning is necessary to accomplish the required end.

There are two common methods of keeping these field notes. One, perhaps the safest, is to write the information in pencil on a slip of paper and drop it into the container in the field. When the collection is put into permanent form the information thus obtained must be copied on the label in (or on) the bottle. The second method is to drop into the container or attach to the specimen a number (written at the time or already prepared), and to record the same number together with the desired information in a field notebook or on cards. This method is easier but involves the risk of loss of the notes (in which case the number means nothing), and the danger of mixing numbers and notes because of haste in the field. Experience will dictate your choice of method.

There is almost no limit to the amount of available information which, if connected with a specimen, might prove valuable to the student of distribution, variation, or ecology. Points commonly noted are: (1) exact *locality*, including county, (2) *date*, (3) exact *location* (as concerns tide level or altitude, for example), (4) *environmental situation* or *habitat* (underside of rock, or the like), (5) any organisms or organic remains apparently associated with it (e.g., "with polychaete worms, no. 63, in hole of boring clam"), (6) anything unusual in the behavior or stage of the animal, and (7) the *collector's name*. The absolute minimum must include locality, date, habitat, tide level, and name of collector.

Field notes may, however, have a usefulness far beyond the proper labeling of specimens in collections. Once the investigator has learned the more important species in the fauna, field trips may be devoted not so much to collecting as to making and recording careful observations on specific organisms or habitats. Learning to produce a worth-while set of field notes in the intertidal takes patience, determination, and practice. The end can be achieved only by limiting the objective, making it sharply definite, and refusing to be deflected from it. Since this in itself requires experience it is generally wise to have earlier field trips devoted to more general objectives, and to confine critical study and recording of ecological features to later trips.

Study and Preservation

On return to the laboratory the first move should be to get material segregated and into dishes of fresh, cool sea water, taking care always to maintain the correlation between specimens and field notes. As a routine it is worth while to observe under the microscope, while they are still living, any small animals or developmental stages not already studied, since this will often bring out much of beauty and interest otherwise lost. If time permits and there is any probability that specimens are to be used for future systematic studies, careful color notes will be of value.

If animals are to be preserved, this should be done while they are still in good condition. Some organisms may be dropped directly into fixative. Others, especially delicate or soft-bodied and contractile organisms, should be first anaesthetized. Isotonic magnesium chloride (about 73 gm. $MgCl_2 \cdot 6\ H_2O$ per liter of tap water) is the best easily obtainable general anaesthetic for marine animals. It may be used straight, or in various proportions with sea water. Other useful anaesthetics are: Epsom salts ($MgSO_4$), of which small amounts may be added directly to dishes of sea water; alcohol, which may be arranged to drip slowly into dishes of sea water or fresh water containing animals; menthol, a few crystals floated on the surface of a dish of sea water; or a few shreds of tobacco added to a small amount of the fluid containing animals. Fixation and preservation of museum specimens may be accomplished with 5-10 per cent formalin or 70-95 per cent alcohol. For detailed methods of preparing and preserving specimens see Bulletins 39-42 of the United States National Museum.

Identifications of animals should not be made lightly (see Introduction). When the species is uncertain the specimen should be labeled with the generic name followed by "sp.?", thus: *Tegula* sp.?. If it is believed to be a certain species but identification is in doubt, a question mark in parentheses should follow the author's name, thus: *Tegula funebralis* Adams (?).

SAMPLE FIELD EXERCISES

Exercises in a course in the natural history of invertebrates normally consist of a number of field trips interspersed with laboratory studies designed to provide preliminary or supplementary knowledge of the animals encountered in the field. Emphasis on field trips will vary with the instructor, but surely in a summer session at the seashore field studies can and should constitute the heart of the course. Trips are usually scheduled to coincide with low tides (tide tables for the year can be obtained at most sporting goods or bait stores, or purchased from the Superintendent of Documents, Government Printing Office, Washington, D.C.) and class programs will therefore vary from year to year.

The following exercises are intended to suggest worth-while areas and problems for investigation, rather than to outline a complete program. The more general exercises come first and are devoted to a survey of the commoner animals and the major types of intertidal habitats. These are followed by intensive studies of more restricted habitats and faunas. The exercises may well need modification for particular local collecting grounds and constellations of organisms in different areas. To be of maximum value the studies should be cumulative and comparative. Each new situation should be compared to others already studied, and similarities and differences should be observed in faunas, environmental conditions, and their interrelationships.

In pursuing field studies it is well to keep in mind certain fundamentals. All animals have similar basic needs or requirements. Ultimately these can be reduced to food, oxygen, protection, and the proper conditions for reproduction, the terms food and protection being used in the broadest possible sense. In situations where an animal occurs regularly and in numbers, we may be sure that these needs are met, though the manner in which they are met is not always obvious.

Food, as used here, includes all substances (except oxygen) from the environment necessary to provide animals with energy and building materials. The intertidal is a region of abundant light and food. In areas of hard substrate there is often much plant life; this includes not only the larger algae and occasional flowering plants such as *Phyllospadix* and *Zostera*, but also the film of microscopic plant life growing on exposed surfaces of rocks and larger plants. A few organisms (e.g., the kelp crab *Pugettia producta*, the sea urchin *Strongylocentrotus franciscanus*, the limpets *Acamea insessa* and *A. paleacea*, and the like) graze directly on the attached larger algae. Seaweeds broken loose and washed into crevices and pools or ashore on beaches provide a rich source of food for other forms (e.g., the purple sea urchin *Strongylocentrotus purpuratus*, the beach amphipods *Orchestia* and *Orchestoidea*, and other scavengers). Finally, plant material pulped into a fine organic detritus by the action of turbulent waters against the substrate provides food for a host of forms that feed in many different ways. Detritus in suspension in moving waters is taken, together with living plankton, by a variety of particle feeders. Some use mucous nets or webs to trap this food (e.g., the echiuroid *Urechis*, the polychaete *Chaetopterus*, the gastropod *Vermetus*, and the like), whereas others use cilia, usually in conjunction with sheets or strands of mucus (e.g., most bivalves, brachiopods, bryozoans, tunicates, serpulid and sabellid polychaetes, sand dollars, sponges, and so on). Still other feeders on suspended detritus and plankton use combs of fine bristles or setae to catch particles (e.g., the anomurous decapods *Petrolisthes* and *Emerita*, barnacles and many other lower crustaceans), and yet others use tentacles (e.g., the sea cucumber *Cucumaria curata*, terebellid and spionid polychaetes, and ctenophores).

Organic detritus of plant and animal origin accumulates on and in soft substrates, and here it forms food for another complex of organisms. Some of these organisms suck up the surface layer of detritus (e.g., bivalves of the genus *Macoma*), others swallow whole mouthfuls of substrate more or less unselectively (e.g., many annelids, sea cucumbers, and so on), and still others burrow and sift the bottom material for edible particles (e.g., the shrimplike anomuran *Callianassa*). Finally, organic detritus tends to cling to exposed surfaces of rocks and plants; here, macroscopic encrusting algae, together with the microscopic plants, bacteria, and protozoa which also occur, serve as food for animals with rasping or scraping organs, particularly the gastropods and chitons.

This great variety of organisms living on plant and animal detritus and plankton provides food in plenty for intertidal predators. Micropredators like the hydroids and *Metridium* feed upon animal plankton brought in by the tides. Predators on larger organisms (e.g., most starfishes—not *Patiria*—, the gastropods *Thais*, *Acanthina*, and *Purpura*, nemerteans, most sea anemones, and others) find a rich diet of clams, snails, barnacles, worms, and other forms. Fishes may be significant predators when the tides are high; shore birds and man when the waters recede. When studying an organism keep in mind the questions: What does it eat? How does it get its food? What is the nature and importance of its role in the food economy of the area?

Oxygen presents less of a problem for most intertidal organisms. In most areas continual movement of the shallow waters insures full oxygenation at all times. The oxygen tension in air is the same as in saturated water, and as long as animals remain damp, atmospheric oxygen can readily diffuse across respiratory membranes. The oxygen requirements of sessile and sedentary animals are not large, and some forms at least are capable of withstanding temporary anaerobic conditions. Some intertidal forms, however, do have special adaptations to meet specialized needs in respect to oxygen. Species living high in the intertidal may show structural modifications enabling them to carry on aerial respiration for prolonged periods. Some species of *Littorina* may have considerable vascularization of the wall of the mantle cavity, which serves at least to some degree as a lung. Crabs such as *Pachygrapsus* may have gills reduced in size and stiff enough for self-support, arrangements for retaining water in the gill cavity, and vascularization of the wall of the cavity itself. Mud-flat forms, stranded in the substrate with a limited water and oxygen supply, may also have a problem. The echiuroid *Urechis* has coelomic corpuscles containing hemoglobin and, under the conditions present when tide flats are submerged, this is fully saturated with oxygen and does not function as a respiratory pigment. But at low tide, as the oxygen in the burrow-water and in the body tissues becomes depleted, the hemoglobin begins to release its stored oxygen to the tissues. The stored supply of oxygen is sufficient to enable the animal to last until the tide again comes in over the mud flats, bringing oxygenated water to the burrow.

Reproduction, another essential for all species, is somewhat more difficult to study in the field. Many intertidal benthic forms shed eggs and sperm into the sea, where developing embryos and feeding larvae lead a pelagic existence for a time. Pelagic stages may be taken in plankton tows but otherwise are seldom seen. However, a large number and variety of marine and brackish-water organisms retain, carry, or brood their eggs (e.g., the sea cucumbers *Cucumaria curata* and *Thyonepsolus nutriens*, the starfish *Leptasterias*, many crustaceans, colonial ascidians, the polychaete *Neanthes lighti*, and the like). Other invertebrates produce characteristic egg masses or capsules within which the larvae develop from yolky eggs (e.g., the snails *Thais* and *Acanthina*, most nudibranchs and cephalopods, some chitons, certain polychaete worms, etc.). These should be observed and if possible connected with the animals that produced them. Still other animals reproduce asexually and may form extensive colonies or aggregations of individuals (e.g., the aggregating anemone *Anthopleura elegantissima*, colonial ascidians, hydroids, bryozoans, the hydrocoral *Allopora*, the alcyonarians *Clavularia* and *Renilla*, and sponges). Observations of such features as can be observed in the field or in individuals collected and brought to the laboratory are important in natural history studies, and may well result in discoveries previously unrecorded in biological literature.

Finally, the question of protection and of the environmental conditions from which organisms require protection, are crucial to these field studies. Animals persistently present at any particular spot must be adapted to survive the most unfavorable extremes of environmental conditions that occur there. All mechanisms—morphological, physiological, and behavioristic—through which an animal preserves internally the conditions for cellular life, are here implied in the term protection. In this sense, protection includes the sheltered places sought by motile forms like crabs, the production of anchored fortresses of lime by barnacles, the capacity of brackish-water (and fresh-water) organisms to osmoregulate, the ability of the brine shrimp to tolerate a warm and hypersaline medium, and a myriad other adaptations that enable functional organisms to withstand conditions *externally* which they could not tolerate *internally*.

Owing to the regular rise and fall of the sea, the intertidal is a region of great periodic fluctuations in environmental conditions. It is a region of transition between sea and land, but the boundary between aquatic and terrestrial conditions is less sharp and far more complex than that at the shore lines of small lakes and ponds. In the intertidal region, animals are subjected to conditions that are alternately marine and semiterrestrial.

Even with their various protective adaptations, animals at the seashore are more or less limited to particular habitats, since each is adapted to a particular way of life under a particular set of conditions. Physical factors of the environment, interacting with specific requirements and adaptations of animals (for food, oxygen, reproduction, and protection), definitely limit the distribution of each species. The most important of the physical factors with which we will be concerned in the field are outlined below.

The nature of the substrate.—The substrate will be found to vary from nearly unbroken cliffs and rocky ledges, through a series of such intergrades as broken rocky reefs, boulders, and pebbles, to the finer substrates of sand, mud, and clay. The nature of the substrate is of cardinal importance in limiting animal and plant distribution, as can be seen at once by comparing the biota of a rocky outcropping with that of an adjacent stretch of sand in a region where both occur. Even the type of rock (e.g., hard granite *vs.* relatively softer sandstone or shale) is of great importance, particularly for rock-boring forms.

Degree of wave shock and current action.—The pounding and abrasive action of waves has many effects on both environment and fauna. Where severe, wave action prevents the accumulation of substrates of of mud and clay; it restricts the distribution of animals that are not either tough or flexible, firmly rooted, or capable of clinging tightly to (or burrowing or boring into) the substrate; and it makes possible the presence of aquatic or semiaquatic forms in a splash zone above the highest level reached by the surface of the sea. Currents may erode soft substrates in one area and redeposit them elsewhere; this shifting of bottom materials and the accompanying scouring action of the particle-laden waters may have important effects on bottom-dwelling organisms present. Ricketts, in *Between Pacific Tides*, clearly recognized the importance of the degree of exposure to wave shock as a factor in limiting animal distribution; his primary ecologic divisions of the Pacific intertidal, based on this factor, are "open coast," "protected outer coast," and "bay and estuary," with appropriate subdivisions of each according to the nature of the substrate.

Degree of tidal exposure to a variety of conditions, particularly the following:

Altered temperatures.—At low tides on sunny summer days and windy winter nights, the temperature extremes for exposed animals markedly exceed those found in the sea. Ability to withstand great changes in temperature, and particularly the higher temperatures, is one important factor in determining the intertidal distribution of organisms.

Desiccation.—Most dwellers in the intertidal have come up from the sea; very few are invaders from the land. It follows that withstanding desiccation is a great problem for many intertidal forms, and surviving prolonged submersion is far less commonly a problem. Exposed intertidal organisms tend to dry out, particularly during low tides on hot, sunny days, and the problem is closely related to that of withstanding higher temperatures. Motile forms may show a protective behavior, shifting to positions under rocks or ledges or seeking the slighter protection of depressions and shallow crevices. Sedentary and sessile forms may have protective shells, adherent layers of sand or gravel, tough and often mucus-covered integuments, or large internal stores of water, etc.

Altered salinity.—On hot days at low tide the salinity of high tide pools may rise owing to evaporation. During rains, direct precipitation and runoff from streams and beaches may greatly reduce the salinity in some areas. Organisms on high rocks, in high pools, and near stream mouths must be able to avoid, regulate against, or tolerate at least temporary brackish and/or hypersaline conditions.

Lowered oxygen availability.—Only in certain instances is this an important factor. It becomes important where intertidal forms are desiccated to a degree that exposed surfaces no longer serve as respiratory membranes. It may also be important for dwellers in mud flats exposed by tides.

Predators—At low tides birds and man, at high tides fishes, may invade the intertidal as predators.

As a result of all these factors and of many others that usually are less accessible to study, intertidal environments, floras, and faunas may vary greatly, even within relatively restricted regions of the coast. Further, at any one locality one will always find a vertical intertidal zonation of organisms. Plants and animals most resistant to vicissitudes live higher in the intertidal; those less tolerant live lower down. Vertical zonation is clearest where the substrate presents uniform, vertical surfaces of rock, as on the walls of deep intertidal channels. It is most complex in areas of irregular and mixed substrates, which provide many different habitats and also conditions under which there is an interdigitation of zones. For an excellent discussion of the general problem of intertidal zonation see Ricketts and Calvin (*Between Pacific Tides*, 1952 ed., chapter 6).

In the sample exercises the term "ecologic niche" is used to refer to the sum total of environmental conditions, physical and biotic, to which a particular species is adapted and which are necessary to its continued existence. "Habitat," on the other hand, refers to a place or region of a certain kind, characterized by particular major environmental features. A habitat such as a tide pool possesses a characteristic fauna, each species of which has its own niche. Since the distribution of ecological conditions within a habitat may spatially limit an ecologic niche we may, for convenience, connect certain niches with places within the habitat, but this is not true of many niches and is not really correct for any niche. To define a niche completely would involve an immense amount of quantitative and experimental investigation of the structure, behavior, and physiology of a species, and a study of the precise physical and biotic conditions of its habitat. However, we are usually able to discern one or more obvious features of each niche which give it objective reality and allow us to give it a preliminary definition and to use it in discussion.

1. Animals of a Protected Rocky Shore with Tidal Flats

(Specifically, such an area as Point Richmond, in San Francisco Bay)

Purpose.—To learn the various common animals of the region as they exist under natural conditions, each in its normal habitat; to determine their distribution with regard to certain physical and biotic factors; to learn their groupings by zones or other environmental situations; and to obtain as clear an idea as possible of their interrelationships in such groupings, especially their food relations.

Procedure.—If this is the first field trip it should be preceded by a laboratory study and practice in the identification of the animals most commonly encountered. Several species of decapods, isopods, amphipods, limpets, and polychaetes form a conspicuous part of this fauna. Preliminary survey of study collections made in the area and the use of unlabeled representative forms put out for identification by means of the keys provide a valuable introduction to the forms which will be seen. In the field collect such animals as are necessary to identify by the use of keys. Make careful field notes with diagrams and sketches.

Report.—The report should be in the form of a synthetic vertical section of the shore, showing the relative locations of all important habitats or associations and accompanied by lists of the species found in each of these different habitats or associations. Discuss some of the reasons for the observed distributions of the different types. Compare the fauna found on rocks here with the fauna of an exposed, wave-swept, rocky shore, if such has been previously visited.

Characteristic species

Polychaetes

Eteone sp. Small, very active, mud-flat dweller.

Boccardia proboscidea In rock crevices in tide pools, with only the paired tentacles showing.

Nephtys caecoides In sand.

Capitellid polychaetes Extremely elongate red worms characteristic of mud flats.

Nereid polychaetes

Isopods

Cirolana harfordi Scavenger, under stones and in mussel beds.

Sphaeroma pentodon Stands prolonged exposure to fresh water.

Ligia occidentalis Lives above high-tide mark.

Amphipods

Hyale pugettensis Conspicuous in high tide pools.

Anisogammarus confervicolus

Corophium spp. Living in tubes in mud.

Orchestia traskiana Burrower, living usually above high-tide mark.

Decapods

Pagurus spp.

Pachygrapsus crassipes Highest intertidal dweller among local decapods.

Hemigrapsus nudus Under boulders resting upon stones.

Hemigrapsus oregonensis Often among stones on mud; can withstand brackish conditions.

Callianassa californiensis In extensive burrows in muddy sand.

Gastropods

Acmaea digitalis

Acmaea scabra

Acmaea pelta

Thais lamellosa A predaceous carnivore.

These forms are merely representative; numerous other species will be found, and a few of the above may not be located.

Sample questions

1. Name three major factors that probably account for the sparse nature of the fauna.

2. Which large and important phylum of animals was not found represented at all? What conditions probably account for its absence?

3. Were oysters present, and if so, where were they largest and most abundant? What differences in physical conditions were correlated with this abundance? What predator was confined to the area of the abundance of oysters?

4. What could be shown to be the usual explanation of the occurrence of shells of gastropods at intertidal levels where the living species were not found?

5. What is the most important predator (or scavenger of freshly killed animals) in each of the following situations: (a) under stones fitting closely against the substrate? (b) in the small, high, rock-bottomed tide pools during low tide? (c) over the higher rock surfaces during high tide?

6. List the animals found in the zone of pebbly beach if such occurred. Which if any of these occur also in sandy beach? How do you explain this great difference?

7. Name a different amphipod characteristic of each of the following locations: (a) high tide pools in rock; (b) under surface of smooth stones resting on substrate, low in tidal zone; (c) felted mass of algae over

surface of rock outcrop in mussel zone; (d) algal debris at or near high-tide line.

8. What biotic reason might you give for the absence of relatively large predators such as the larger sea anemones and starfishes? What two physical conditions may also be concerned?

9. Which two species are diagnostic of the higher splash zone? Which two of the lower splash zone or the highest tide line?

10. What three vicissitudes must be met by animals living in small high-tide pools?

11. What is the extent of mussel beds? Judging from the situations in which you found them, name as many as you can of the more than a dozen species whose numbers would probably be greatly increased if the mussel beds were fully developed.

12. On a line representing the beach line, indicate the habitats present and their position in the intertidal; then indicate by brackets where you would expect to find each of the following (a) *Hemigrapsus oregonensis*; (b) *Hemigrapsus nudus*; (c) *Thais lamellosa*; (d) chitons.

13. Choose four species whose habitats seem best defined by our study and attempt to state the factors involved.

14. Which one species had the widest intertidal range? What makes possible this wide range?

15. Which species common here are absent on exposed rocky intertidal? Attempt to explain the absence of each species.

2. Animals of a Horizontal Shale Exposure at Mid-Tide Level on Open Coast

(As on rocks at the northern end of the beach at Dillon Beach, Marin County, or at Moss Beach, San Mateo County)

Purpose.—Familiarity with the names, normal habitats, and distributions of members of the fauna; experience in the observation of animals in the field in terms of the effects of ecological factors on the distribution and success of species; increased understanding of faunal associations.

Physical conditions.—A rocky substrate (consisting at least in part of fairly hard slaty shale), with relatively few crannies but cut at intervals by crevices or gullies. These may be intertidal channels at intermediate tides, but contain tide pools at low tides, often with bottoms of sandy gravel. In some regions a precipitous seaward face of rock, rising some feet above the low-tide level, receives the full force of the waves and is cut at intervals by tidal channels which may not penetrate far. In places the reef rises to high-tide level or above. Shoreward are

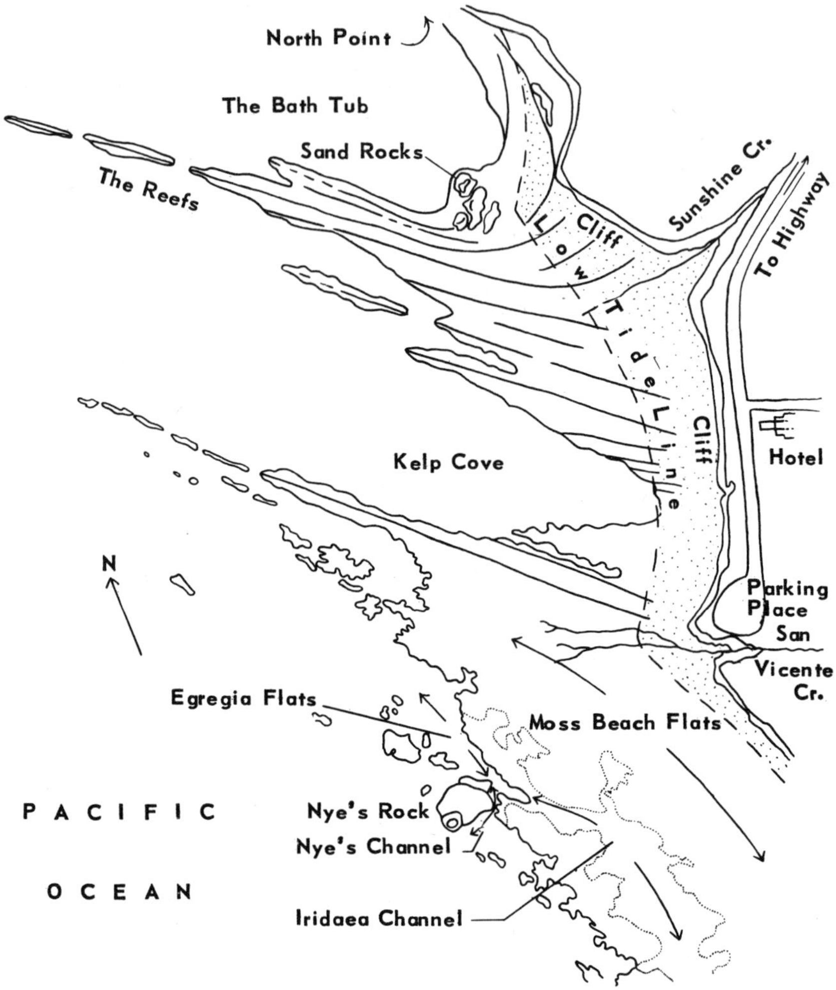

Fig. 134. Map of the Moss Beach area, San Mateo County.

protected channels and tide pools with a sloping rocky or sandy beach, sometimes backed by a sheer bluff.

Physical habitats of diverse types are found. The factor of exposure, varying primarily with vertical height above the low-tide line, is variously affected by the nature and arrangement of the substrate. Thus we have tide pools, shallow gravel-filled crevice pools, deep protected tidal channels, outer exposed tidal channels, ledges, small caves, exposed rock surfaces with varying relations to wave impact, and so on.

Special biotic features.—Living seaweed often forms a close growth over much of the horizontal rock surface, expecially at Moss Beach.

Close-set aggregations of the sea anemone *Anthopleura elegantissima*, whose individuals cover themselves with pieces of shell and gravel, form beds often several feet across. A *Mytilus-Mitella* association occurs on high, fully exposed rocks.

Procedure.—If this is the first field trip it should be preceded by laboratory study and practice identification of the animals most commonly encountered. In the field, go out to the sea front to take advantage of the low tide. Study the animals at the lowest tidal exposure first. Look under ledges and under movable stones (replacing the latter afterward). Break off small sections of overhanging ledges (where rock is soft) to get a more intimate knowledge of the dwellers on the surface, in the crannies, and in burrows (if any); distinguish between cranny dwellers, nestlers, and borers. Work shoreward across the horizontal or broken surface, observing characteristic associations in different habitats. Observe faunal differences, if any, (1) between seaward and landward sides of rock formations, or (2) correlated with different algal types. Compare the animals (3) of narrow crevice pools, (4) of gravel-bottomed higher pools, (5) of the lowest intertidal pools, and (6) of the rocky pools in the splash zone. Contrast the faunal assemblages (7) of the higher shoulders of the outer reef rocks, (8) of the rocks between outer reef and beach slope, (9) of higher rocks at the level of the beach, and (10) of the crevices at the foot of the cliff, if one is present.

If this follows a trip to the protected rocky shore, differences between the two faunas should be observed and an attempt made to explain these differences. Some of the differences are those to be expected between bay and open-shore faunas; others are owing to special conditions in the two situations. Attempt to distinguish between these two types of difference.

Characteristic species

Anthopleura elegantissima On rocks, often partly buried in sand.

Epiactis prolifera Larger specimens often with young (which develop from larvae, *not* buds), sitting near margins of basal disc.

Leptoplana or *Notoplana* Under large upper mid-tide level rocks.

Paranemertes peregrina

Halosydna brevisetosa

Nereid polychaetes Under stones, among seaweed, and in mussel beds.

Dexiospira spirillum (formerly *Spirorbis*) On and under rocks in upper mid-tide pools.

Phascolosoma agassizi In crevices and under protected rocks.

Balanus glandula

Chthamalus fissus and *C. dalli* Abundant but easily overlooked on upper mid-tide rocks with *Balanus glandula*. Parasitized by a highly modified isopod.

Mitella polymerus Leaf barnacle; occurs in dense clusters in mussel beds.

Cirolana or *Exocirolana* Under rocks and in mussel beds.

Spirontocaris spp. The common tide-pool shrimps, excellent for demonstrating internal organs in action during life.

Lophopanopeus spp.

Pagurus spp. Important tide-pool scavengers.

Petrolisthes cinctipes Well adapted for life under upper mid-tide rocks.

Hemigrapsus nudus

Pachygrapsus crassipes

Cancer antennarius

Mytilus californianus Beds formed by mussels shelter many animals which could not otherwise survive.

Pholadidea penita A common mechanical borer in shale.

Botula falcata and *B. californiensis* The "pea-pod" borers.

Kellia laperousii Often a nestler in crevices and old pholad holes.

Saxicava (= *Hiatella*) *arctica* A nestler in holes and crevices.

Littorina planaxis Characteristic of the splash zone.

Littorina scutulata Found in the upper mid-tidal.

Tegula funebralis Also in upper mid-tidal. Shells often bear *Crepidula adunca* and *Acmaea asmi*. After death the shells are used by hermit crabs and with them a fascinating series of commensals.

Tegula brunnea Larger than *T. funebralis* and characteristic of lower mid-tidal.

Acmaea digitalis and *A. scabra* In the upper and splash zones.

Acmaea pelta, A. scutum, A. limatula Mid-tidal.

Nuttallina californica Common on exposed rock near mussel beds.

Ischnochiton heathiana Under rocks partly buried in sand; fine for dissection.

Tonicella lineata In low, protected places, on corallines into which its color and pattern blend.

Platyodon cancellatus A mechanical borer.

Pisaster ochraceus An important predator on *Mytilus* and *Tegula*.

Amphipholis squamata A small, ovoviviparous brittle star.

Amaroucium californicum The commonest colonial ascidian, forming yellowish encrusting sheets on and under rocks.

Sample questions

1. What phylum well represented here was absent on protected bay shores? Differences in what two important ecologic factors might be thought to explain this difference?

2. Which two phyla are represented most extensively in the fauna of horizontal algae-covered flats? Can you give any reason for the absence of large individuals and species here?

3. Contrast the assemblages found in crevice pools and surface growth. What explains the differences?

4. Compare the limpets (*Acmaea*) of the flat and of the inner shore rocks and high shoulder. How would you explain this situation?

5. Attempt to explain the occurrence of the patches of *Anthopleura elegantissima*. Why not more extensive?

6. What seems to restrict the occurrence of *Anthopleura xanthogrammica* in this area?

7. Describe the home habits of *Lophopanopeus* in the larger, more shoreward, crevice pools.

8. What characteristic assemblage of animals occurs under rocks with *Petrolisthes*?

9. What differences do you find in the distribution of the species of *Pagurus*?

3. Fauna of an Exposed Rocky Low Intertidal Reef

(Based on Frenchman's Reef, south of Moss Beach, San Mateo County)

Purpose.—Familiarity with the species present and the associations characteristic of different habitats on a low and exposed reef; further experience in attempting to explain the distribution of animals on the basis of environmental differences in the different habitats of a given area.

Physical features.—A shale reef, with a relatively flat top, extending into the open sea, exposed on its outer side to direct wave action and cut by numerous deep, narrow, tidal channels with steep, often overhanging, walls; numerous tide pools; ledges under exposed horizontal rock strata; a protected shallow lagoon between the shoreward end and the shore on the north side. This habitat, though having a good many features in common with the shale rock exposure discussed above, differs in having a long flat part in the low intertidal rather than in the mid-tidal zone, and a more pronounced lagoon behind the reef.

Biotic features.—A very great abundance of algae except on the higher shoreward part; beds of *Phyllospadix* and *Laminaria*, and the like on the protected side; sea urchin beds near the outer end of the reef.

Procedure.—Since unprotected, exposed reefs on the open coast are visited with real profit only during the lowest tides, pay special attention to the fauna of the lowest intertidal regions here. Study first the deep tidal channels, then the low beds of eelgrass and seaweed of the protected sides. Compare the assemblages of animals on outer and inner edges of the reef at particular levels. On return, study and attempt to

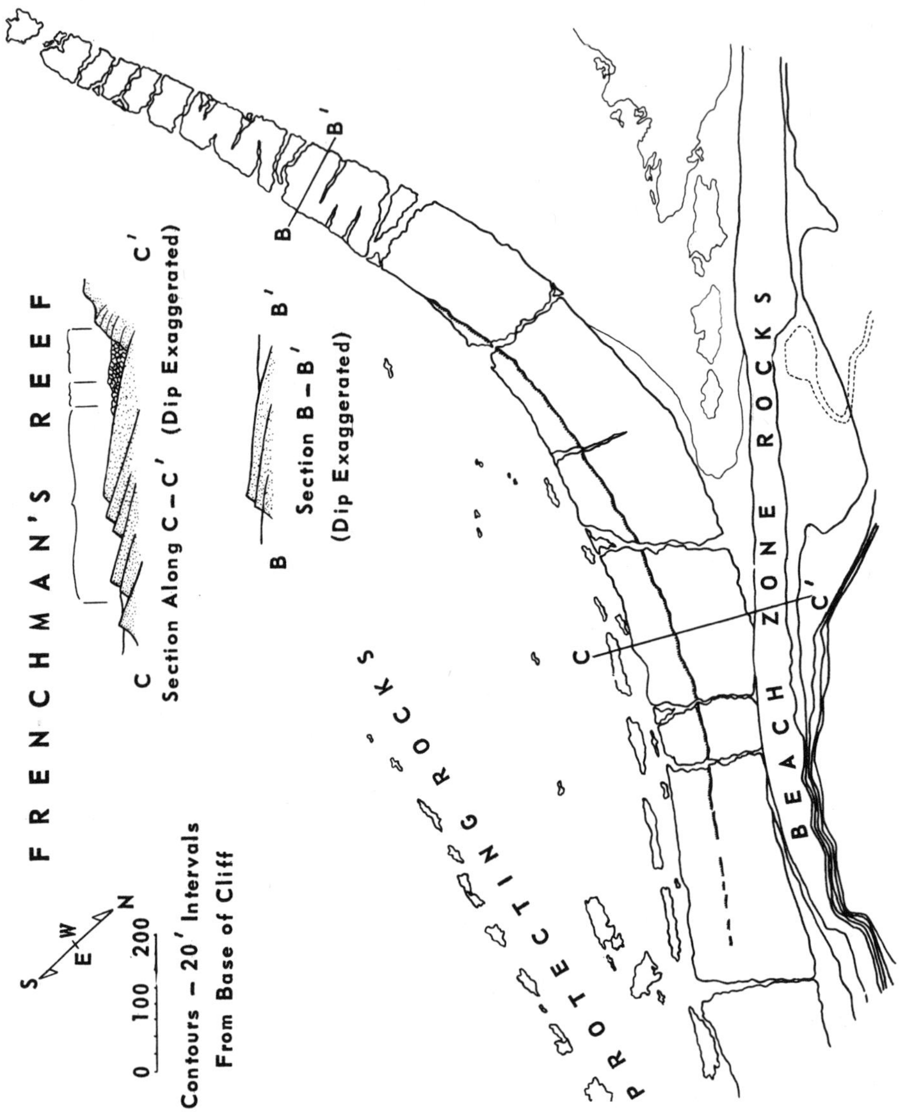

Fig. 135. Map of Frenchman's Reef, San Mateo County.

explain shoreward changes in fauna on the reef surface. Some attention should be paid to the under-rock fauna of the lagoon, as well.

Special notice should be given to three species of limpets, each confined to a single plant species: *Acmaea paleacea* on the eelgrass *Phyllospadix*, *A. insessa* on *Egregia*, and *A. instabilis* on *Laminaria*.

Characteristic species.—The low intertidal zone in rocky stretches of open coast contains an appalling wealth of species. The beginner is at once entranced by the quantity, variety, and beauty of the rich display, and confused and discouraged by the seemingly endless list of names to learn and the complexity of the ecological picture. By concentrating efforts on the commonest and most conspicuous forms and the major habitats and faunal assemblages, however, order soon begins to appear out of apparent chaos, and a complete mastery of the fauna, even if this were possible at the present time, is by no means necessary for attaining ecological insight. Most of the forms common here may have been encountered in the localities described above. Among these, species almost certain to be seen include *Anthopleura elegantissima, Epiactis prolifera, Pisaster ochraceus, Amaroucium californicum, Halosydna brevisetosa, Dexiospira spirillum, Balanus glandula, Chthamalus, Mitella polymerus, Pagurus, Hemigrapsus nudus, Pachygrapsus crassipes, Mytilus californianus, Littorina planaxis* and *scutulata, Tegula funebralis* and *brunnea, Thais* spp., *Acmaea* spp. (*digitalis, scabra,* and *pelta*), and *Nuttallina californica.* Additional forms, some undoubtedly encountered previously but not listed in earlier exercises, will probably include:

Rhabdodermella nuttingi Small vase-shaped sponges under ledges.
Haliclona permollis Purple encrusting sponge in illuminated crevices.
Garveia annulata Hydroid with orange hydranths.
Abietinaria spp.
Aglaophenia spp. "Ostrich-plume" hydroids.
Sertularella sp.
Anthopleura xanthogrammica The great green anemone of tide pools.
Balanophyllia elegans Solitary orange coral.
Eudistylia polymorpha Large "feather-duster" sabellid polychaete.
Tetraclita squamosa rubescens Red barnacle.
Pugettia producta The kelp crab; adapted for clinging to seaweeds.
Katharina tunicata Leathery black chiton; in exposed situations below mussel beds.
Cryptochiton stelleri Giant chiton; in low channels.
Diodora aspera Keyhole limpet; often with commensal polynoid worm in mantle cavity.
Acmaea spp. (*mitra, insessa, instabilis,* and *paleacea*)
Haliotis cracherodii and *rufescens* Black and red abalones.
Nudibranchs
Bugula californica
Leptasterias spp. Eggs are brooded beneath the parent.
Henricia leviuscula Also broods its eggs.
Patiria miniata The bat star; largely a vegetarian.
Strongylocentrotus purpuratus In great beds on horizontal strata; each urchin in an individual pocket in the shale.
Strongylocentrotus franciscanus Found singly in deep pools and channels.

Eudistoma ritteri
Eudistoma psammion
Distaplia occidentalis Highly variable in color and pattern.
Styela montereyensis Simple stalked tunicate; beneath ledges.

Sample questions

1. List nine or more habitats or subhabitats, each with an essentially distinct fauna. Give one to three characteristic species for each.
2. Name five extremely stenotopic species of sessile gastropods.
3. At many points a fauna characteristic of the lowest littoral or the sublittoral was encountered at relatively high levels. How could you explain this in terms of physical factors?
4. What biotic factor (associated with a physical feature of the subhabitat) operates with regard to hydroids?
5. What factors operate to determine the distribution of *Strongylocentrotus purpuratus*?
6. What major difference with regard to *S. franciscanus*?
7. Name four types (belonging to as many phyla) which are found competing for space in certain favorable spots? What makes these spots favorable?
8. Name five genera of hydroids encountered. Which was most abundant?
9. What arthropods seem adapted for life on this particular hydroid?
10. Be able to list (a) the in-rock dwellers, as borers, nestlers, or cranny dwellers; (b) the on-rock dwellers, both those of exposed and of under surfaces; (c) the under-rock dwellers; and (d) the dwellers in the under-rock substratum.
11. Have in mind the food habits of each and, so far as possible, the reasons for their presence there.
12. For each of the following name one or more species which are conspicuous in, or characteristic of, the fauna of exposed rocky intertidal:

Brachyurous anomurans	Limpets
Anomurous anomurans	Barnacles
Brachyurans	Isopods
Chitons	Rock-boring pelecypods
Errant polychaetes	Nestling pelecypods
Sedentary polychaetes	Surface-dwelling pelecypods
Small snail-like gastropods	Cranny-dwelling pelecypods
Large snail-like gastropods	Rock-dwelling brachyurous anomurans
Asteroids	Ophiuroids
	Holothuroids

13. What was the largest animal? What correlation between its size and its location?
14. What important predators were not seen?

4. Fauna of an Exposed Coast of Granitic Ledges and Boulders

(As at Point Pinos, Pacific Grove; or Mission Point, Carmel; both on Monterey peninsula)

Purpose.—To study a coast comparable in exposure to the areas discussed, under (2) and (3), but composed of granitic rock rather than shale and provided with flats of heavy boulders.

Physical features.—Shore very broken, with prominent ledges receiving full impact of surf, lacking the flat bench structure and long horizontal fissures of the shale coasts described above; ledges broken by deep crevices of every possible size, shape, and exposure to sea and light; flat areas composed of large boulders too heavy to be moved by waves, with a protected understory of other boulders and coarse gravel in the lower tidal zones, but bedded in sand in higher and more protected areas; material of rock too hard to bore into; numerous tide pools.

Biotic features.—Algae extremely abundant; a drapery of *Egregia* over boulders in lower mid-tidal; laminarians and *Phyllospadix* on rocks below tidal "zero"; *Postelsia* on exposed ledges receiving full impact of surf at mid-tidal height; *Ulva* in high, protected pools. The fauna has much in common with that on other exposed coasts, but there will be noticed a complete absence of the distinctive rock-boring bivalves of shale reefs, a restriction of such forms as sea urchins to sheltered pools and crevices, and a marked increase in the number of large under-rock forms such as crabs and abalones, correlated with the great abundance of sizable protected crannies provided by boulder shores.

Procedure.—Ledges, crevices, and channels may be explored as were shale ledges. In the boulder zone turn aside one or more large rocks and examine carefully the undersurfaces, which are covered with attached animal life. Then remove underlying rocks so as to excavate a deep cavity, studying each rock in turn down to the level of bedded sand and gravel. Such collecting is extremely rich and varied. Be sure to replace the rocks.

Characteristic species.—Many forms listed in the previous exercises will also be found here. The following list includes additional species likely to be prominent in this area.

On exposed faces and in illuminated crevices

Halichondria panicea A greenish sponge in well-lighted situations.

Archidoris montereyensis Associated with, and feeding upon, *Halichondria.*

Paranemertes peregrina Creeping about among mussels and corallines.

Cucumaria curata Small "tar-spot" sea cucumbers among mussels and corallines.

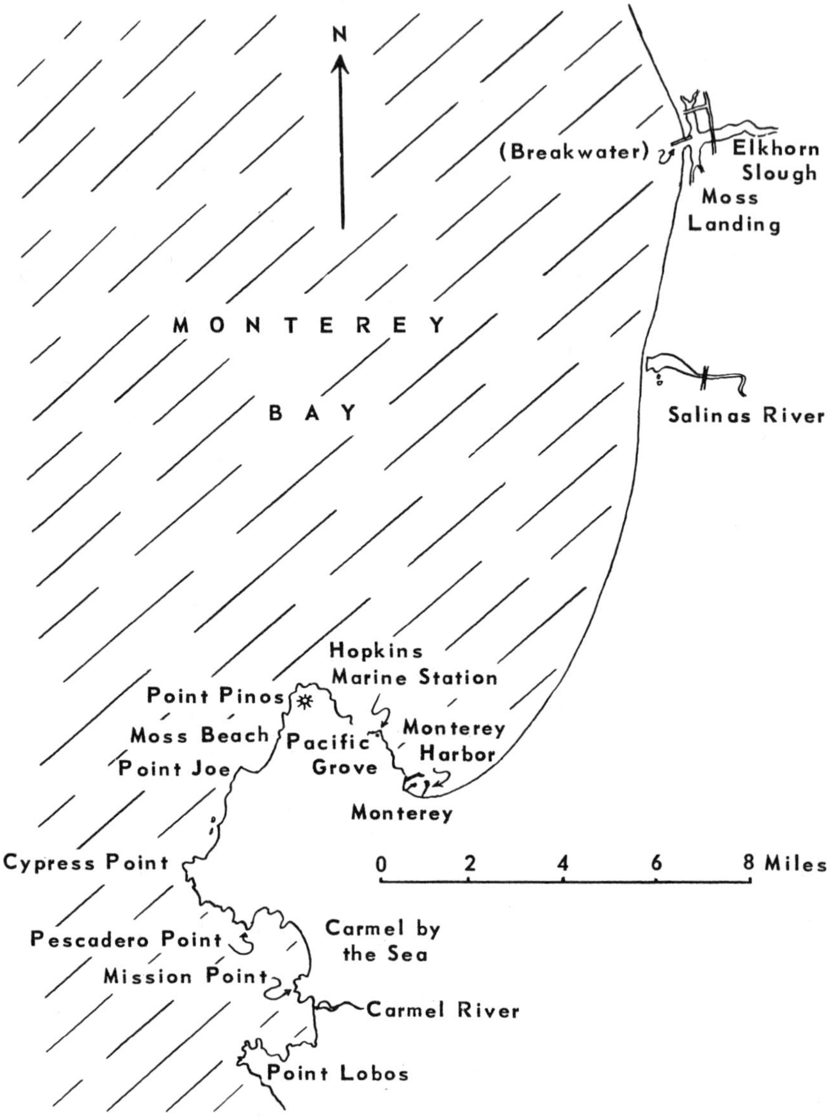

Fig. 136. Map of the Monterey area.

Allopora porphyra Lower down, in semiexposed crevices.
Haliotis cracherodii Black abalone, mostly young.

Under ledges, often well back into cracks

Hinnites multirugosus Cemented as adult; young resemble scallops, attached by a byssus.
Cucumaria sp. Extremely difficult to remove undamaged.

Undersurface of large boulders

Sabellaria cementarium In very firm sand tubes.

Serpula sp. In large calcareous tubes.

Salmacina tribranchiata In fine white tubes, ramifying or in erect masses.

Vermetid snails

Encrusting bryozoans

Numerous sponges

Tunicates

Hydractinia sp.

Corynactis californica

In spaces formed by layers of boulders

Crangon spp. The snapping shrimps.

Paraxanthias taylori

Loxorhynchus crispatus The masking crab.

Hapalogaster cavicauda

Pachycheles rudis

Petrolisthes spp.

In upper mid-tidal gravelly pools

Polychoerus carmelensis Small red-orange acoelous flatworm, on *Ulva* or under stones partly embedded in gravel.

Amphiodia occidentalis Long-armed brittle star, buried in sand.

Leptosynapta albicans Wormlike holothurian, also in sand.

Lumbrinereis zonata Beneath rocks in sand.

5. Fauna of a Semi-protected Sandy Beach and Elevated Rocks in Sand

(As on rocks at the northern end of the beach at Dillon Beach, Marin County, or rocks at the north end of Moss Beach near Asilomar, Pacific Grove, Monterey County)

Purpose.—To gain familiarity with the members of the fauna, their names, their ecological niches, and their interrelationships.

Physical conditions.—Shore semi-protected by configuration of local geography, but subjected to wave action which may be violent in times of storm. The beach is moderately long, with a gentle, more or less even slope from low-tide level to the dune or bluff at the upper margin of the beach. The rocks are relatively hard, without many crevices and crannies, irregular in shape, generally broadest at the base, sloping upward to a narrower top, and are immovably fixed in position. Sand is washed out about the bases of some rocks, leaving temporary tide pools. The two major habitats are rocks and sand.

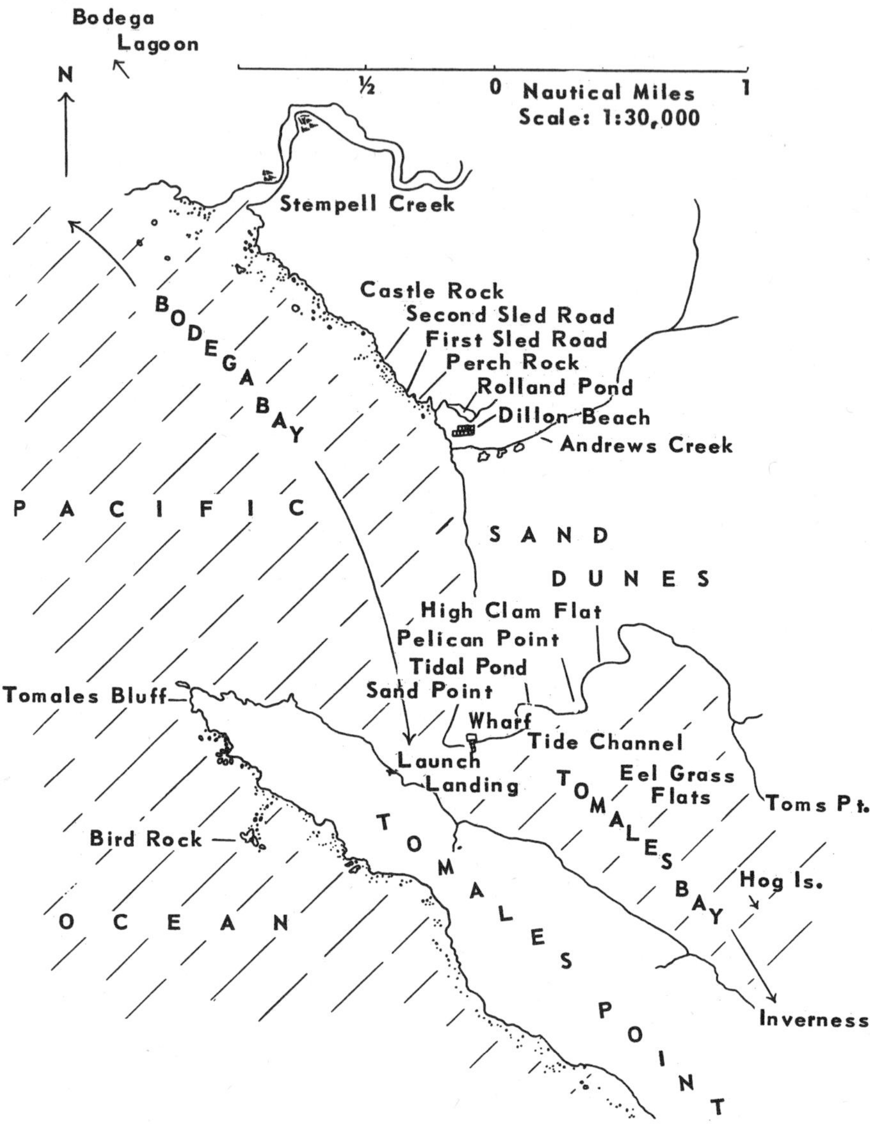

Fig. 137. Map of Dillon Beach area.

Biotic conditions.—Living seaweed is variable in quantity; often very little on the rocks. Scattered clumps of *Egregia* may be rooted to rocks just buried below the surface of the sand. Dead seaweed may be present in windrows on the beach. The upper rocks may seemingly lack an algal film. Two striking rock communities are present, one dominated by the California sea mussel, *Mytilus californianus*, the other by the sea anemone, *Anthopleura elegantissima*.

Procedure.—Examine the surfaces of the rocks for clinging forms. Pry off a small number of mussels to sample the rich association of forms that find protection around the bases of attached mussels. Inspect the attached algae and carefully turn over masses of dead algae on the beach to search for scavengers. Dig in the sand at the water level and below and, at selected areas higher on the beach, particularly investigate any holes or tubes appearing in the sand at any level. Keep careful notes on distribution and associations of each species, and on physical and biotic factors which seem to explain them. Collect forms not identified in the field for later study in the laboratory.

Characteristic species.—The following animals are among those usually present. Identify as many of the animals as possible in the field, and use the names until they become familiar.

Rock fauna

Anthopleura elegantissima
Balanus glandula
Chthamalus sp.
Mitella polymerus
Ligia occidentalis
Pachygrapsus crassipes
Hemigrapsus nudus
Nuttallina californica
Katharina tunicata
Mopalia spp.
Mytilus californianus
Tegula funebralis
Thais emarginata
Littorina planaxis
Littorina scutulata
Acmaea digitalis
Acmaea scabra
Acmaea pelta
Acmaea scutum
Pisaster ochraceus

In Mussel Beds (where *Mytilus* and *Mitella* predominate)

Arabella
Neanthes brandti
Cirolana
Amphipods
Petrolisthes
Pachycheles
Saxicava arctica
Thais emarginata
Barnacles and limpets

Sand fauna

Nainereis
Nephtys californiensis
Orchestoidea spp.
Emerita analoga

Sample questions

1. What is the dominant animal in the most conspicuous animal association of the area?

2. What other animals are present in this association?

3. What is their relation to it?

4. What other sedentary animal is present in considerable concentration?

5. Does it seem to compete with other species?

6. What evidence do you observe of differences in the niches of the species of *Acmaea*?

7. What is the most important sand-dwelling organism?
8. What plant scavengers were observed?
9. Where are the different polychaetes to be found?
10. Do you find any clams? Reasons?

6. Fauna of an Exposed Sandy Beach

(As the beach seaward of Elkhorn Slough, Monterey County, or the open coast between Point Reyes and Tomales Point, Marin County)

Purpose.—To examine the fauna and its zonation in the specialized habitat provided by an exposed sandy beach, and to see some of the ways in which the animals present are adapted for life in this area.

Physical conditions.—A long and wide sandy beach, sloping gently upward from the water line to dunes above the high-water mark; wave action varying with the weather, from a slight surge to great breakers; substrate fine to medium sand, shifting to some degree with the action of the waves; absence of rocks or other hard surfaces for attachment of animal or plant life.

Biotic conditions.—No attached plant life, but large algae may at times wash up in windrows on the beach together with dead animal material; large populations of a small number of animal species, as the sand crab *Emerita,* the pismo clam *Tivela,* and the high beach amphipods *Orchestoidea* spp.

Procedure.—If the sea is calm, wade out into the water and probe with a shovel, turning over the substrate for sand crabs and clams. If dip nets are available, these may be dragged along the bottom like a trawl for shrimp. Work slowly back to the water's edge and up to the high beach, digging, sieving, and examining the substrate at intervals of a few feet. Be sure to investigate any holes or protruding tubes noticed. Look under masses or pieces of algae washed ashore, observing carefully for small amphipods, insects, and small red mites. Look under driftwood. Pieces of timber recently cast ashore may bear populations of the goose barnacle *Lepas.* At the levels above high watermark notice the transition from beach to dune conditions and an essentially terrestrial fauna.

List possible food sources available to dwellers on a sandy beach. What feeding mechanisms observed in the animals collected appear especially adapted for obtaining these materials? If possible, visit the same or a similar beach at night with a flashlight, and notice the large numbers of nocturnal forms out on the higher beach surface.

Characteristic species

Low beach

Nephtys californiensis

Archaeomysis maculata Small transparent mysids, living in the sand; best found by digging a shallow pool in the low beach and puddling the sand. Sometimes only their shadows reveal the swimming animals.

Emerita analoga Very effective digger in wet sand; strains food from water with long, bristly antennae; parts of population may migrate up and down beach with the tide.

Blepharipoda occidentalis Usually lower in the intertidal than *Emerita*; larger but similar in general appearance.

Crago sp. In shallow water along and in the sandy bottom.

Tivela stultorum The pismo clam; burrows fairly rapidly in a complex manner; foot wiggles rapidly, loosening sand, then jets of water ejected anteriorly displace this up along the sides of the shell, and finally the heavy weight of the shell and the pull of the foot together drag the animal down.

Siliqua patula One of the razor clams, a very rapid burrower.

Olivella biplicata More common on and just below the surface of protected beaches and flats, but may be found here in small numbers.

Dendraster excentricus The sand dollar; usually subtidal, but some small specimens may be found at low tide, or washed up after rough weather.

High beach

Orchestoidea californica The two species of *Orchestoidea* often occur together in medium sand, but *O. californica* seems to prefer a finer grade, whereas *O. corniculata* is commoner on sheltered beaches of coarser composition.

Orchestoidea corniculata

Orchestia traskiana

Alloniscus perconvexus A high beach isopod, common under boards on dry sand.

Staphylinid beetles

Sand flies

In addition to the above, a variety of pelagic forms may be found cast ashore in fair condition. Among these are jellyfish, hydromedusae, siphonophores (especially the purple *Velella*), ctenophores, and goose barnacles (*Lepas*) on floating timbers.

7. Fauna of Flats in a Protected Bay or Slough

(Based on tidal flats at the north end of Tomales Bay, Marin County, and Elkhorn Slough near Moss Landing, Monterey County)

Purpose.—To gain familiarity with the fauna, the environment, and the organization of organisms into particular associations, in a general habitat very different from the rocky intertidal.

Physical conditions.—The substrate consists of broad, gently sloping tidal flats of mixed sand, mud, and clay. Wave shock is absent except during prolonged storms, but tidal currents during spring tides, especially when augmented by increased drainage during winter rains, may cause considerable shifting of sediments from place to place. Tidal currents are prolonged, bringing an immense amount of water over a given area of bottom. The soft substrate offers an enormous surface for a variety of burrowing forms.

Biotic features.—Aside from patches of plant life (*Zostera* flats in Tomales Bay, *Enteromorpha*, and occasional clumps of *Zostera* in Elkhorn Slough), and certain tidal channels or pools, the gently sloping tidal flats do not show clearly delimited habitats. However, rather clearly marked animal assemblages occur, and the area can be roughly zoned according to the location of these. Three important communities are to be seen, each at a different intertidal level and each named according to its dominant species. The *Macoma-Schizothaerus* community, in some spots overlain with patches of the broad-leaved eelgrass *Zostera* (not abundant in Elkhorn Slough), occurs near the low low-tide level; it is easily recognized by the holes and siphons of the large clam *Schizothaerus nuttallii.* Above this lies the *Macoma-Phoronopsis* community, easily recognized by the presence of thousands of small green lophophores belonging to *Phoronopsis viridis.* Where these are withdrawn, one turn of a shovel will reveal the elongate tubes of cemented sand formed by this species. Highest of the major communities is the *Macoma-Callianassa* community, well up in the mid-tide zone, and characterized by the holes and tunnels of the ghost shrimp *Callianassa californiensis.* The roots and leaves of the eelgrass *Zostera* offer a specialized habitat for a special assemblage of forms not found elsewhere. Groups best represented are the bivalve molluscs and the polychaetes.

Procedure.—Locate, in turn, each of the major communities listed above. With shovel, trowel, and sieve investigate each community, observing not only the dominant species but others also abundant in these biotic formations. Try to find communities or associations other than the three enumerated. Collect animals from any vegetation present, distinguishing between forms found on the leaves and those occurring in the mud about the roots. In a preliminary way, try to map the distribution and range for the commonest species on a transect running from the high to the low water marks.

Characteristic species.

Species occurring both in Tomales Bay and in Elkhorn Slough

Cerebratulus californiensis

Glyceridae (esp. *Glycera* and *Hemipodus*)

Capitellidae (esp. *Notomastus tenuis*)

Polynoidae (esp. *Halosydna* and commensal *Hesperonoë*)

Urechis caupo Builds U-shaped burrow in mud, containing commensal crabs, fish, polychaetes, clams. Hemoglobin acts as an oxygen reservoir during low tides.

Callianassa californiensis Tunnels through muddy sand, sifting substrate for edible detritus.

Upogebia pugettensis A mud burrower feeding on plankton and suspended detritus; inhabits much softer mud than *Callianassa.*

Hemigrapsus oregonensis A scavenger.

Scleroplax granulata Commensal in *Urechis* and *Callianassa* burrows; legs often bear colonies of bryozoan *Triticella.*

Pinnixa spp. Commensal in clams, and in tubes of tubicolous polychaetes.

Macoma nasuta The commonest, most widely distributed mud clam; feeds on detritus at surface of mud, using incurrent siphon like a vacuum cleaner.

Macoma secta

Schizothaerus nuttallii With siphons up to three feet long when extended.

Cryptomya californica Extends siphons into burrows of *Urechis* and *Callianassa.*

Saxidomus nuttallii

Protothaca staminea Ranges from rocky intertidal to mud flats.

Polinices lewisii Predaceous on bivalves.

Phoronopsis viridis Sometimes occurring in enormous dense beds.

Leptosynapta albicans Same species as that found in sand under rocks along open coast, but seemingly more resistant to brackish conditions.

Additional species common at Tomales Bay but not Elkhorn Slough

Nephtyidae (*Nephtys caecoides*)

Terebellidae (*Eupolymnia crescentis* and *Pista pacifica*)

Maldanidae (*Axiothella rubrocincta*) Very long metameres—the "bamboo worm."

Cirratulidae (*Cirriformia spirabrancha*)

Sabellidae (*Eudistylia vancouveri*)

Spionidae (*Boccardia* and *Polydora*)

Exocirolana

Tellina bodegensis

Panope generosa The geoduck; less common than *Schizothaerus*.

Nassarius On surface of mud.

Olivella biplicata Burrows in sand surface.

Acteon punctocoelata On surface of mud.

Hermissenda crassicornia Sometimes also common in Elkhorn Slough in the winter.

Phyllaplysia taylori On leaves of *Zostera*.

Amphiodia occidentalis In mud among roots of *Zostera*.

Additional species common or conspicuous at Elkhorn Slough but not at Tomales Bay.

Arenicola sp. In burrows with a mound marking one end, a crater the other. Abundant in occasional years.

Mya arenaria Occuring high in the intertidal mud flats, near *Salicornia*.

Macoma irus In very muddy areas, such as *Upogebia* inhabits.

Cardium nuttallii A shallow burrower with an extremely active foot.

Zirphaea pilsbryi A mechanical borer occurring in stiff, sticky clay.

Epinebalia pugettensis Small leptostracan, sometimes abundant in *Enteromorpha*.

Sample questions

1. Which species were found to occur on the surface of the substrate?

2. What common problems do burrowing forms face in such a habitat as this? How are they similar to those faced by rock borers? How do they differ?

3. Distinguish between deposit-feeding bivalves and those that feed on plankton and detritus in suspension.

4. What species present are predators? On what do they feed? How?

5. What differences in distribution do you find for the commoner polychaetes?

6. What examples of commensalism were observed?

7. Describe three different methods of detritus feeding found in animals seen.

8. Compare the three species of *Macoma* in:

a.) Position assumed in the substrate.

b.) Relative muddiness or sandiness of substrate inhabited.

8. Comparison of Faunas and Environments of an Exposed Rocky Shore, a Partly Protected Rocky Shore, and Protected Tidal Flats

(Based on Tomales Point, Bodega Bay, and Tomales Bay, respectively, but comparable triads of habitats can be found in and around San Francisco Bay and Monterey Bay)

Purpose.—To bring the knowledge gained from preceding field trips and laboratory studies to a focus in an intensive study of three faunal assemblages not far apart spatially, but ecologically distinct; to bring out more clearly the interrelationships between physical factors of the environment on one hand, and faunas and biotic communities on the other.

Procedure.—The period of the low tide on one day will be devoted to the fauna of the rocks, pools, and tidal channels of an exposed shore. Observe the height of the splash zone, as indicated by the distribution of littorines, limpets, and barnacles; notice the forms occurring on fully exposed rock faces at all levels. Among other things, make certain you investigate fully the association of animals in a well-developed mussel bed such as occurs here on higher rocks. Try to determine the role of each type of animal in the association, or the advantages it receives by the association.

The following day a trip will be made to a protected rocky shore exposed at low tide. Here, studies will be made of the relatively rich and diversified fauna to be found in tide pools, on rock surfaces, under rocks, among algae, in gravel and sand about the roots of eelgrass, and other special habitats. Keep in mind faunal differences between this area and the open coastal rocks, and attempt to correlate these with differences in environmental conditions found.

Finally, the third day will be given over to the fauna to be found at various levels of the tidal flats of muddy sand. Here the number of species is not so great as in the faunas studied on preceding days, but there is a wealth of individuals. Notice particularly the distribution of populations of the giant clam *Schizothaerus nuttallii*, the bent-nosed clam *Macoma nasuta*, the phoronid *Phoronopsis viridis* whose green lophophores may practically cover the ground over large areas, the ghost shrimp *Callianassa californiensis* burrowing in great numbers at higher levels, and polychaetes of several interesting types (especially the maldanids, glycerids, capitellids, nephtyids, and cirratulids).

Sample questions

1. (a) What are two extremely important physical features in which the tide-flat environment differs from that of the rocky shore? (b) What seems the most obvious over-all biotic difference between the two situations?

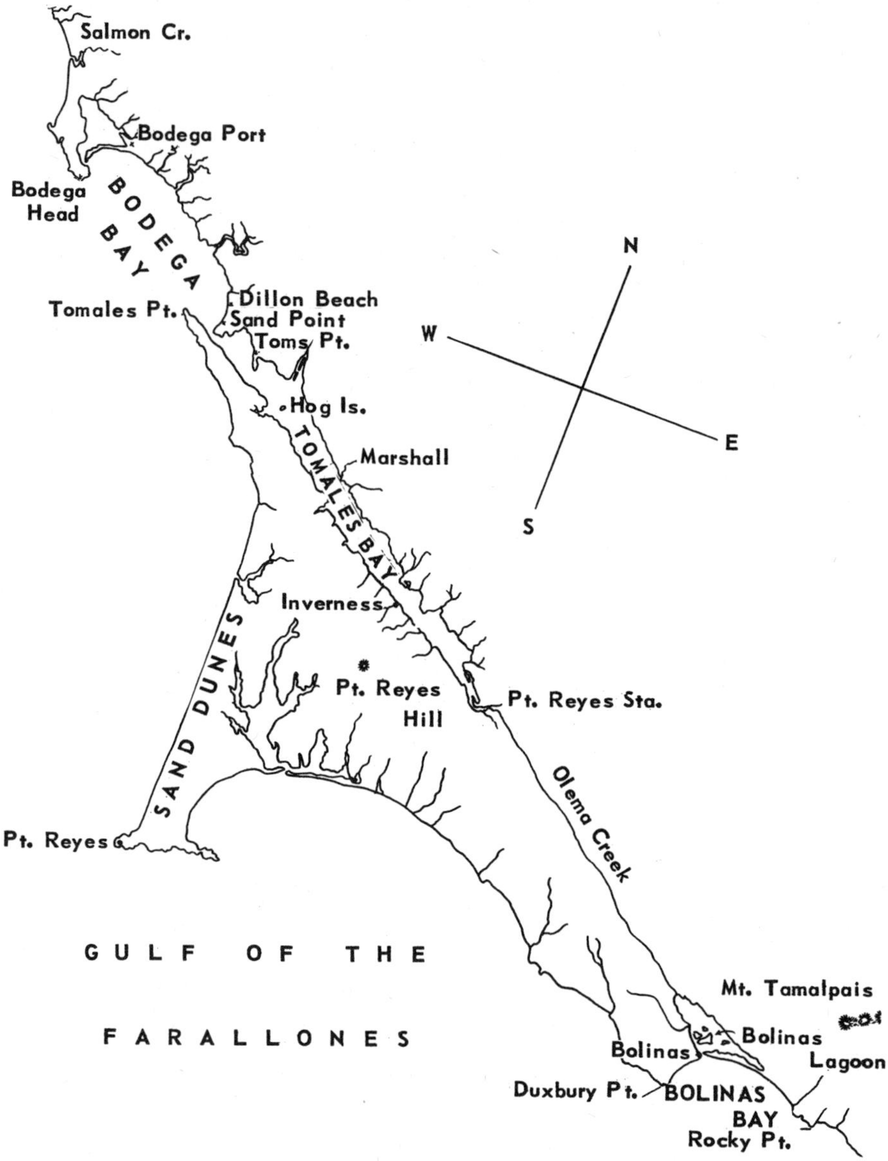

Fig. 138. Map of Point Reyes, Bolinas, Bodega, and Tomales bays.

2. (a) What species occur in all three, or at least in two, of the environments studied here? (b) In which fauna is each most abundant? (c) For each species, state the conditions common to all three (or two) situations which make possible its existence where it was found. (d) Did you observe any differences in individuals which would enable you to

distinguish between specimens taken on tidal flats from those taken from rocky shores?

3. What were the five or six most abundant and widespread species of the intertidal fauna of the mud flats? Do any of these occur on protected or exposed rocky shores? For each that does not, give the conditions favorable to it on the tidal flats, and explain why conditions on the rocky shores are unfavorable.

4. What five or six abundant and widespread species occur on rocky shore? Were they found on the tidal flats? What conditions in the rocky intertidal favor these species. If they were absent from the tidal flats, explain this absence in terms of environmental factors.

5. Contrast the important predators of the rocky intertidal with those of the tidal flats. What predaceous species occur in one area but not the other? Can you explain this distribution on ecological grounds?

6. Ignoring *Phoronopsis*, (a) What were the two dominant *classes* of animals found in the tidal flats? (b) What *class* of animals forms an important feature of both exposed rocky shores and tidal flats? (c) What *class*, not dominant on the mud flats, forms a highly important element in the fauna of rocky shores? (d) Explain the findings in a-c above in terms of differences in the physical environment, and in terms of the structural adaptations of the animals involved.

7. Draw two cross-sectional diagrams, one representing a transect from highest to lowest tide marks through a tidal flat, the other representing a similar transect through a broken reef formation in an exposed rocky intertidal (in the latter include the splash zone). On each, mark four locations which represent equivalent degrees of tidal exposure. For each location on each diagram, list the animals found most abundantly at this approximate level.

8. What animals found (a) on rocky shores, and (b) on tidal flats, seemed to show the most striking structural adaptations for life?

9. Fauna of Sandstone Rocks

(As at Moss Beach, San Mateo County)

Purpose.—Experience in thinking in terms of ecologic factors as determining general habitats, and of the conditions required by common species.

Physical conditions.—An area of varied physical conditions involving several very distinct habitats (and their intergrades), such as (1) high tide pools: (2) sheer, soft-rock surfaces exposed to tidal and wave action; (3) a rock surface sloping upward toward the shore; (4) tiny tide pools contained in the surface of this rock layer; (5) a shaly low

low-tide bottom exposure; (6) the inner ends of nearly upright reefs of hard rock extending out some distance; (7) a region of rocks and small boulders near low-tide line; and (8) a sandy inner beach of coarse sand.

Biotic conditions.—These are correspondingly varied, from areas practically devoid of life to those with heavy growths of large seaweeds.

Procedure.—Study each of these habitats and its characteristic fauna. Be able to discuss the distribution of the different groups and species in this area in terms of the ecologic factors involved.

Sample questions

1. What species not found or rare on sandstone rocks were found on granite or other types of rocks?
2. Name species found abundantly on sandstone that were rare or not found on granite rock.
3. Suggest one general reason for the fact that the former list is longer than the latter.
4. What were the most important plant feeders, scavengers, predators, and plankton feeders in the sandstone area?
5. Name animals characteristic of sandstone that were not found among smaller rocks around its base.
6. List differences in fauna between loose rocks on sandy substratum closer in and farther out in the intertidal zone. Which of the three major factors seems most important here? Illustrate.
7. How would you distinguish the niches of the three shore crabs, *Pachygrapsus crassipes*, *Hemigrapsus nudus*, and *H. oregonensis*?
8. What evidences did you note of vertical zonation of the species of *Acmaea*?
9. Which polychaete was the most abundant? What two features of the situation probably account for its abundance?
10. What are the advantages and disadvantages for *Acmaea scabra* of sandstone as contrasted with granite rock?
11. Why so few barnacles and no mussels on sandstone?
12. Why so few littorines?
13. What structural character of *Cirriformia luxuriosa* not found in *C. spirabrancha* is correlated with its habit of life?
14. Name three genera of isopods encountered. Characterize the very distinct ecologic niche of each.
15. With what other animal did the presence of *Acanthina* seem to be correlated?
16. Why were borers not abundant? Nestlers?

10. Under-Rock Faunas

Purpose.—To visualize the importance of the factor of protection, and to see the characteristic assemblages of organisms living in this protected habitat.

Procedure.—Choose a rock, preferably one in the upper mid-tide zone, that is resting on or partly embedded in sand or gravel. The rock should be as large as can conveniently be overturned and be so situated as to be relatively stable. Notice carefully the physical site of the rock with regard to tide level, the degree to which it is protected from wave action, and so on. Notice whether it is embedded in, resting on, or partly elevated above, the substrate.

Overturn the rock and make immediate notes of the active population that almost at once seeks protection elsewhere. Among the first to escape will be the small crabs and anomurans, followed quickly by the isopods. Carefully list the remaining fauna of the lower surface of the rock and of its crevices and crannies, if any. Distinguish between sessile and motile forms.

Attempt to arrange the total fauna thus listed on various bases: (1) reasons for their presence there, (2) their food relations and so on.

Repeat the process with another rock, this time low in the intertidal zone. Observe faunal differences and environmental differences and attempt to correlate the two. *Do not forget to replace the rocks.*

11. Boring, Nestling, and Cranny-dwelling Animals, Particularly Pelecypods

A number of pelecypods and a few other organisms are able to bore into relatively hard substrata, or they occur in cavities in substrata so hard as to suggest that they have bored into it. Mechanical wood borers (e. g., the bivalves *Teredo, Bankia,* and the isopod *Limnoria*), are generally confined to pilings, floats, and boats in protected waters, and are not generally encountered in the exposed rocky intertidal, where the important borers are other bivalves and the sea urchin *Strongylocentrotus purpuratus.*

As a result of the action of the borers, many of the softer intertidal rocks (but not granite) are full of holes and crannies in addition to those which might normally arise from fracturing and weathering. These holes and crannies furnish protected sites for a large array of species of various groups. Some of these live within the shells left by preceding species. These have been termed nestlers in distinction to the cranny dwellers which live in crevices or cavities of one kind or another in rock. Sometimes we may find a series of successive nestlers; for example, a tiny *Kellia* in a *Saxicava* shell, itself in a *Protothaca* shell which lies within the shell of *Pholadidea*, the original borer. The distinction between borers, nestlers, and cranny dwellers is not sharp. Relatively feeble borers (e.g., *Petricola carditoides*) may occupy and slightly enlarge a hole made originally by another more effective borer. *Protothaca staminea* may nestle or live in crannies, but it also flourishes in some protected sand and mud flats in bays. However, the terms are useful for ecological studies.

Local rocks in many areas are soft enough to permit their being occupied by borers, and conditions favorable for the study of cranny dwellers are equally common.

Procedure.—At low tide in a region where rock dwellers occur, choose a small rock, or break a small piece from a shoreward bank or ledge. Collect and make a careful record of the surface-dwelling animals, then carefully break up the rock for the in-rock fauna. Identify and list (1) the crevice and cranny dwellers, (2) the nestlers, and (3) the true rock borers. Attempt to determine the food habits and to explain the presence of each species in its particular location.

Common boring, nestling, and cranny-dwelling organisms, especially bivalves

Borers (never all found in any one habitat)

Botula falcata This and the following species are mechanical rock borers equipped with a very heavy periostracum.

Botula californiensis

Lithophaga plumula A chemical borer, using acids applied directly to the surface of calcareous rocks by mantle edges and siphons; no free acid occurs in the burrow.

Saxicava (= *Hiatella*) spp. Mechanical rock borers but also found in mussel beds, on pilings, embedded in sponges, and so on.

Pholadidea penita A true rock borer that anchors the body by attaching the foot, and rasps the burrow not by opening and closing the valves but by alternate contractions of anterior and posterior adductor muscles which rock the valves over a fulcrum formed by the hinge. Essentially the same mechanism is used by *Parapholas*, *Teredo*, and *Bankia*.

Parapholas californica Mechanical borer in soft rock.

Platyodon cancellatus Mechanical borer in mudstones.

Zirphaea pilsbryi A mechanical borer in stiff and sticky clay; in bays.

Teredo navalis A mechanical wood borer, capable of digesting some wood.

Bankia setacea Similar to *Teredo.*

Strongylocentrotus purpuratus Bores shallow, cuplike depressions in soft rock, primarily by means of teeth and Aristotle's lantern, but also by rotating body and scraping the ventral spines against the rock.

Limnoria spp. Forms shallow tunnels in wood, boring with asymmetrical mouth parts. Female apparently does most of the work.

Common nestlers

Petricola carditoides Commonly a nestler, though some members of this genus are rock borers.

Irus lamellifer

Saxicava (= *Hiatella*) spp. Sometimes a borer.

Kellia laperousii A very common and characteristic nestler in old pholad holes.

Protothaca staminea Sometimes a nestler, but also living in soft substrates.

Mytilimeria nuttallii Completely embedded in tests of compound ascidians.

Lasaea cistula Among barnacles and byssal threads of mussels.

Crepidula nummaria (?) Often a nestler in old pholad holes.

Crevice and cranny-dwellers

Pododesmus macroschismus Also found under rocks and protected ledges.

Septifer bifurcatus

Glans carpenteri

And many other forms, such as polychaetes (esp. polynoids), isopods (esp. *Cirolana*), sipunculids (esp. *Phascolosoma agassizi*), chitons (e. g., *Ischnochiton heathiana*), anomurans (esp. *Pachycheles* spp.), and numerous others.

12. Fauna of Seaweed Holdfasts

With a knife or geologist's pick separate from the rock a holdfast of one of the larger brown algae (*Macrocystis, Laminaria, Egregia,* or the like) found at the lower tide levels. The tangled holdfasts of *Macrocystis* are especially favorable. In the laboratory, carefully dismantle the holdfast and list the species found and the numbers in which they occur. Examination of holdfasts from one area was found to yield sixty-two different species, representing ten phyla. Attempt to explain the presence of the different species found, and evaluate the specialized habitat in terms of the protection it offers from wave shock and the tendency for organic debris to accumulate in interstices between the rootlets. Were any animals present found to be feeding directly on the tissues of the holdfast?

13. Fauna of Small Tide Pools, with Emphasis on Behavior and Feeding Habits

Choose a small, permanent tide pool, relatively high in the intertidal zone but not in the splash zone. Do not disturb or remove the inhabitants of the pool and avoid casting a shadow over the pool during observations. Make a careful survey and census of the pool so far as this is possible without removing the animals, and indicate the distribution of the inhabitants in a diagram. If specimens are needed for laboratory examination, try to take similar forms from adjacent pools if available. Record all activities observed during a period of an hour's observation, and attempt to explain these movements of the animals.

Devise simple experiments to determine the reactions of different animals to various types of stimuli. Introduce a small piece of mussel (previously prepared to avoid disturbances). By this method seek (1) to determine the numbers and locations of carnivorous animals, (2) to understand and record their feeding behavior, and (3) to determine whether sight, touch, or chemical sense enables them to find their food.

By varying this device, seek to determine which are facultative carnivores and which obligatory. Use tiny fragments of dead seaweed to detect plant scavengers, if any. By using tiny whole organisms, seek to determine which, if any, of the carnivores are predators. Different animal bodies or tissues may be used to determine whether different organisms show marked food preferences. Interpret negative results with caution.

Consider also the relations of the inhabitants of the pool with other species present only during high tide, and attempt thus to construct the food chains involved. If feasible, visit the pool several times under different conditions of tides and try to relate these conditions to differences, if any, in the presence, positions, and activities of various inhabitants of the pool. Attempt to determine the degree to which the fauna changes from day to day, and the extent to which the permanent inhabitants have permanent locations.

14. Fauna of High Tide Pools and the Problem of the Distribution of the Red Copepod *Tigriopus*

High tide pools, ranging from slightly below the high-water line to well up into the splash zone, may vary in size from puddles a few inches in depth to deep potholes. The substratum, too, may vary from soft shale or sandstone through conglomerate to granite. Finally, the pools show differences in other physical factors such as exposure to sunlight and so on.

The dominant animals of these pools are three: the crab *Pachygrapsus crassipes*, the annelid *Boccardia proboscidea* whose waving tentacles are readily seen on close observation, and the small red copepod *Tigriopus californicus*, present in varying numbers or absent. In addition, some of the pools are inhabited by the cirratulid polychaete *Cirriformia luxuriosa*, whose dark tentacles form conspicuous patches.

Study the distribution of the copepods to determine why they are absent or nearly so in some pools, and present in fair to large numbers in others. Important among the factors to be considered are predatory animals and food requirements. Previous studies have shown the copepod adapted to withstand extreme ranges in temperature and salinity as would be expected of an animal living in small high tide pools. The problem of *Tigriopus* distribution has never been very satisfactorily settled, however.

Make observations on a series of pools, and propose a hypothesis. If possible, check the pools at several different dates and criticize the hypothesis in the light of new findings.

15. Vertical Distribution of Rock Dwellers in Relation to Wave Action

A characteristic assortment of animals inhabits vertical granite surfaces in the upper intertidal and splash zones on exposed and partly protected rocky shores. The assortment includes the limpets *Acmaea digitalis*, *A. scabra*, *A. pelta*, and sometimes others, the snails *Littorina planaxis* and *L. scutulata*, and the barnacles *Balanus glandula*, *Chthamalus* sp. (either *C. dalli* or *C. fissus*), and *Tetraclita squamosa rubescens*. These animals are favorable objects for a study whose purpose is to determine the effect of wave action and height of splash upon vertical distribution.

Select two or three nearly vertical rock surfaces as much alike as possible, which extend from mid-tide level to above the splash limits. One of these should face the sea and be exposed to breaking surf and spray. A second should be situated on the inside of a point, or be otherwise protected from the surf. If possible, a third should be in relatively quiet waters. Study these rocks at both low and high tide, and record the extent of surf and spray upon them. Then at a favorable tide, mark off vertical strips one foot wide (wax crayon), divided vertically into one foot square sections. Count the individual limpets, snails, and barnacles (or any one species if time is short) in as many of the squares as possible, selecting squares that furnish fair samples of all levels. Make a chart or series of graphs comparing the vertical pattern of population density for each species on each of the rocks examined. Correlate with the observed height of wave action and splash. Notice how the vertical height of a given intertidal zone may be modified by local topography and its effect upon wave action.

16. Distribution and Habits of Limpets

The genus *Acmaea* offers an unusual opportunity for the study of numerous species of a single genus living in the same general habitat. Fourteen species occur in this area, of which ten are fairly common. The purpose of this study is to consider the structural differences of these species, their habitats and behavior, so far as time allows, with a view to discovering what correlation, if any, is to be found between these structural characters and the habitats and behavior of the different species. The following ten species may be used for the study, all of them common except *Acmaea persona* and *A. instabilis*:

Acmaea asmi
Acmaea digitalis
Acmaea insessa
Acmaea mitra
Acmaea pelta
Acmaea paleacea
Acmaea persona
Acmaea scabra
Acmaea scutum
Acmaea instabilis

(Other species occurring but not so favorable for such study are: *Acmaea fenestrata*, *A. limatula*—common in the Monterey area, *A. ochracea*, and *A. triangularis*. A comparable study can be carried out upon the common chitons or on barnacles.)

Make careful notes as to the characteristic location of these species as seen during low tides. Look for characters of general shape, color, and marginal ornamentation that seem to have an adaptive relation to the particular niche which each occupies. Prepare charts showing the vertical range of each species, and their positions in the intertidal relative to each other.

17. Ecological Distribution of Polychaete Worms

A large number of species of polychaetes belonging to many genera and families are to be found in any ecologically diversified area. Polychaetes should be collected whenever found and, after microscopic observation and identification, be properly preserved and labeled. If it is not possible to identify them at the time, preserve them for later study or transmittal to a specialist. There is a distinct value in identifying and becoming familiar with a wide range of types in a single group. It gives a picture of the range of structural diversity, resulting from a long evolution, and in the realm of taxonomy a rare species is just as important as a common one, sometimes more so.

To the student of ecology, however, the species that occur in large numbers are the important ones, since they have a proportionately much larger role in the communities of which they form part. Make certain, therefore, that you collect and identify properly all of the common, abundant species of polychaetes, keeping as complete a record as possible of their occurrence and ecologic distribution. Be prepared to report on the distribution and ecology of at least a dozen such common polychaetes.

18. Structural and Ecological Comparison of *Callianassa, Upogebia,* and *Emerita*

Purpose.—To compare the structural modifications of body and appendages in three different burrowing anomuran decapods, and to correlate differences found here with differences in habitat, behavior, and general way of life in the three species.

Laboratory procedure.—Remove and arrange in parallel series all the appendages on the right side of specimens of the three species. Care must be taken in removing the pereiopods to obtain the gills with them; each walking leg and cheliped is provided with a gill attached near its articulation with the body. The mandibles of *Emerita* are very small and fused to the epistome and are easily overlooked. Make a table for these appendages, with separate columns for the three species in the order *Callianassa, Upogebia, Emerita,* and compare each appendage with the others to bring out differences.

Observe the behavior of living specimens of the three species in laboratory aquaria in which the bottoms have been covered with sand (for *Emerita*) and muddy sand or mud (for *Upogebia* and *Callianassa*). Observe swimming, walking, and digging operations, paying special attention to the use of various appendages. *Upogebia* and *Callianassa* may feed under aquarium conditions, and this too should be observed and compared. Attempt to correlate differences in structure with differences in behavior. Is *Upogebia* more like *Callianassa* in structure? If so, why? Which of the two is most like *Emerita*? Is this to be explained in terms of phylogenetic relationship? If not, on what basis do you explain it?

Field studies.—Sandy beaches abound with *Emerita*, the adults scattered but numerous, the young present in countless thousands. These should be observed. On the basis of aquarium and field observations, be able to describe their method of walking, swimming, and digging in, the type of trail they make, and the part played in these different activities by the different body appendages. If possible, observe feeding in the area where waves are breaking on the beach. Refer to MacGinitie (1938, 1949). See if you can discern by your own observations an animal that preys on sand crabs; this will best be discovered when the tide is coming in.

Investigate a series of stations along a transect running from high to low water lines across a protected tidal flat. Pay special attention to the distribution of the ghost shrimp *Callianassa californiensis* and the mud shrimp *Upogebia pugettensis*. Become familiar with the holes and burrows of the ghost shrimp and see the various organisms associated with them, especially the gobies, pinnotherid crabs, and *Cryptomya*. If *Upogebia* is not encountered at the lower levels of the transect, seek out areas of soft, sticky mud and clay. Attempt to state how, spatially and ecologically, it is isolated from *Callianassa*. Refer to MacGinitie (1930 and 1934, or 1949).

19. Fauna of Flats of the Broad-Leaved Eelgrass (*Zostera*)

Sand flats, such as those near the mouth of Tomales Bay, separated from the shore flats by a tidal channel and exposed by the lowest tides, are often covered to a varying extent with the broad-leaved eelgrass *Zostera*. Such flats may be subject to fluctuations in height from time to time, and consequently show changes as well in fauna and flora. Whatever the condition of the eelgrass beds, they are well worth investigation.

The fauna of this association is rich but relatively monotonous. There are really three important habitats or zones here, each with its characteristic animals: (1) the mud below the beds, especially around the roots of the eelgrass; here occur the brittle star *Amphiodia occidentalis*, several polychaetes, but especially *Nephtys caecoides*, *Notomastus tenuis*, *Eupolymnia*, and *Hemipodus*, and in smaller numbers the sea cucumber *Leptosynapta albicans*, as well as a scattering of bivalves; (2) the luxuriant growth of eelgrass and larger algae (*Ulva*, and the like); here may be found the tectibranch *Phyllaplysia taylori*, caprellids, creeping copepods, isopods, the gastropod *Lacuna*, and sometimes bryozoans, tunicates, hydroids, sea anemones, and sea cucumbers. Most of these species require a substrate other than the mud and would not be here except for the plant life to give them sites above the bottom; (3) the water; here there are such forms as the shrimp *Spirontocaris*.

Collect and make notes on the fauna found in the habitats outlined above. Do not neglect the fauna of the mud. Material should be collected and brought back to the laboratory for a study of the submacroscopic fauna of each habitat.

20. *Phoronopsis* Associations

(As in Tomales Bay or Elkhorn Slough)

Laboratory study.—*Phoronopsis viridis* is a representative of a small group now given phylum rank, the Phoronidea (see p. 283). Familiarize yourself with the more important features of the anatomy, particularly the lophophore, which is here a feeding and respiratory organ and the only part of the animal ordinarily exposed to view. On living individuals demonstrate the ciliary action of the lophophore by adding a few drops of carmine suspension to the dish containing the animal.

Field studies.—Where the substrate has been built up (by ordinary deposition or mud transportation) to the level of low tides, plant growth is inhibited; here one finds luxuriant beds of *Phoronopsis*, with *Macoma* spp. less abundant.

Although a crude measure of population density and size can be obtained by counting the phoronids in one small sample and multiplying this by the size of the total area of the bed, the procedure outlined in the following paragraphs will yield more reliable figures. Students should work together in teams of two, each team provided with a shovel and a metal box six inches square, open at both ends.

1. Each team should select a small area of the bed where only phoronids occur, and clams appear to be absent. Push the sampling box down into the substrate to outline an area six by six inches; count all the phoronid tubes seen from the surface and check the count. Then remove the earth on all sides of the box, lift out the boxed sample, spread it out, and count all phoronids found within it. Comparison of this figure with the surface count will give the correction necessary to make surface counts reasonably accurate, when the class figures are averaged.

2. Next it will be necessary to determine the population of clams in the area and the extent to which they displace *Phoronopsis* where they occur together. Each team should select a six-inch square containing both phoronids and clams, insert the sampling box, and count the clam openings visible from the surface (remembering that *Macoma* has two separate siphons, one smaller than the other and sometimes buried). After counting clam openings, dig out the boxed sample and count both the clams and phoronids in it. By comparing the number of clams revealed in the surface count with that obtained in digging out the sample, one may obtain a correction factor to make surface counts reasonably accurate. At the same time, comparison of the phoronid and clam count of this sample with the phoronid count made in (1) above, will show about how many phoronids are displaced in the beds by an average clam.

3. Having obtained correction factors that make surface counts reasonably reliable, the teams should now sample the phoronid beds much more extensively. Half the teams should measure off areas a yard square at intervals over the bed, count the number of clam holes or siphon pairs visible, and apply the correction factor [obtained in (2) above] for clam counts. This will give a fair picture of the density of clam distribution for the bed as a whole when class figures are averaged. The other teams should use the six-inch metal sampling boxes to make surface counts of phoronids at several different points on the bed using the correction factor for surface counts obtained in (1) above.

4. The approximate dimensions of the bed can be measured and the total phoronid and clam populations calculated, or the population figures may be calculated for the average numbers of clams and phoronids per unit area of the bed.

Exercises of this type offer a simple introduction to the approach of population ecology. They introduce methods of critical sampling and reveal in an impressive manner the enormous abundance of some species in a limited habitat. Somewhat similar studies may be made of mussel beds, barnacle beds, or polychaete populations.

21. Piling Communities

(As found at bridges, wharves, and yacht basins)

Protected pilings offer a specialized habitat often richly populated with marine life. Pilings at the shore may not be as richly inhabited as those in deeper water, and studies in some areas may have to be made from boats. Collecting equipment and a yardstick or folding carpenter's rule are needed, and specimens taken should be studied and identified in the laboratory on the same day.

First try to get a general picture of the common forms present and of their vertical and horizontal distribution. What factor usually determines vertical distribution? You will find some types here at higher levels than would ordinarily be expected; what are these types? Do you find any difference in this condition between central and marginal piles? What is the second factor thus modifying the action of the primary factor that usually determines vertical distribution?

Observe the hydroids and anemones and attempt to distinguish different genera of hydroids and species of anemones in the field; collect specimens on which to check field identifications. Observe also the highly modified amphipods (Caprellidea), which are often abundant on the hydroids, and collect some for study in the laboratory. Are the caprellids different on different genera of hydroids? Where do you find nudibranchs and their coiled egg masses?

Some of the piles may be covered with muddy debris, consisting of the tubes of great numbers of small amphipods. Many species of animals seek shelter in and about the mud tubes. Collect some for later study. Does the location of these piles give you any clue as to why they, and not others, bear this association? Barnacles and mussels may form a conspicuous element in the fauna of some pilings. Do specimens of the commonest piling barnacle, *Balanus glandula*, show any differences from members of this species taken from rocks on the open coast? If so, how would you account for the differences?

Select a pile that shows zonation especially clearly. Working in teams of three or four, study the zonation carefully, designating various zones and measuring their extent and relation to the low-tide level. Keep notes on the animals found to be common and characteristic of each zone. Seek to determine the factors, physical and biotic, influencing the zonation. Try to determine what factors might operate to set lower limits on the distribution of various species, as well as upper limits. As time allows, compare findings on the first pile with conditions to be seen on neighboring piles, and attempt to account for any differences noticed.

If any floats or buoys are available for study in the immediate vicinity, investigate the fauna along the sides of these. Observe differences in the fauna and its distribution between these and the pilings. Here, forms may live within an inch of the surface yet never be exposed more than momentarily. Do you find a zonation here?

22. Plankton

A surprising number of littoral marine invertebrates get at least part of their food in the form of plankton. No complete understanding of the food relationships, food chains, and food cycles can be obtained if we depend for information solely on the low-tide investigations to which we are largely confined. One exercise, at least, should be devoted to the floating population, the organisms concerned, and something of their relations to the animals already studied.

Plankton may be collected with a fine-meshed tow net from a boat, or from a bridge or pier over a tidal current, or simply by towing a throw net from a suitable spot on shore. Study of the material collected is necessarily confined to the laboratory where microscopes are available. Sample the collected material and identify common forms to major systematic groups. For this, Johnson, Scott, and Chadwick, *The Marine Plankton*, is very helpful. The organisms in the haul will be found to vary greatly from place to place, and from season to season for hauls made in the same spot. Those most conspicuous or commonly encountered will probably be diatoms, protozoa, medusae, ctenophores, rotifers, chaetognaths, pelagic molluscs, larvacean tunicates, and a great variety of crustaceans. The larvae of many groups may be present, particularly young stages of echinoderms, bryozoans, molluscs, annelids, and crustaceans. Through observations and outside reading seek to understand the roles of some plankters in the food economy of the inshore waters.

23. A Brackish Estuarine Lake

A brackish-water estuarine lake is exemplified by Lake Merritt in Oakland. Its level is controlled by tide gates; hence the factor of the tides is eliminated. During the rainy months the salinity is very low. During the summer and fall it reaches a concentration (30 parts per 1,000) not far below that normal for sea water (34 parts per 1,000). An abundant algal growth, chiefly green algae and diatoms, furnishes a rich food supply. As a result, a few species, adjusted to low and varying salinity and not adapted to withstand exposure or wave action, flourish and occur in great numbers except during periods of high salinity.

Such species are the tube worm *Mercierella enigmatica*, the associated amphipods *Corophium* spp. and *Anisogammarus confervicolus*, the worm *Neanthes succinea*, the crab *Rhithropanopeus harrisii*, all of which live among the masses of tubes of *Mercierella*; the barnacle *Balanus improvisus*; the entoproct *Pedicellina*, common as a mossy growth over the tube-worm masses, especially near the exit of the lake; and the ribbed mussel *Modiolus demissus*.

Several other species, though typically marine, occasionally invade the lake at times of high salinity but maintain a precarious existence during the remainder of the year. Such are the soft-shelled clam *Mya*, barnacles, the encrusting bryozoan *Membranipora*, the mud snail *Nassarius obsoleta*, the oyster drill *Urosalpinx*, the tunicate *Ciona intestinalis*, and various other species.

Procedure.—Study the fauna to be found on the sides of a pier. If possible, see a boat lifted and observe the enormous numbers of gammarids leaving it as it rises out of the water, and the thick encrustation of *Mercierella* tubes. Break apart one of the tube-worm masses and study the relations among the different animals in this association. In the time allotted, work to gain a knowledge of this fauna, the food habits of its chief elements, and the chief physical and biotic features responsible for the abundance of each.

Observe *Modiolus* and *Mya* in life in the mud. Mud from the bottom may also show great numbers of polychaete worms, chiefly of the genus *Streblospio*. Plankton tows should be made and these studied on return to the laboratory. They will probably contain very many larvae of *Mercierella* in various stages of development: trochophores, polytrochs, and young worms. Veliger larvae, probably of *Mya*, possibly also of *Modiolus*, may be found in great numbers. In late spring or summer, entomostracans (especially copepods) and rotifers may be abundant.

24. Fauna of a Permanent Fresh-water Pond or Lake

Purpose.—To get a comprehensive picture of the environment, fauna, and food chains of a fresh-water pond or lake.

Preliminary laboratory study.—Laboratory preparations for such a trip should include introductory studies of some of the so far neglected invertebrate groups whose members, if not exclusively inhabitants of fresh water, are at least important in fresh-water environments. Such are the rotifers, nematodes, cladocerans, ostracods, copepods, and insects (both larvae and adults). Preparations may well include preliminary study of some of the other forms listed at the end of this exercise.

Procedure.—Collect as thoroughly as possible from the various habitats in the pond. Especially important are: (1) floating plants, (2) the lower surface of floating wood and debris, (3) underneath the bark of floating branches and logs, (4) the fine roots of trees, (5) plants growing in the water at various levels, (6) the bottom mud, and (7) the open water. Plankton tows should be made near the surface, some inches below, and near the bottom. Mud should be taken from the bottom for examination.

Identify your collections with the aid of references provided. Those most useful are Ward and Whipple, Needham and Needham (1951), and works dealing with individual groups which are listed in the bibliography. Be sure to sample the fauna of the plankton tows and bottom mud, as well as organisms collected in other areas. Insects need not be identified beyond order or family groups, thus, belostomid bugs, dytisicid beetles, damsel-fly larvae, dragon-fly larvae, chironomid larvae, phantom larvae (*Corethra*), and so on. The common, small, somewhat transparent leeches found in central California fresh waters are *Glossiphonia stagnalis*. They are predators on snails rather than bloodsuckers, as are typical leeches exemplified by the larger species.

Consider carefully the food habits of each species or group found. Fit them into food chains in a diagram showing food interrelationships in the pond.

Some animals of a permanent pond, listed according to food habits

Plankton feeders: Protozoa, sponges, rotifers, cladocerans, bryozoans, *Musculium* (pelecypod), copepods.

Predators: *Hydra, Dugesia* (=*Planaria*), leeches, water mites, spiders, dragonfly larvae, water bugs (back swimmers, water boatmen, Belostomatidae), water striders (Gerridae), beetles and their larvae (Dytiscidae, Hydrophilidae), ghost larvae (*Corethra*), fishes, amphibians, reptiles, birds.

Plant feeders: Isopods, snails, aphids, amphibian larvae, some fishes.

Plant scavengers (including detritus feeders):Nematodes, oligochaetes, ostracods, isopods, amphipods, May-fly nymphs, bloodworms (larval Chironomidae), mosquito larvae, gnat larvae (Simuliidae).

Animal scavengers: Ostracods, isopods, amphipods, water striders (Gerridae), gyrinid beetles.

25. Fauna of a Temporary Pond

Life in all small bodies of water is subject to relatively rapid seasonal changes in temperature and other factors. The permanent pond or lake also undergoes a vastly slower series of changes known as ecological succession, as each body of water gradually becomes more and more sedimented and ultimately passes through a marsh or bog stage to become a meadow. The temporary or seasonal pond undergoes each year profound changes in its biota, since it is not only subject to the influence of climatic factors but also to the change from dry land to pond and back again through marshy stages to dry land. Only those organisms will be found in it that have a resistant stage to carry them over the dry period, or that may be introduced either by visiting animals or by wind-borne resting stages.

In temporary ponds the shifts of population are very rapid. First is a short period of rapid development of phytoplankton; this furnishes the food basis for a rapidly pyramiding population of plankton feeders (Protozoa, phyllopods, Cladocera, copepods, rotifers) which develop in enormous numbers; these are followed somewhat later by a horde of predators and scavengers. The predators are largely insects, such as waterbugs and water beetles, both larvae and adults. Important as scavengers at this stage are the ostracods, which oftentimes occur in very great numbers in such ponds.

To get a picture of the seasonal cycle of life in such a pond, a series of three or four visits, beginning early in the spring, should be made to one or two selected ponds, and records of the catches on each occasion kept for comparison.

26. Fauna of a Small Permanent Stream

Streams differ from lakes and ponds chiefly in having a continuous, unidirectional current; the swifter the current the more conditions and fauna are likely to differ from those of a pond. However, the current is not uniform for the whole stream. The rate of flow is lowest along the banks, through vegetation, along the bottom, in areas protected by fallen branches, under stones, in accumulations of dead leaves, and in semi-isolated pools.

Plankton may be present; it develops principally in protected areas and is continually being washed away and renewed. The majority of the stream inhabitants belong to two phyla, the molluscs and arthropods, and nearly all of them are benthic forms. Crustaceans may be represented by crayfish, amphipods, and sometimes shrimp; molluscs by gastropods and fresh-water clams. The predominant forms are generally the larval insects. Commonly found are the larvae of caddis flies, stone flies, May flies, crane flies, and dragon flies. Planarians may occur under stones, and annelids along the banks and in the bottom mud.

Sample the faunas of (1) vegetation along the shore, (2) accumulations of dead leaves trapped by branches, (3) exposed surfaces of stones, (4) under stones, (5) the bottom mud or sand, (6) the open running waters, and other likely spots. Pay special attention to adaptations in structure or behavior which appear to fit the animals for life in an environment of flowing water. Identify forms as in the study of a fresh-water lake and compare your findings here with those in nonflowing bodies of fresh water.

27. Special or Individual Problems

In addition to field trips and other projects that are carried out as group efforts, we have found that individual student problems arouse interest, diversify the work of a course, provide much useful information, and test the student's potentialities as a zoölogist. The exact nature of the problem chosen is not of too great consequence, as long as it is compatible with the general program and activities of the class. But it is extremely important that the scope of the problem be clearly defined and sufficiently limited to permit it to be studied in a thorough fashion within the available time. Ordinarily the student problems fall into one of the following categories.

Taxonomic problems.—These usually involve making a well-preserved, identified, and properly labeled collection of animals of some particular taxonomic group, supplemented by an annotated check list and suggestions for the improvement of the available key to the group. The student is expected to consult all original taxonomic literature pertinent to the group.

Ecological-faunistic-distributional problems.—Problems here may be of several different sorts.

A survey of the fauna of a spatially or geographically defined area.

A survey of the fauna of a particular habitat or association (e.g., a kelp bed, a *Pelvetia* bed, a *Macrocystis* holdfast, a series of high splash pools).

A study of the distribution of one or a few selected species, and an attempt to define the characteristic habitat-niche(s) and to discover some of the factors restricting the organism(s) thereto.

A comparison of two different areas with special attention to physical factors and faunas, and an attempt to correlate differences in factors and faunas found.

Anatomical problems.—These may involve dissection and an investigation of the internal or external anatomy of one form, or a comparison of similar structures in a series of related forms. Particularly valuable here are problems in functional anatomy, in which an attempt is made to demonstrate the relation of structure to use. The best structures for such purposes are those which are exposed to the external environment and are involved in some visible activity (e.g., the palps and ctenidia of various bivalves, particular crustacean appendages, the radulae and related structures of gastropods, the setae and parapodia of various polychaetes).

Natural history problems.—Problems here may involve a study of one or two species in relation to such things as food and feeding behavior, reproduction, some phases of development, locomotion, commensals and parasites, and so on.

Problems in physiological ecology.—Attempts may be made to determine the resistance of selected species to such environmental vicissitudes as salinity changes, desiccation, higher temperatures, and the like, and to relate tolerances found to the observed position and behavior of these species in the intertidal.

The problem should be chosen with the advice of the instructor, and work initiated early in the course. A certain part of the laboratory time should be allotted to it. It is to be carried out and reported on as a research problem. Every effort should be made to make it as scientific as possible (i.e., clear, accurate, logically sound, and intellectually honest). All statements should be supported by the author's own observations or by definite reference to authority. Reference should be made to sources found to be useful, such as previous reports on the same subject, general treatises, and especially articles in scientific journals. The following is suggested as a standard form for these reports:

Brief table of contents
Statement of problem
Brief survey of literature and previous reports
Materials and methods
General findings, including lists of species, keys, and detailed presentation of results
Discussion of results (if needed)
Summary of your findings
Bibliography

Illustrations should be placed in the body of the paper where they will be of most value. Both *The Zoological Record* and *Biological Abstracts* may be used in making up the bibliography. The standard bibliographical form should be used, as in the example below, and references made by name and year; thus, Oldroyd (1924):

Oldroyd, I. S., 1924. The marine shells of the west coast of North America. I. Pelecypoda. Stanford Univ. Publ. Geol. Sci., 1:1-247, pls. 1-57.

A rough draft, or at least an outline, should be turned in for criticism and suggestions on a date designated by the instructor.

In our courses the finished reports are held for use of future students and have proved of great value. Duplicates should be made if the student desires to retain a copy. Carefully labeled examples of the species dealt with should be turned in with each report.

Bibliography

GENERAL REFERENCE SOURCES

Biological Abstracts. A monthly journal containing brief abstracts of articles in current United States and foreign biological periodicals; comprehensive coverage of the literature.

The Zoological Record. Volumes published yearly by the Zoological Society of London; references arranged according to taxonomic groupings; no abstracts, but excellent and comprehensive in coverage; follows about two years behind current literature.

Most of the volumes listed below contain more or less extensive bibliographies which are well worth consulting; of these, the bibliography in Ricketts and Calvin (1952 edition), brought up to date by J. W. Hedgpeth. is the most generally useful and renders superfluous a large list of ecological references in the present volume. The *Treatise on Marine Ecology and Paleoecology*, now in preparation under the auspices of the National Research Council and interested scientific societies, should be of tremendous value to students of marine life when it appears.

General Anatomy and Classification: Texts and Reference Works

Borradaile, L. A., F. A. Potts, L. E. S. Eastham, and J. T. Saunders, 1935. *The Invertebrata.* 2d ed. London: Macmillan. 725 pp., illus. Usually known as "Borradaile and Potts." One of the standard invertebrate texts, particularly good on the crustacea.

H. G. Bronn. 1866—. *Klassen und Ordnungen des Tierreichs.* Leipzig: Winter. Many volumes, each a monographic account by a specialist of a taxonomic group, covering anatomy, development, general biology, and classification, and with an extensive bibliography. New issues still being published.

Brown, F. A., Jr. (ed.). 1950. *Selected Invertebrate Types*. New York: Wiley. 597 pp., illus. Primarily a guide for the laboratory study and dissection of common inland and East Coast marine invertebrates; numerous notes on living animals are included; some parts are excellent.

Buchsbaum, R. 1948. *Animals without Backbones*. Rev. ed., University of Chicago Press. 405 pp., illus. Excellent elementary and semi-popular work; profusely illustrated with diagrams and photographs.

Bullough, W. S. 1950. *Practical Invertebrate Anatomy*. London: Macmillan. 463 pp., illus. Covers somewhat the same ground as Brown (1950); emphasis on anatomy and classification.

Carter, G. S. 1948. *A General Zoology of the Invertebrates*. Rev. ed., London: Macmillan. 509 pp., illus. A refreshing elementary treatment of invertebrates from the point of view of principles and function rather than anatomy and taxonomy.

Delage, Y., and E. Hérouard, (eds). 1896-1903. *Traité de Zoologie Concrète*. Paris: Schleicher. Vols. 1-8. A comprehensive treatise, old but still very useful; many of the colored plates and diagrams are unequaled in modern works; covers mainly anatomy and taxonomy.

Grassé, P. (ed.). 1948- . *Traité de Zoologie. Anatomie, Systématique, Biologie*. Paris: Masson. Another comprehensive treatise, containing a proposed series of 17 volumes, of which only volumes dealing with echinoderms, protochordates, and some of the arthropod groups among the invertebrates have so far appeared; excellent and modern.

Harmer, S. F., and A. E. Shipley (eds.). 1895-1909. *The Cambridge Natural History*. London: Macmillan. Vols. 1-6. A large work dealing with the morphology, systematics, and biology of invertebrate groups; still useful for many groups.

Hesse, R. (present ed.). 1896- . *Das Tierreich*. " . . . attempts to describe all known species of animals in the world. Begun in 1896, the work at present comprises about 70 large volumes, but these cover only a small part of the animal kingdom. R. Hesse, in charge of the undertaking, reported in 1929 that at the present rate of publication, 750 years would be required to complete the task of describing the species of animals now known, leaving out of consideration the new species that would be discovered in the interim. These facts may give . . . some idea of the magnitude of the task of classifying and naming animals." (Hyman, *The Invertebrates*, Vol. 1, p. 25).

Hyman, L. H. 1940- . *The Invertebrates*. New York: McGraw-Hill. Illus. Vols. 1-3, covering Protozoa through the pseudocoelomate phyla, have appeared, and others are in preparation or contemplated. indispensable.

Kükenthal, W. G., and T. Krumbach (eds.). 1923- . *Handbuch der Zoologie.* Leipzig: De Gruyter. Several volumes; not yet completed. An excellent great treatise, each animal group being treated from all points of view by a specialist; well illustrated and with very extensive bibliographies.

Lang, A. 1891-1896. *Text-book of Comparative Anatomy.* Trans. by Bernard and Bernard, London: Macmillan. 2 vols., illus. Old, but with much good material on anatomy of hard and soft parts; one of the few books which emphasizes the "comparative" aspect of invertebrate anatomy.

Lankester, E. R. (ed.) 1900-1909. *A Treatise on Zoology.* London: Macmillan. Several volumes, illus., never completed. Excellent for reference, particularly on anatomy; volumes on molluscs and crustaceans are particularly good.

Liverpool Marine Biological Committee *Memoirs* (L.M.B.C. Memoirs) to 1947. Liverpool University Press. 34 vols., illus. Each of the volumes presents a detailed study of the anatomy of a particular organism, many closely related to forms found on the Pacific Coast. Very valuable for reference in dissections.

Parker, T. J., and W. A. Haswell. 1940. *Textbook of Zoology.* 6th ed., London: Macmillan. Vols. 1-2, illus. Rather out of date in several places, but still the most useful invertebrate text.

Ecology, Natural History, and the Marine Environment: Books

Allee, W. C., A. E. Emerson, O. Park, T. Park, and K. P. Schmidt. 1949. *Principles of Animal Ecology.* Philadelphia: Saunders. 837 pp., illus. A comprehensive treatment of ecology, more a reference than a text, with an excellent bibliography.

Flattely, F. W., and C. L. Walton. 1922. *The Biology of the Seashore.* London: Macmillan. 336 pp., illus. Indispensable.

Hesse, R., W. C. Allee, and K. P. Schmidt. 1951. *Ecological Animal Geography.* 2d ed. New York: Wiley. 715 pp., illus. Excellent treatment of ecology, adaptations, and distribution of animals in all parts of the marine environment.

Johnson, M. E., and H. J. Snook, 1927. *Seashore Animals of the Pacific Coast.* New York: Macmillan. 659 pp., illus. Exceedingly useful for identifications, though many parts are out of date, and others never were at all complete. Out of print; now being revised.

Johnstone, J., A. Scott, and H. C. Chadwick. 1924. *The Marine Plankton.* Liverpool University Press. 194 pp., illus. The best single introduction to planktonic forms ever published for student use; the pictures enable one to place within major groups nearly anything small that he is likely to find in a plankton haul.

MacGinitie, G. E., and N. MacGinitie. 1949. *Natural History of Marine Animals.* New York: McGraw-Hill. 473 pp., illus. Contains a great many interesting and useful observations on the biology of West Coast invertebrates and should be read by all students.

Marmer, H. A. 1926. *The Tide.* New York: Appleton. 282 pp., illus. Excellent discussion of tidal causes and effects.

Ricketts, E. F., and J. Calvin. 1952. *Between Pacific Tides.* 3d ed., rev. by J. W. Hedgpeth, Stanford University Press. 502 pp., illus. We urge every user of the present manual to buy and read Ricketts and Calvin, since in many essential respects it complements and supplements this work. Treatment of intertidal animals is primarily from the standpoint of ecology and natural history; the book is less useful for the identification of any but relatively distinctive forms. An excellent chapter on intertidal zonation has been added by Hedgpeth to the new edition, and the bibliography expanded and brought up to date.

Russell, F. S., and C. M. Yonge. 1936. *The Seas.* Rev. ed., London: Warne. 379 pp., illus. Still the finest general and semipopular account of the life in the sea, written by two outstanding authorities; highly informative and readable.

Shelford, V. E. 1929. *Laboratory and Field Ecology.* Baltimore: Williams & Wilkins. 608 pp., illus. Ecological methods.

Sverdrup, H. U., M. W. Johnson, and R. H. Fleming. 1942. *The Oceans, Their Physics, Chemistry, and General Biology.* New York: Prentice-Hall. 1087 pp., illus. An indispensable reference of technical information on the sea.

Yonge, C. M. 1949. *The Sea Shore.* London: Collins. 311 pp., illus. "This is the finest book of this *genre* yet published; a *sine qua non* for the bookshelf of all who go to the shore." (Hedgpeth, in Ricketts and Calvin, 1952, p. 475.) We agree wholeheartedly; beautifully written and illustrated.

Fresh Water Biology

Morgan, A. H. 1930. *Field Book of Ponds and Streams.* New York: Putnam. 448 pp., illus.

Needham, J. G., and J. T. Lloyd. 1937. *The Life of Inland Waters.* Ithaca, New York: Comstock. 438 pp., illus.

Needham, J. G., and P. R. Needham. 1941. *A Guide to the Study of Fresh Water Biology*. Ithaca, New York: Comstock. 89 pp., illus.

Pennak, R. W. 1953. *Fresh-water Invertebrates of the United States*. New York: Ronald. 769 pp., illus. A work comparable in scope to Ward and Whipple; it has appeared so recently that we cannot predict whether it will prove more or less useful than the older work, but it is up to date, at least.

Pratt, H. S. 1935. *A Manual of the Common Invertebrate Animals*. Rev. ed., Philadelphia: Blakiston. 854 pp., illus. Keys to many common forms.

Ward, H. B., and G. C. Whipple. 1918. *Fresh-water Biology*. New York: Wiley. 1111 pp., illus. This old stand-by is now in process of revision by W. T. Edmondson of the University of Washington.

Welch, P. S. 1952. *Limnology*. 2d ed. New York: McGraw-Hill. 538 pp., illus.

GENERAL METHODS AND EQUIPMENT

Dall, W. H. 1907. Instructions for Collecting Mollusks, and Other Useful Hints for the Conchologist. Part G, Bull. 39, U. S. Nat. Mus., 1-56 (2d ed.).

Lo Bianco, S. 1899. The Methods Employed at the Naples Zoological Station for the Preservation of Marine Animals. Trans. from the Italian by E. O. Hovey. Part M, Bull. 39, U. S. Nat. Mus., 1-42.

McGinitie, G. E. 1947. Sea Water Systems at Marine Laboratories. Science, 106:171-173.

———. 1948. Dredges for Use at Marine Laboratories. Turtox News, 26: No. 12, Dec.

———. 1948. Choice, Operation, and Care of Boats at Marine Laboratories. Turtox News, 26: No. 5, May.

Needham, J. G., P. S. Galtsoff, F. E. Lutz, and P. S. Welch. 1937. *Culture Methods for Invertebrate Animals*. A compendium prepared for section F, American Assoc. for the Advancement of Science. Ithaca, New York: Comstock. 590 pp.

Pantin, C. F. A. 1946. *Notes on Microscopical Technique for Zoologists*. Cambridge University Press. 75 pp. Concise, practical, and invaluable.

Riley, C. V. 1892. Directions for Collecting and Preserving Insects. Part F, Bull. 39, U. S. Nat. Mus., 1-147.

Smithsonian Institution. 1944. *A Field Collector's Manual in Natural History*. (Prepared by members of the staff of the Smithsonian Institution). Publ. 3766, 118 pp.

Wards, Inc. 1950. How to Embed in Bio-plastic. Rochester, New York: Ward's Natural Science Establishment. 19 pp.
Welch, P. S. 1948. *Limnological Methods*. Philadelphia: Blakiston. 381 pp.

PROTOZOA

Arnold, Zach M. 1951. Occurrence of *Gromia oviformis* Dujardin in California (Testacea). Wasmann Jour. Biol., 9: 351-353.
Jepps, Margaret W. 1926. Contributions to the Study of *Gromia oviformis* Dujardin. Quart. Jour. Micr. Sci., 70: 701-719.
Myers, E. H. 1940. Observations on the Origin and Fate of Flagellated Gametes in Multiple Tests of *Discorbis* (Foraminifera). Jour. Mar. Biol. Assoc., 24: 201-226.

PORIFERA

Bergmann, W. 1949. Comparative Biochemical Studies on the Lipids of Marine Invertebrates, with Special Reference to the Sterols. Jour. Mar. Res., 8 (2): 137-176. (A discussion of the possible application of biochemical characteristics to systematic problems in the Porifera.)
Bidder, G. P. 1923. The Relation of the Form of a Sponge to Its Currents. Quart. Jour Micr. Sci., 67: 293-323.
Burton, M, 1949. Notes on the Ecology of Sponges. British Science News, 2 (15): 83-85.
———. 1949. Observations on Littoral Sponges, Including the Supposed Swarming of Larvae, Movement and Coalescence in Mature Individuals, Longevity and Death. Proc. Zool. Soc. London, 118 (4): 893-915.
Duboscq, O., and O. Tuzet. 1937. L'ovogénèse. la fécondation, et les premiers stades du développement des éponges calcaires. Arch. Zool. gén. exp., 79 (2): 157-316.
Herlant-Meewis, H. 1949. Contribution a l'étude histologique des spongiaires. Ann. Soc. Roy. Zool. Belg., 79:5-36. (Studies on the distribution of elastin and spongin fibers in several groups of sponges.)
Hyman, L. H. 1940. *The Invertebrates: Protozoa through Ctenophora.* New York: McGraw-Hill. pp. 284-364.

Jepps, M. W. 1947. Contribution to the Study of the Sponges. Proc. Roy. Soc. London, B, 134: 408-417. (Studies on contractile vacuoles in sponges.)

Jewell, M. E. 1935. An Ecological Study of the Fresh-water Sponges of Northern Wisconsin. Ecol. Mon., 5: 461-504.

———. 1939. II. The Influence of Calcium. Ecology, 20 (1): 11-28. (Two excellent studies of the environmental factors limiting the distribution of spongillids, including information on the relationship between silicon content of the water and spicule form.)

Jørgensen, C. B. 1944. On the Spicule-formation of *Spongilla lacustris* (L.) 1. The Dependence of the Spicule-formation on the Content of Dissolved and Solid Silicic Acid of the Milieu. Kgl. Danske Vidensk. Selsk. Biol. Meddel., 19 (7): 1-45.

———. 1947. 2. The Rate of Growth of the Spicule. *Ibid.*, 20 (10):1-21.

Killian, E. F. 1952. Wasserströmung und Nahrungsaufnahme beim Süsswasserschwamm *Ephydatia fluviatilis.* Z. vergl. Physiol., 34:407-447. (The best description of sponge feeding and digestion.)

Laubenfels, M. W. de. 1932. Morphology and Physiology of Porifera Exemplified by *Iotrochota.* Carneg. Inst. Wash. Papers. Tortugas Lab., 28: 37-66. (Interesting observations of bispecific conglomerations.)

———. 1932. The Marine and Fresh-water Sponges of California. Proc. U. S. Nat. Mus., 82 (4): 1-140.

———. 1936. A Discussion of the sponge fauna of the Dry Tortugas, with Material for a Revision of the Families and Orders of the Porifera. Carneg. Inst. Wash. Publ. no. 467. 225 pp.

———. 1948. The Order Kerotosa of the Phylum Porifera—A Monographic Study. Allan Hancock Found. Publ., Occ. Pap. no. 3: 1-217.

Parker, G. H. 1910. The Reactions of Sponges, with a Consideration of the Origin of the Nervous System. Jour. Exp. Zool., 8 (1): 1-41.

Renouf. L. 1937. The Importance of Field Notes to the Study of Porifera. C. R. XIIe Congr. Internat. Zool., Lisbonne, 1935, pp. 831-840. (Observations on variability in sponge species as related to environmental factors.)

Tuzet, O. 1948. La place des Spongiaires dans la classification. Rapport, Section VI, XIIIe Congr. Internat. Zool., Paris, 1948. (A discussion of the relationship of the sponges to other Metazoa.)

Van Weel, P. B. 1949. On the Physiology of the Tropical Fresh-water Sponge, *Spongilla proliferans* Annand. I. Ingestion, Digestion, and Excretion. Physiol. Compar. et Oecol., 1 (2): 110-126.

Vosmaer, G. C. J., 1928. Bibliography of Sponges 1551-1913. Ed. by G. P. Bidder and C. S. Vosmaer-Röell. Cambridge University Press. 234 pp.

Vosmaer, G. C. J. 1932-1935. Sponges of the Bay of Naples. Porifera Incalcaria. Capita Zoologica, 3 and 5. 875 pp. (An excellent study of variation in sponge species. Includes a review of literature on the clionids.)
Wilson, H. V., and J. T. Penney. 1930. The Regeneration of Sponges (*Microciona*) from Dissociated Cells. Jour. Exp. Zool., 56: 73-147.

COELENTERATA

General

Hyman, L. H. 1940. *The Invertebrates: Protozoa through Ctenophora.* New York: McGraw-Hill. pp. 365-661.
Weil, R. 1934. Contribution à l'étude des cnidaires et de leurs nématocysts. Trav. Sta. Zool. Wimereux, Vols. 10 and 11.

Hydrozoa

Fisher, W. K. 1938. Hydrocorals of the North Pacific Ocean. Proc. U. S. Nat. Mus., 84: 493-554.
Fraser, C. McL. 1937. *Hydroids of the Pacific Coast of Canada and the United States.* University of Toronto Press. 207 pp., 44 plates.
Hand, Cadet, and J. R. Hendrickson. 1950. A Two-tentacled, Commensal Hydroid from California (Limnomedusae, *Proboscidactyla*). Biol. Bull., 99: 74-87.
Hyman, L. H. 1947. Two New Hydromedusae from the California Coast. Trans. Amer. Micr. Soc., 66:262-268.
Mayer, A. G. 1910. *Medusae of the World, Vols. I and II. The Hydromedusae.* Carneg. Inst. Wash., Publ. no. 109.
Skogsberg, T. 1948. A Systematic Study of the Family Polyorchidae (Hydromedusae). Proc. Calif. Acad. Sci., ser. 4, 26: 101-124.

Scyphozoa

Mayer, A. G. 1910. *Medusae of the World. Vol. III. The Scyphomedusae.* Carneg. Inst. Wash., Publ. no. 109.

Anthozoa

Carlgren, O. 1949. A Survey of the Ptychodactaria, Corallimorpharia and Actiniaria, with a Preface by T. A. Stephenson. K. Svenska Vetenskapakad. Handl., Ser. 4, 1(1): 1-121.

———. 1952. Actiniaria from North America. Arkiv för Zoologi, Ser. 2, 3(30): 373-390.

Hand, Cadet, 1951. The Intertidal Sea Anemones of the Central Californian Coast. Unpublished Ph. D. dissertation, University of California, Berkeley. 306 pp.

Kükenthal, W. 1913. Über die Alcyonarienfauna Californiens und ihre tiergeographischen Beziehungen. Zool. Jahrb., Abt. f. Syst., 35: 219-270.

Nutting, C. C. 1909. Alcyonaria of the California Coast. Proc. U. S. Nat. Mus., 35: 681-727. (Kükenthal is considered more reliable.)

Stephenson, T. A. 1925, 1928. *The British Sea Anemones.* London: Ray Society. Vols. 1, 148 pp., and II, 426 pp., illus.

PLATYHELMINTHES

Boone, E. S. 1929. Five New Polyclads from the California Coast. Ann. Mag. Nat. Hist. (10), 3: 33-46.

Costello, H. M., and D. P. Costello. 1938. A New Species of *Polychoerus* from the Pacific Coast. Ann. Mag. Nat Hist. (11), 1:148-155.

Heath, H., and E. A. McGregor 1912. New Polyclads from Monterey Bay, California. Proc. Acad. Nat. Sci., Philadelphia, 64: 455-488.

Hyman, L. H. 1931. . . . Recent European Revisions of the Triclads, and Their Application to the American Forms, with a Key to the Latter . . . Trans. Amer. Micr. Soc., 50: 316-335.

———. 1939. . . . The Priority of *Dugesia* Girard 1950 over *Euplanaria* Hesse 1897 with Notes on the American Species of *Dugesia. Ibid.*, 58: 264-275.

———. 1943. Endemic and Exotic Land Planarians of the United States . . . Amer. Museum Novitates, no. 1241: 1-21.

———. 1951. *The Invertebrates. Vol. II. Platyhelminthes and Rhynchocoela.* New York: McGraw-Hill. 550 pp., illus.

———. 1953. The Polyclad Flatworms of the Pacific Coast of North America. Bull. Amer. Mus. Nat Hist., 100 (2): 269-392.

Lehman, H. E. 1946. A Histological study of *Syndisyrinx franciscana*, gen. et sp. nov., an Endoparasitic Rhabdocoel of the Sea Urchin, *Strongylocentrotus franciscanus*. Biol. Bull., 91: 295-311.

Stunkard, H. W., and J. O. Corliss. 1951. New Species of *Syndesmis* and a Revision of the Family Umagillidae Wahl, 1910 (Turbellaria: Rhabdocoela). Biol. Bull., 101: 319-334.

NEMERTEA

Coe, W. R. 1901. Papers from the Harriman Alaska Expedition, 20, The Nemerteans. Proc. Wash. Acad. Sci., 3:1-111.

———. 1904. The Nemerteans. Harriman Alaska Exped. 11:1-220.

———. 1905. Nemerteans of the West and Northwest Coasts of America. Bull. Mus. Comp. Zool. Harvard, 47:1-319.

———. 1940. Revision of the Nemertean Fauna of the Pacific Coast of North, Central, and Northern South America. Allan Hancock Pacific Exped., 2 (13):247-323.

———. 1943. Biology of the Nemerteans of the Atlantic Coast of North America. Trans. Conn. Acad. Arts and Sci., 35:129-328.

Hyman, L. H., 1951. *The Invertebrates, Vol. II. Platyhelminthes and Rhynchocoela.* New York: McGraw-Hill. 550 pp., illus.

ASCHELMINTHES

Hyman, L. H. 1951. *The Invertebrates, Vol. III. Acanthocephala, Aschelminthes, and Entoprocta.* New York: McGraw-Hill. 572 pp., illus.

ANNELIDA

Polychaeta

The literature on polychaetes is very great. Fortunately the more important references have recently been assembled and published by Dr. Olga Hartman in an extremely useful book, *The Literature of the Polychaetous*

Annelids, 1951*b* (see below). The following list is selected to include examples of studies on diverse aspects of polychaete biology as well as works most generally useful in the identification of West Coast species (Ed.).

Ashworth, J. H. 1912. *Catalogue of the Chaetopoda in the British Museum.* (Nat. Hist.). *A. Polychaeta. Part 1. Arenicolidae.* London: William Clowes and Sons. 175 pp., 15 pl.

Benham, W. B., 1910. Archiannelida, Polychaeta, and Myzostomaria, in *The Cambridge Natural History*, Vol. II. London: Macmillan.

Berkeley, Edith, and Cyril Berkeley, 1948. Annelida, Polychaeta Errantia. Canadian Pacific Fauna, no. 9*b* (1). Fish. Research Board, Canada. University of Toronto Press. pp. 1-100.

———. 1952. Annelida, Polychaeta Sedentaria. *Ibid.*, no. 9*b* (2), 1-139. Both of these works by Mr. and Mrs. Berkeley are extremely useful, especially to the north of California. See numerous earlier references in Hartman, 1951*b*.

Bullock, T. H. 1948. Physiological Mapping of Giant Nerve Fiber Systems in Polychaete Annelids. Physiol. Compar. et Oecol., 1:1-14.

Goodrich, E. S. 1945. The study of Nephridia and Genital Ducts since 1895. Quart. Jour. Micr. Sci., 86:113-392.

Fauvel, P. 1923. *Polychètes Errantes.* Faune de France, Paris; 5:1-488.

———. 1927. *Polychètes Sedentaires. Addenda aux Errantes, Archiannèlides, Myzostomaires. Ibid.* 16:1-494. The two volumes by Fauvel are extremely useful references, even on this coast.

Fox, D. L., S. C. Crane, and B. H. McConnaughey. 1948. A Biochemical Study of the Marine Annelid Worm, *Thoracophelia mucronata.* Jour. Mar. Research, 7:567-585. Excellent quantitative study of nutrition and natural history.

Gravier, C. 1923. La ponte et l'incubation chez les "Annélides Polychètes." Ann. Sci. Nat. Zool., sér. 10, 6:153-248.

Hanson, Jean. 1949. The Histology of the Blood System in Oligochaeta and Polychaeta. Biol. Rev., 24:127-173.

Hartman, O. 1936*a*. Nomenclatorial Changes Involving California Poly chaete Worms. Jour. Wash. Acad. Sci., 26:31-32.

———. 1936*b*. New Species of Polychaetous Annelids of the Family Nereidae from California. Proc. U. S. Nat. Mus., 83:467-480.

———. 1936*c*. New Species of Spionidae (Annelida polychaeta) from the Coast of California. Univ. Calif. Publ. Zoöl., 41:45-52.

———. 1936*d*. A Review of the Phyllodocidae (Annelida polychaeta) of the Coast of California, with Descriptions of Nine New Species. *Ibid.*, 41:117-132.

———. 1938*a*. Brackish and Fresh-water Nereidae from the Northeast

Pacific with the Description of a New Species from Central California. *Ibid.*, 43:79-82.

———. 1938*b*. Descriptions of New Species and New Generic Records of Polychaetous Annelids from California of the Families Glyceridae, Eunicidae, Stauronereidae, and Opheliidae. *Ibid.*, 43:93-112.

———. 1938*c*. Review of the Annelid Worms of the Family Nephtyidae from the Northeast Pacific, with Descriptions of Five New Species. Proc. U. S. Nat. Mus., 85:143-158.

———. 1939a. Polychaetous Annelids. Part 1. Aphroditidae to Pisionidae. Allan Hancock Pacific Exped., 7:1-156.

———. 1939*b*. New Species of Polychaetous Annelids from Southern California. *Ibid.*, 7:157-172.

———. 1940*a*. Polychaetous Annelids. Part 2. Chrysopetalidae to Goniadidae. *Ibid.*, 7:173-287.

———. 1940*b*. *Boccardia proboscidea*, a New Species of Spionid Worm from California. Jour. Wash. Acad. Sci., 30:382-387.

———. 1941*a*. Some Contributions to the Biology and Life History of Spionidae from California. Allan Hancock Pacific Exped., 7:289-324.

———. 1941*b*. Polychaetous Annelids. Part 4. Pectinariidae, with a Review of All Species from the Western Hemisphere. *Ibid.*, 7:325-345.

———. 1944*a*. Polychaetous Annelids. Part 5. Eunicea. *Ibid.*, 10:1-238.

———. 1944*b*. Polychaetous Annelids from California, Including the Description of Two New Genera and Nine New Species. *Ibid.*, 10: 239-307.

———. 1944*c*. Polychaetous Annelids. Part 6. Paraonidae, Magelonidae, Longosomidae, Ctenodrilidae, and Sabellariidae. *Ibid.*, 10:311-389.

———. 1947*a*. Polychaetous Annelids. Part 7. Capitellidae. *Ibid.*, 10: 391-481.

———. 1947*b*. Polychaetous Annelids. Part 8. Pilargiidae. *Ibid.*, 10: 482-523.

———. 1948. The Polychaetous Annelids of Alaska. Pacific Science, 2:1-58.

———. 1950. Goniadidae, Glyceridae, and Nephtyidae. Allan Hancock Pacific Exped., 15:1-181.

———. 1951*a*. Fabricinae (Feather-duster Polychaetous Annelids) in the Pacific. Pacific Science, 5:379-391.

———. 1951*b*. *The Literature of the Polychaetous Annelids. Vol. I. Bibliography*. Los Angeles, California. Privately published, 290 pp. The bibliography of this work contains more than 4,000 titles, alphabetically arranged by author, followed by a subject-analysis summarizing references under twenty selected headings. See p. 279, Faunal Studies, California, for literature on this area.

———. 1952. On the Identity of *Stylarioides inflata* (Treadwell) and Its Extended Distribution (Annelida). Pacific Science, 6:71-74.

Hartman, Olga, and D. J. Reish. 1950. The Marine Annelids of Oregon. Oregon State College Press. Oregon State Monographs, Studies in Zoology, No. 6, 64 pp.

Hempelman, F. 1931. Archiannelida and Polychaeta. *In* Kukenthal und Krumbach, *Handbuch der Zoologie,* Bd. 2:2:12 and 13: pp. 1-213.

———. 1937. Polychaeta. *In* Bronn, *Klassen und Ordnungen des Tierreichs,* Bd. 4: 3: 2: 1: pp. 1-106. (Another very extensive and useful bibliography, although not having a subject index.)

Johnson, H. P. 1897. A Preliminary Account of the Marine Annelids of the Pacific Coast. . . . Proc. Calif. Acad. Sci., 1:153-190.

———. 1901. The Polychaeta of the Puget Sound Region. Proc. Boston Soc. Nat. Hist., 29:381-437.

Johnson, M. W. 1943. Studies on the Life History of the Marine Annelid, *Nereis vexillosa.* Biol. Bull., 84:106-114.

Lillie, F. R., and E. E. Just. 1913. Breeding Habits of the Heteronereis Form of *Nereis limbata* at Woods Hole, Mass. Biol. Bull., 24:147-168. (Probably this is a synonym of *Neanthes succinea.*)

MacGinitie, G. E. 1939. The Method of Feeding of *Chaetopterus.* Biol. Bull., 77:115-118.

Moore, J. P. 1909. Polychaetous Annelids from Monterey Bay and San Diego, California. Proc. Acad. Nat. Sci., Philadelphia: 61:235-295.

———. 1923. The Polychaetous Annelids Dredged by the U.S.S. *Albatross* off the Coast of Southern California in 1904. Spionidae to Sabellariidae. Proc. Acad. Nat. Sci., Philadelphia, 75:179-259.

Nicol, E. A. T. 1930. The Feeding Mechanism, Formation of the Tube, and Physiology of Digestion in *Sabella pavonina.* Trans. Roy. Soc. Edinburgh, 56:537-598.

Nicol, J. A. C. 1948. The Giant Axons of Annelids. Quart. Rev. Biol., 23:291-323.

Okuda, S. 1947. On an Ampharetid Worm, *Schistocomus sovjecticus* Annenkova, with Some Notes on Its Larval Development. Jour. Fac. Sci., Hokkaido Imperial University, Japan, ser. 6, 9:321-329.

Pettibone, M. H., 1953. Some Scale-Bearing Polychaetes of Puget Sound and Adjacent Waters. Univ. Washington Press, Seattle. 89 pp., illus.

Potts, F. A. 1911. Methods of Reproduction in the Syllids. Ergebnisse und Fortschritte der Zoologie, 3:1-72.

———. 1914. Polychaeta from the Northeast Pacific. The Chaetopteridae. . . . Proc. Zool. Soc. London, for 1914:955-994.

Shearer, C. 1911. On the Structure and Development of the Trochophore of *Hydroides uncinatus* (*Eupomatus*). Quart. Jour. Micr. Sci., 56:543-591.

Smith, R. I. 1950. Embryonic Development in the Viviparous Nereid Polychaete, *Neanthes lighti* Hartman. Jour. Morph., 87:417-466.

Wells, G. P. 1945. The mode of life of *Arenicola marina* L. Jour. Mar. Biol. Assoc., 26:170-207.
Wells, G. P., and R. P. Dales, 1950. . . . The irrigation of the burrow in the polychaetes *Chaetopterus variopedatus* Renier and *Nereis diversicolor* O. F. Müller. Jour. Mar. Biol. Assoc., 29:661-680.
Wilson, D. P. 1948. The Relation of the Substratum to the Metamorphosis of *Ophelia* Larvae. Jour. Mar. Biol. Assoc., 27:723-760. See Hartman, 1951*b*, bibliography for other papers.
Wilson, E. B. 1892. The Cell-lineage of *Nereis*. Jour. Morph., 6:361-480.

Oligochaeta and Hirudinea

Altman, L. C. 1931. *Enchytraeus pugetensis* (n. sp.), a New Marine Enchytraeid from Puget Sound. Trans. Amer. Micr. Soc., 50:154-163.
———. 1936. Oligochaeta of Washington. Univ. Wash. Publ. Biol., 4:1-137.
Moore, J. P., and M. C. Meyer, 1951. Leeches (Hirudinea) from Alaskan and adjacent waters. Wasmann Jour. Biol., 9(1):11-77.

SIPUNCULOIDEA

Balzer, F. 1931. Sipunculida. *In* Kükenthal and Krumbach, *Handbuch der Zoologie*, Bd. 2: Halfte 2:14 (9): pp. 15-61.
Fisher, W. K. 1952. The Sipunculid Worms of California and Baja California. Proc. U. S. Nat. Mus., 102:371-450.
Peebles, F., and D. L. Fox. 1933. The Structure, Functions, and General Reactions of the Marine Sipunculid Worm *Dendrostoma zostericola*. Bull. Scripps Inst. Oceanog., tech. ser., 3:201-224.
Pickford, G. 1947. Sipunculida. *Encyclopedia Brittanica*.

ECHIUROIDEA

Balzer, F. 1931. Echiurida. *In* Kükenthal and Krumbach, Bd. 2:2:14(9): pp. 62-168.
Fisher, W. K. 1946. Echiuroid Worms of the North Pacific Ocean. Proc. U. S. Nat. Mus., 96:215-292.
Fisher, W. K., and G. E. MacGinitie, 1928*a*. A New Echiuroid Worm from California. Ann. Mag. Nat. Hist., ser. 10, 1:199-204.
———. 1928*b*. The Natural History of an Echiuroid Worm (*Urechis*). Ann. Mag. Nat. Hist., ser. 10, 1:204-214.

Pickford, G. E. 1947. Echiurida. *Encyclopedia Brittanica.*
Redfield, A. C., and M. Florkin. 1931. The Respiratory Function of the Blood of *Urechis caupo.* Biol. Bull., 61:185-210.

PRIAPULIDA

Balzer, F., 1931. Priapulida. *In* Kükenthal and Krumbach, Bd. 2:2:14(9): pp. 1-14.
Hyman, L. H. 1951. *The Invertebrates: Acanthocephala, Aschelminthes, and Entoprocta.* Vol. III., New York: McGraw-Hill. 572 pp. illus.
Lang, K., 1948*a*. Contribution to the Ecology of *Priapulis caudatus* Lam. Arkiv f. Zool., 41A (no. 5):1-12.
———. 1948*b*. On the Morphology of the Larva of *Priapulus caudatus* Lam. *Ibid.*, 41A (no. 9):1-8.

ARTHROPODA

General

Calman, W. T. 1909. Crustacea, *in* Lankester, *Treatise on Zoology,* Part VII, Third Fascicle. London: Adam and Charles Black. 346 pp.
(Ward and Whipple, *Fresh-water Biology,* contains excellent general accounts of fresh-water groups.)

Branchiopoda

Baker, H. M. 1938. Studies on the Cladocera of Monterey Bay. Proc. Calif. Acad. Sci., ser. 4: 23: 311-365.
Kuenen, D. J. 1939. Systematical and Physiological Notes on the Brine Shrimp *Artemia.* Arch. néerl. Zool., 3:365-449.
Linder, Folke. 1952. Contributions to the Morphology and Taxonomy of the Branchiopoda Notostraca, with Special Reference to the North American Species. Proc. U. S. Nat. Mus., 102:1-69.

Ostracoda

Hoff, C. C. 1942. The Ostracods of Illinois. Illinois Biol. Monographs, 19 (1, 2):1-196.

Juday, C. 1906. Ostracoda of the San Diego Region. I. Halocypridae. Univ. Calif. Publ. Zoöl., 3:13-38.

——— . 1907. II. Littoral forms. *Ibid.*, 3:135-157.

Müller, G. W. 1894. Die Ostracoden des Golfes von Neapel. Monogr. 21, Fauna und Flora des Golfes von Neapel.

Skogsberg T. 1928. Studies on marine ostracods. Part II. External Morphology of the Genus *Cythereis* with Descriptions of Twenty-one New Species. Occ. Papers Calif. Acad. Sci., no. 15:155 pp.

——— . 1950. Two New Species of Marine Ostracoda (Podocopa) from California. Proc. Calif. Acad. Sci., ser. 4, 25:483-505.

Tressler, W. L. 1940. Ostracoda from Beaufort, North Carolina, Sand Beaches. Am. Midl. Nat., 24:365-368.

——— . 1947. A Checklist of the Known Species of North American Fresh-water Ostracoda. Amer. Midl. Nat., 38:698-707.

Tressler, W. L., and E. M. Smith, 1948. An Ecological Study of Seasonal Distribution of Ostracoda, Solomons Island, Maryland, Region. State of Maryland, Bd. of Nat. Resources, Publ. No. 71, Chesapeake Biological Laboratory.

Copepoda

Esterly, C. O. 1924. The Free-swimming Copepoda of San Francisco Bay. Univ. Calif. Publ. Zoöl., 26:81-129.

Coker, R. E. 1934. Contribution to Knowledge of North American Harpacticoid Freshwater Copepod Crustacea. Jour. Elisha Mitchell Sci. Soc., 50:75-141.

Davis, C. C. 1949. The Pelagic Copepoda of the Northeastern Pacific Ocean. Univ. Wash. Publ. Biol., 14:1-118. (Key)

Fraser, J. H. 1936*a*. The Occurrence, Ecology, and Life History of *Tigriopus fulvus* (Fischer). Jour. Mar. Biol. Assoc., 20:523-536.

——— . 1936*b*. The Distribution of Rock Pool Copepods According to Tidal Level. Jour. Animal Ecol., 5: 23-38.

Illg, P. L. 1949. A Review of the Copepod Genus *Paranthessius* Claus. Proc. U. S. Nat. Mus., 99:391-428.

Johnson, M. W. 1939. *Pseudodiaptomus* (*Pseudodiaptallous*) *euryhalinus*, a New Subgenus and Species of Copepoda, with Preliminary Notes on its Ecology. Trans. Amer. Micr. Soc., 58:349-355.

Johnson, M. W. 1948. The Postembryonic Development of the Copepod, *Pseudodiaptomus euryhalinus* Johnson, and Its Phylogenetic Significance. *Ibid.*, 67:319-330.

Lang, Karl. 1948. *Monographie der Harpacticiden.* Stockholm: H. Ohlsson. 2 vols., 1682 pp.

Light, S. F. 1938. New Subgenera and Species of Diaptomid Copepods from the Inland Waters of California and Nevada. Univ. Calif. Publ. Zoöl., 43:67-78.

———. 1939. New American Subgenera of *Diaptomus* Westwood (Copepoda, Calanoida). Trans. Amer. Micr. Soc., 58:473-484.

Light, S. F., and O. Hartman, 1937. A Review of the Genera *Clausidium* Kossmann and *Hemicyclops* Boeck (Copepoda, Cyclopoida). Univ. Calif. Publ. Zoöl., 41:173-188.

Marsh, C. D. 1929. Distribution and Key of the North American Copepods of the Genus *Diaptomus*. Proc. U. S. Nat. Mus., 75:1-27.

———. 1933. Synopsis of the Calanoid Crustaceans, Exclusive of the the Diaptomidae, Found in Fresh and Brackish Waters, Chiefly of North America. *Ibid.*, 82:1-58.

Monk, C. R. 1941. Marine Harpacticoid Copepods from California. Trans. Amer. Micr. Soc., 60:75-99.

Wilson, C. B. 1932. The Copepods of the Woods Hole Region, Massachusetts. Bull. 158, U. S. Nat. Mus., 635 pp. (Includes both marine and fresh-water forms.)

———. 1935. Parasitic Copepoda from the Pacific Coast. Amer. Midl. Nat., 16:776-797.

———. 1944. Parasitic Copepods in the United States National Museum. Proc. U. S. Nat. Mus., 94:529-582. (Contains bibliography of earlier work.)

Yeatman, H. C. 1944. American Cyclopoid Copepods of the *viridis-vernalis* Group (including a description of *Cyclops carolinianus*, n. sp.). Amer. Midl. Nat., 32:1-90. (Excellent key.)

Branchiura

Meehean, O. L. 1940. A Review of the Parasitic Crustacea of the Genus *Argulus* in the Collections of the United States National Museum. Proc. U. S. Nat. Mus., 88:459-522.

Wilson, C. B. 1944. Parasitic Copepods in the United States National Museum. *Ibid.*, 94:529-582. (Includes key to species of *Argulus*.)

Cirripedia

Boschma, H. 1927. On the Larval Forms of Rhizocephala. Proc. Koninklijke Akademie Van Wetenschappen, Amsterdam, 30 (2) .

Cornwall, I. E. 1935. On the Nervous System of Four British Columbian Barnacles. Jour. Biol. Board Canada, 1:469-475.

———. 1937. A New Species of Barnacle from the Coast of California. Ann. Mag. Nat. Hist., ser. 10, 20:232-235.

———. 1951. The Barnacles of California (Cirripedia). Wasmann Jour. Biol., 9:311-346.

Darwin, C. *A Monograph of the sub-class Cirripedia. Part I, Lepadidae,* 1851; *Part II, Balanidae,* 1854. London: Ray Society.

Henry, D. P. 1940. Notes on some Pedunculate Barnacles from the North Pacific. Proc. U. S. Nat. Mus., 88 (308):225-236.

———. 1940*a*. The Cirripedia of Puget Sound with a Key to the Species. Univ. Wash. Publ. Oceanogr., 4 (1):1-48.

———. 1942. Studies on the Sessile Cirripedia of the Pacific Coast of North America. *Ibid.*, 4 (3):95-134.

Krüger, P. 1940. Cirripedia. *In* Bronn, *Klassen und Ordnungen des Tierreichs*, 5:1:3:3:pp. 1-560.

Pilsbry, H. A. 1916. The Sessile Barnacles (Cirripedia) Contained in the Collections of the United States National Museum; Including a Monograph of the American Species. Bull. 93, U. S. Nat. Mus. 366 pp.

Reinhard, E. G. 1944. Rhizocephalan Parasites of Hermit Crabs from the Northwest Pacific. Jour. Wash. Acad. Sci., 34:49-58.

Rice, L. 1930. Peculiarities in the Distribution of Barnacles in Communities and Their Probable Causes. Puget Sound Biol. Sta. Publ. 7:249-257.

Shelford, V. E. 1930. Geographic Extent and Succession in Pacific North American Intertidal (*Balanus*) Communities. Puget Sound Biol. Sta. Publ. 7:217-224.

Tomlinson, J. T. 1953. A Burrowing Barnacle of the Genus *Trypetesa* (Order Acrothoracica) Jour. Wash. Acad. Sci. (In press).

Towler, E. D. 1930. An Analysis of the Intertidal Barnacle Communities of the San Juan Archipelago. Puget Sound Biol. Sta. Publ. 7:225-232.

Worley, L. G. 1930. Correlation between Salinity, Size, and Abundance of Intertidal Barnacles. Puget Sound Biol. Sta. Publ. 7:233-240.

Leptostraca

Cannon, H. G. 1927. On the Feeding Mechanism of *Nebalia bipes*. Trans. Roy. Soc. Edinburgh, 55:355-369.

Clark, A. E. 1932. *Nebaliella caboti* n. sp. with Observations on Other Nebaliacea. Trans. Roy. Soc. Canada, ser. 3, 26:217-235, pls. 1-6. (Description of *Epinebalia pugettensis*.)

Stomatopoda (Hoplocarida)

Schmitt, W. L. 1940. The Stomatopods of the West Coast of America. Allan Hancock Pacific Exped., 5(4):129-225.

Mysidacea

Banner, A. H. 1947. A Taxonomic Study of the Mysidacea and Euphausiacea (Crustacea) of the Northeastern Pacific. Part I. Mysidacea, from Family Lophogastridae through Tribe Erythropini. Trans. Roy. Canad. Inst., 26:345-399; pls. 1-9. 1948. Part II. Mysidacea, from Tribe Mysini through Subfamily Mysidellinae. *Ibid.*, 27:65-125; pls. 1-7.

Tattersall, W. M. 1932. Contributions to a Knowledge of the Mysidacea of California. Univ. Calif. Publ. Zoöl., 37:301-347.

———. 1951. A review of the Mysidacea of the United States National Museum. Bull. 201, U. S. Nat. Mus., 1-292.

Tattersall, W. M., and Olive S. Tattersall. 1951. *The British Mysidacea*, London: Ray Society. 460 pp., illus.

Cumacea

Calman, W. T. 1907. On New or Rare Crustacea of the Order Cumacea from the Collection of the Copenhagen Museum. Part I. Trans. Zool. Soc. London, 18:1-58.

Zimmer, C. 1936. California Crustacea of the Order Cumacea. Proc. U. S. Nat. Mus., 83:423-439.

Isopoda and Chelifera

Arcangeli, A. 1932. Isopodi terrestri raccoltidal. Prof. Silvestri nel Nord-America. Boll. Lab. Zool. Gen. Agr., Portici, 26:121-141.

Gurjanova, E. 1933. Contributions to the Isopoda fauna of the Pacific Ocean. I. New species of Valvifera and Flabellifera. Explor. des Mers d'U.R.S.S., Institut Hydrobiologique, Leningrad, no. 17:87-106. II. New Species of Gnathiidea and Asellota. *Ibid.*, no. 19: 79-91. (Russian and English texts.)

Hansen, H. J. 1905. On the Propagation, Structure and Classification of the Family Sphaeromidae. Quart. Jour. Micr. Sci., 49(1): 69-135.

Hatch, M. H. 1947. The Chelifera and Isopoda of Washington and Adjacent Regions. Univ. Wash. Publ. Biol., 10: 155-274.

Holmes, S. J., and M. E. Gay. 1909. Four New Species of Isopods from the Coast of California. Proc. U. S. Nat. Mus., 36: 375-379.

Howard, A. D. 1952. Molluscan shells occupied by Tanaids. The Nautilus, 65(3): 75-76.

Hult, J. 1941. On the Soft-bottom Isopods of the Skager Rak. Zool. Bidrag från Upsala, 21: 3-223. (Excellent ecological-distributional treatment.)

Johnson, M. W. 1935. Seasonal Migrations of the Wood-borer *Limnoria lignorum* (Rathke) at Friday Harbor, Washington. Biol. Bull., 69: 427-438.

Koepcke, H. W. 1948. Über das Zeichnungsmuster einiger Idotea-Arten (Isopoda). Zool. Jahrb., 61(4): 413-450.

Maloney, J. O. 1933. Two New Species of Isopod Crustaceans from California. Jour. Wash. Acad. Sci., 23: 144-147.

Menzies, R. J. 1949. A New Species of Apseudid Crustacean of the Genus *Synapseudes* from Northern California (Tanaidacea). Proc. U. S. Nat. Mus., 99: 509-515.

———. 1950. The Taxonomy, Ecology, and Distribution of Northern California Isopods of the Genus *Idothea* with the Description of a New Species. Wasmann Jour. Biol., 8: 155-195.

———. 1950. Notes on California Isopods of the Genus *Armadilloniscus*, with the Description of *Armadilloniscus coronacapitalis* n. sp. Proc. Calif. Acad. Sci., ser. 4, 26: 467-481.

———. 1951. A New Species of *Limnoria* . . . Bull. So. Calif. Acad. Sci., 50: 86-88.

———. 1951. New Marine Isopods, chiefly from Northern California, with Notes on Related Forms. Proc. U. S. Nat. Mus., 101: 105-156.

———. 1952. Some Marine Asellote Isopods from Northern California, with Descriptions of Nine New Species. *Ibid.*, 102:117-159.

———. 1953. The Apseudid Chelifera of the Eastern Tropical and North Temperate Pacific Ocean. Bull. Mus. Comp. Zool., Harvard, 107(9): 443-496.

Menzies, R. J., and J. L. Barnard. 1951. The Isopodan Genus *Iais* (Crustacea). Bull. So. Calif. Acad. Sci., 50: 136-151.

Menzies, R. J., and R. J. Waidzunas. 1948. Post-embryonic Growth Changes in the Isopod *Pentidotea resecata* (Stimpson), with Remarks on their Taxonomic Significance. Biol. Bull., 95: 107-113.

Miller, M. A. 1936. California Isopods of the Genus *Porcellio* with Descriptions of a New Species and a New Subspecies. Univ. Calif. Publ. Zoöl., 41: 165-172.

———. 1938. Comparative Ecological Studies on the Terrestrial Isopod Crustacea of the San Francisco Bay Region. *Ibid.*, 43: 113-142.

Monod, Th. 1931. Tanaidacés et isopodes aquatiques de l'Afrique occidentale et septentrionale. 3(1), Sphaeromatidae. Mem. Soc. Nat. Maroc, No. 29: 7-91.

Peabody, E. B. 1939. Pigmentary Responses in the Isopod, *Idothea.* Jour. Exp. Zool., 82: 47-83.

Richardson, Harriet. 1905. Monograph on the Isopods of North America. Bull. 54, U. S. Nat. Mus. 727 pp.

———. 1909. Isopods Collected in the Northwest Pacific by the U. S. Bureau of Fisheries Steamer "Albatross" in 1906. Proc. U. S. Nat. Mus., 37: 75-129.

Suneson, S. 1947. Color Changes and Chromatophore Activators in *Idothea.* Lunds Univ. Arsskrift. N. F., Avd. 2, 43: 1-34.

Van Name, W. G. 1936. The American Land and Fresh-water Isopod Crustacea. Bull. Amer. Mus. Nat. Hist., 71: 1-535.

———. 1940. A Supplement to the American Land and Fresh-water Isopod Crustacea. *Ibid.*, 77: 109-142.

———. 1942. A Second Supplement to the American Land and Fresh-water Isopod Crustacea. *Ibid.*, 80: 299-329.

Amphipoda (Gammaridea)

Alderman, A. L. 1936. Some New and Little Known Amphipods of California. Univ. Calif. Publ. Zoöl., 41:53-72.

Barnard, J. L. 1950. The Occurrence of *Chelura terebrans* Philippi in Los Angeles and San Francisco Harbors. Bull. So. Calif. Acad. Sci., 49:90-97.

———. 1952. Some Amphipods from Central California. Wasmann Jour. Biol., 10:9-36. (This deals with the Morro Bay region of the coast, but is useful in the area covered in our key.)

Holmes, S. J. 1908. The Amphipoda Collected by the . . . "Albatross" off the West Coast of America. . . . . Proc. U. S. Nat. Mus., 35: 489-543.

Shoemaker, C. R. 1925. The Amphipoda Collected by the . . "Albatross" in 1911, Chiefly in the Gulf of California. Bull. Amer. Mus. Nat. Hist., 52:21-61.

———. 1941*a*. A New Genus and a New Species of Amphipoda from the Pacific Coast of North America. Proc. Biol. Soc. Wash., 54:183-186.

———. 1941*b*. On the Names of Certain California Amphipods. *Ibid.*, 54:187-188.

———. 1949. The Amphipod Genus *Corophium* on the West Coast of America. Jour. Wash. Acad. Sci., 39:66-82.

Skogsberg, T., and G. H. Vansell. 1928. Structure and Behavior of the Amphipod, *Polycheria osborni*. Proc. Calif. Acad. Sci., ser. 4, 17:267-295.

Stebbing, T. R. R. 1888. Report on the Amphipoda Collected by H.M.S. Challenger . . . Repts. Sci. Results Voyage Challenger, 1873-1876, Zoology, 29:i-xxiv, 1-1737; pl. I-CCX.

——— 1906. *The Amphipoda, I. Gammaridea.* In *Das Tierreich* (Prussian Academy of Sciences, Berlin; not to be confused with Bronn's *Tierreich*), vol. 21: 806 pp. (in English)

Thorsteinson, E. D. 1941. New or Noteworthy Amphipods from the North Pacific Coast. Univ. Wash. Publ. Oceanogr., 4:50-94.

Amphipoda (Caprellidea)

Boeck, A. 1872. Bidrag til Californiens Amphipodefauna. Forhandlinger i Videnskabs-Selskabet i Christiania aar 1871: 32-51.

Dougherty, E. C., and Joan E. Steinberg. 1953. Notes on the Skeleton Shrimps (Crustacea:Caprellidae) of California. Proc. Biol. Soc. Wash., 66:39-50.

La Follette, R. 1914. Caprellidae from Laguna Beach. Pomona College, Jour. Entom. and Zool. 6(4):222-232.

———. 1915. Caprellidae from Laguna Beach. II. *Ibid.*, 7(1):55-63.

Mayer, P. 1882. Die Caprelliden des Golfes von Neapel . . . Fauna and Flora des Golfes von Neapel. 6:i-x, 1-201.

———. 1890. Nachtrag zur Monographie derselben. *Ibid.*, 17:i-viii,1-157.

———. 1903. Die Caprellidae der Siboga-Expedition. Siboga Expeditie, 34:1-160.

Say, T. 1818. An Account of the Crustacea of the United States. Jour. Acad. Nat. Sci., Philadelphia, 1:374-401.

Stebbing, T. R. 1888 (see under Gammaridea)

Stimpson, W. 1857. The Crustacea and Echinodermata of the Pacific Shores of North America. Boston Jour. Nat. Hist., 6:444-532.

———. 1864. Descriptions of New Species of Marine Invertebrates from Puget Sound Collected by the Naturalists of the North-West Boundary Commission. . . . Proc. Acad. Nat. Sci., Philadelphia, 16:153-161.

Euphausiacea

Banner, A. H. 1949. A Taxonomic Study of the Mysidacea and Euphausiacea (Crustacea) of the Northeastern Pacific. Part III. Euphausiacea. Trans. Roy. Canad. Inst., 28:1-63.

Decapoda

Brown, F. A., Jr. 1944. Hormones in Crustacea: Their Sources and Activities. Quart. Rev. Biol., 19:32-46 and 118-143.

Chace, F. A., Jr. 1951. The Grass Shrimps of the genus *Hippolyte* from the West Coast of North America. Jour. Wash. Acad. Sci., 41:35-39.

Goodwin, D. G., 1952. Crustacea Collected during the 1950 Bottom-fish Investigations of the M. V. *N. B. Scofield*. Calif. Fish and Game, 38:163-181.

Gurney, Robert. 1939. *Bibliography of the Larvae of Decapod Crustacea.* London: Ray Society. 123 pp.

———. 1942. *Larvae of Decapod Crustacea.* London: Ray Society. 306 pp.

Hart, J. L. 1935. The Larval Development of British Columbia Brachyura. Canad. Jour. Research, 12:411-432.

———. 1937. Larval and Adult Stages of British Columbia Anomura. *Ibid.*, D, 15:179-220.

Hay, W. P. 1899. Synopses of North-American Invertebrates. VI. The Astacidae of North America. Amer. Nat., 33:957-966.

Hiatt, R. W. 1948. The Biology of the Lined Shore Crab, *Pachygrapsus crassipes* Randall. Pacific Science, 2:135-213.

Israel, H. R. 1936. A Contribution toward the Life Histories of Two California Shrimps, *Crago franciscorum* (Stimpson) and *Crago nigricauda* (Stimpson). Fish Bull. no. 46, Calif. Div. of Fish and Game.

Johnson, M. W. 1940. The Correlation of Water Movements and Dispersal of Pelagic Littoral Animals, especially the Sand Crab, *Emerita.* Jour. Marine Research, 2:236-245.

Johnson, M. W., F. A. Everest, and R. W. Young. 1947. The Role of Snapping Shrimp (*Crangon* and *Synalpheus*) in the Production of Underwater Noise in the Sea. Biol. Bull., 93:122-138.

Johnson, M. W., and W. M. Lewis, 1942. Pelagic Larval Stages of the Sand Crabs *Emerita analoga* (Stimpson), *Blepharipoda occidentalis* Randall, and *Lepidopa myops* Stimpson. Biol. Bull., 83:67-87.

Jones, L. L. 1941. Osmotic Regulation in Several Crabs of the Pacific Coast of North America. Jour. Cell. Comp. Physiol., 18: 79-92.

MacGinitie, G. E. 1930. The Natural History of the Mud-shrimp, *Upogebia pugettensis*. Ann. Mag. Nat. Hist., ser. 10, 6:36-44.

———. 1934. The Natural History of *Callianassa californiensis* Dana. Amer. Midl. Nat., 15: 166-177.

———. 1938. Movements and Mating Habits of the Sand Crab, *Emerita analoga*. *Ibid.*, 19: 471-481.

Menzies, R. J. 1948. A Revision of the Brachyuran Genus *Lophopanopeus*. Allan Hancock Found. Publ., Occ. Pap. No. 4:1-44.

Phillips, J. B. 1939. The Market Crab of California and Its Close Relatives. Calif. Fish and Game, 25: 18-29.

Rathbun, M. J. 1917. The Grapsoid Crabs of America. Bull. 97, U. S. Nat. Mus. 444 pp.

———. 1925. The Spider Crabs of America. Bull. 129, U. S. Nat. Mus. 613 pp.

———. 1930. The Cancroid Crabs of America of the Families Euryalidae, Portunidae, Atelecyclidae, Cancridae, and Xanthidae. Bull. 152, U. S. Nat. Mus. 609 pp.

———. 1937. The Oxystomatous and Allied Crabs of America. Bull. 166, U. S. Nat. Mus. 278 pp.

Schmitt, W. L. 1921. The Marine Decapod Crustacea of California. Univ. Calif. Publ. Zoöl., 23:1-470.

———. 1935. The West American Species of Shrimps of the Genus *Penaeus*. Proc. Biol. Soc. Wash., 48:15-24.

Snodgrass, R. E. 1952. The Sand Crab *Emerita talpoida* (Say) and Some of Its Relatives. Smithsonian Misc. Coll., 117(8):1-34.

Wells, W. W. 1940. Ecological Studies on the Pinnotherid Crabs of Puget Sound. Univ. Wash. Publ. Oceanogr. 2: 19-50.

Myriapods

Attems, C. G. *et al.* 1926. Progoneata and Chilopoda. *In* Kükenthal and Krumbach, *Handbuch der Zoologie*, Vol. 4: part 1.

Bollman, C. H. 1893. The Myriapoda of North America. Bull. 46, U. S. Nat. Mus. 210 pp. (Not a monograph, but the collected papers of Charles Bollman, who did remarkable work before his death at the age of 20.)

Buck, J. B., and M. L. Keister. 1950. *Spirobolus marginatus* (millipede). *In* F. A. Brown (ed.), *Selected Invertebrate Types*. New York: Wiley. pp. 462-475.

Chamberlin, R. V. 1910-1912. The Chilopoda of California. Part I. Pomona College Jour. Entom. and Zool., 2:363-374, Part II, *Ibid.*, 4:470-479.

———. 1912. New North American Chilopods and Diplopods. Ann. Entom. Soc. Amer., 5:141-176.

Hansen, H. J. 1903. The Genera and Species of the Order Symphyla. Quart. Jour. Micr. Sci. 47:1-101.

Insecta

Brues, C. T., and A. L. Melander. 1932. Classification of Insects. A Key to the Known Families of Insects and Other Terrestrial Arthropods. Bull. Mus. Comp. Zool., Harvard, 73:1-672.

Chu, H. F. 1949. *How to Know the Immature Insects*. Dubuque, Iowa: W. C. Brown. 234 pp.

Essig, E. O. 1926. *Insects of Western North America*. New York: Macmillan. 1035 pp.

Saunders, L. G., 1928. Some marine insects of the Pacific Coast of Canada. Ann. Ent. Soc. Amer., 21: 521-545.

Usinger, R. L., I. La Rivers, H. P. Chandler, W. W. Wirth. 1948. Biology of Aquatic and Littoral Insects (Entomology 133 Syllabus). University of California Syllabus Series. 244 pp. (mimeographed). Contains additional references.

Wirth, W. W. 1949. A Revision of the Clunionine midges. . . . Univ. Calif. Publ. Entom., 8:151-182.

Arachnida

Baker, E. W., and G. W. Wharton. 1952. *An Introduction to Acarology*. New York: Macmillan. 465 pp.

Buck, J. B., and M. L. Keister. 1950. *Argiope aurantia*. In *Selected Invertebrate Types*, F. A. Brown (ed.). New York: Wiley. pp. 382-394.

Chamberlin, J. C. 1921. Notes on the genus *Garypus* in North America (Pseudoscorpionida-Cheliferidae). Canad. Entomologist, 53:186-191.

———. 1931. The Arachnid Order Chelonethida. Stanford Univ. Publ. Biol. Sci., 7:1-284.

Comstock, J. H. 1940. *The Spider Book*. 2d ed., Ithaca, New York: Comstock. 729 pp.

Gertsch, W. J. 1949. *American Spiders*. New York: Van Nostrand. 285 pp.

Grassé, P. (ed.). 1949. *Traité de Zoologie, Tome* VI, *Onychophora, Tardigrades, Arthropodes (généralités), Trilobitomorphs, Chélicérates*. Paris: Masson. 979 pp.

Newell, I. M. 1947. A Systematic and Ecological Study of the Halacaridae of Eastern North America. Bull. Bingham Oceanogr. Coll.,10(3):1-232. (Methods of handling marine mites and keys to subfamilies and genera are included.)

———. 1949. New Genera and Species of Halacaridae (Acari). Amer. Museum Novitates, 1411:1-22

———. 1950. New Species of *Copidognathus* from the Aleutians. *Ibid.*, 1476:1-19.

———. 1951. New Species of *Agaue* and *Thalassarachna* from the Aleutians. *Ibid.*, 1489:1-19.

———. 1951. *Copidognathus curtus* Hall, 1912, and Other Species of *Copidognathus* from Western North America. *Ibid.*, 1499:1-27.

———. 1952. Further Studies on Alaskan Halacaridae. *Ibid.*, 1536:1-56.

Savory, T. H. 1935. *The Arachnida.* London: Arnold and Co. 218 pp.

Pycnogonida

Cole, Leon J. 1904. Pycnogonida of the West Coast of North America. Harriman Alaska Exped., 10:249-298.

Dohrn, Anton. 1881. Die Pantopoden des Golfes von Neapel und der angrenzenden Meeres Abschnitte. Fauna und Flora Golfes von Neapel, 3:1-252.

Hall, H. V. M. 1913. Pycnogonida from the Coast of California. Univ. Calif. Publ. Zoöl., 11:127-142.

Hedgpeth, Joel W. 1940. A New Pycnogonid from Pescadero, California, and Distributional Notes on Other Species. Jour. Wash. Acad. Sci., 30(2):84-87.

———. 1941. A Key to the Pycnogonida of the Pacific Coast of North America. Trans. San Diego Soc. Nat. Hist., 9(26):253-264.

———. 1947. On the Evolutionary Significance of the Pycnogonida. Smithsonian Misc. Coll., 106(18):1-53.

———. 1951. Pycnogonids from Dillon Beach and Vicinity, California, with Descriptions of Two New Species. Wasmann Jour. Biol., 9: 105-117.

Helfer, H., and E. Schlottke. 1935. Pantopoda. Bronn, *Klassen und Ordnungen der Tierreichs*, 5, Abt. IV, Buch 2, Lief. 1:1-160.

Hilton, W. A. 1916. The Life History of *Anoplodactylus erectus* Cole. Pomona College Jour. Entom. and Zool., 8(1):25-34.

MOLLUSCA

General

Burch, J. Q. (ed.). 1944-1946. Distributional List of the West American Marine Mollusks from San Diego, California, to the Polar Sea. Minutes Conch. Club So. Calif., nos. 45-63, mimeographed. (Issued separately as parts I and II, the latter with index.)

Coe, W. R. 1943. Sexual Differentiation in Mollusks. I. Pelecypods. Quart. Rev. Biol., 18:154-164. II. Gastropods, Amphineurans, Scaphopods, and Cephalopods. *Ibid.*, 19:85-97.

Dall, W. H. 1907. Instructions for Collecting Mollusks, and Other Useful Hints for the Conchologist. Part G, Bull. 39, U. S. Nat. Mus., pp. (2d ed.).

———. 1921. Summary of the Marine Shell-bearing Mollusks of the Northwest Coast of America. . . . Bull. 112, U. S. Nat. Mus. 217 pp., pls. 1-22.

Grant, U. S., IV, and H. R. Gale. 1931. Catalogue of the Marine Pliocene and Pleistocene Mollusca of California. . . . Mem. San Diego Soc. Nat. Hist., 1:1-1036. (Paleontological but useful for taxonomy.)

Keen, A. Myra. 1937. *An Abridged Checklist and Bibliography of West American Marine Mollusca.* Stanford University Press. 87 pp.

Keep, Josiah, rev. by J. L. Baily, Jr. 1935. *West Coast Shells.* Stanford University Press. 350 pp. (A charmingly written little book, useful especially to the beginner and the general student. Illustrations and nomenclature a little old-fashioned.)

Morris, P. A., 1952. *Field Guide to Shells of the Pacific Coast and Hawaii.* Boston: Houghton Mifflin. 220 pp.

Newell, I. M. 1948. Marine Molluscan Provinces of Western North America: a Critique and a New Analysis. Proc. Amer. Phil. Soc., 92: 155-166.

Oldroyd, Ida S. 1924-27. The Marine Shells of the West Coast of North America. I. Pelecypoda. Stanford Univ. Publ. Geol. Sci., 1:1-247, 57 pl. II. Gastropoda and Amphineura. *Ibid.*, 2: Parts 1-3, pp. 1-941, 108 pls.

Packard, E. L. 1918. Molluscan Fauna from San Francisco Bay. Univ. Calif. Publ. Zoöl., 14:199-452.

Pelseneer, P. 1906. Mollusca. Part V of Lankester, *Treatise on Zoology*, London: Adam and Charles Black. 355 pp. (Old, but still the best available single textbook on molluscs.)

Schenck, H. G., and A. M. Keen. 1936. Marine Molluscan Provinces of Western North America. Proc. Amer. Phil. Soc., 76:921-938.

———. 1937. An Index-method for Comparing Molluscan Faunules. *Ibid.*, 77:161-182.

Smith, A. G., and M. Gordon, Jr. 1948. The Marine Mollusks and Brachio-Pods of Monterey Bay, California, and Vicinity. Proc. Calif. Acad. Sci., ser. 4, 26:147-245.

Thiele, J. 1929-1935. *Handbuch der systematische Weichtierkunde.* Parts I-IV, 1154 pp. Jena: Gustav Fischer. (Classification of Mollusca down to genera, but without keys.)

Tryon, G. W., and H. A. Pilsbry. 1879-1898. *Manual of Conchology, Structural and Systematic.* 17 volumes, Acad. Nat. Sci. Philadelphia.

Yonge, C. M. 1947. The Pallial Organs in the Aspidobranch Gastropoda and their Evolution throughout the Mollusca. Phil. Trans. Roy. Soc., Lond., B, 232: 443-518.

Scaphopoda

Yonge, C. M. 1937. Circulation of Water in the Mantle Cavity of *Dentalium entalis*. Proc. Malacol. Soc., 22: 333-337.

Cephalopoda

Berry, S. S. 1912. A Review of the Cephalopods of Western North America. Bull. U. S. Bur. Fisheries, 30(1910): 269-336, pls. 32-56.

Fisher, W. K. 1923. Brooding Habits of a Cephalopod. Ann. Mag. Nat. Hist., ser. 9, 12: 147-149.

———. 1925. On the Habits of an Octopus. *Ibid.*, 15: 411-414.

Fox, D. L. 1938. An Illustrated Note on the Mating and Egg-brooding Habits of the Two-spotted Octopus. Trans. San Diego Soc. Nat. Hist., 7: 31-34.

Pickford, G. E., and B. H. McConnaughey, 1949. The *Octopus bimaculatus* Problem: A Study in Sibling Species. Bull. Bingham Oceanogr. Coll., 12, art. 4: 1-66.

Robson, G. C. 1929. *A Monograph of the Recent Cephalopoda, Part I,* 236 pp. 1932, *Part II,* 359 pp. British Museum (Nat. Hist.)

Amphineura

Berry, S. S. 1946. A Re-examination of the Chiton *Stenoplax magdalenensis* (Hinds), with Description of a New Species [*Ischnochiton heathiana* (Berry)]. Proc. Malacol. Soc., 26: 161-166.

Berry, S. S. 1948. Two Misunderstood West American Chitons [*Cyanoplax dentiens* and *Lepidochitona keepiana*]. Leaflets in Malacology, 1(4): 13-15.

Fretter, V. 1937. The Structure and Function of the Alimentary Canal of Some Species of Polyplacophora (Mollusca). Trans. Roy. Soc. Edinburgh, 59: 119-164.

Heath, H. 1899. *The Development of Ischnochiton* [*heathiana*]. Jena; Gustav Fischer. 90 pp.

———. 1905*a*. The Breeding Habits of Chitons of the Californian Coast. Zool. Anz., 29: 390-393.

———. 1905*b*. The Excretory and Circulatory Systems of *Cryptochiton stelleri* Midd. Biol. Bull., 9: 213-225.

Pilsbry, H. A. *See* Tryon, G. W., and H. A. Pilsbry, *Manual of Conchology*, volumes 14 (*Polyplacophora*) 1892, and 15 (*Polyplacophora and Tectibranchiata)* 1893, Acad. Nat. Sci., Philadelphia.

Yonge, C. M. 1939. On the Mantle Cavity and Its Contained Organs in the Loricata (Placophora). Quart. Jour. Micr. Sci., 81: 367-390.

Lamellibranchia

Bonnot, P. 1940. The Edible Bivalves of California. Calif. Fish and Game, 26:212-239.

Coe, W. R. 1943. Sexual Differentiation in Mollusks. I. Pelecypods. Quart. Rev. Biol., 18:154-164.

———. 1947. Nutrition, Growth and Sexuality of the Pismo Clam (*Tivela stultorum*). Jour. Exp. Zool., 104:1-24.

Coe, W. R., and D. L. Fox. 1942. Biology of the California Sea-mussel (*Mytilus californianus*). I. Influence of Temperature, Food Supply, Sex and Age on the Rate of Growth. Jour. Exp. Zool., 90:1-30.

———. 1944. III. Environmental Conditions and Rate of Growth. Biol. Bull., 87:59-72.

Fitch, J. E. 1953. Common Marine Bivalves of California. Calif. Dept. Fish and Game, Fish Bull. no. 90, 102 pp. illus.

Fox, D. L. (ed.). 1936. The Habitat and Food of the California Sea Mussel. Bull. Scripps Inst. Oceanogr., tech. ser., 4:1-64. (A group of eight papers by various co-authors.)

Fox, D. L., and W. R. Coe. 1943. II. Nutrition, Metabolism, Growth and Calcium Deposition. Jour. Exp. Zool., 93:205-249.

Fox, D. L., H. U. Sverdrup, and J. P. Cunningham. 1937. The Rate of Water Propulsion by the California Mussel. Biol. Bull., 72:417-438.

Fraser, C. M., and G. M. Smith. 1928. Notes on the Ecology of the Little Neck Clam, *Paphia staminea* Conrad. Trans. Roy. Soc. Canada, ser. 3, 22:249-271.

Fraser, C. M., and G. M. Smith. 1928. Notes on the Ecology of the Butter Clam, *Saxidomus giganteus* Deshayes. *Ibid.*, 22:271-286.

Hewatt, W. G. 1935. Ecological Succession in the *Mytilus californianus* Habitat as Observed in Monterey Bay, California. Ecology, 16: 244-251.

Hill, C. L., and C. A. Kofoid, (eds.). 1927. *Marine Borers and Their Relation to Marine Construction on the Pacific Coast.* San Francisco Bay Marine Piling Committee. 357 pp.

Hunter, W. R. 1949. The Structure and Behaviour of *Hiatella gallicana* (Lamarck) and *H. arctica* (L.), with Special Reference to the Boring Habit. Proc. Roy. Soc. Edinburgh, B, 63:271-289.

Ingram, W. M. 1948. The Larger Freshwater Clams of California, Oregon and Washington. Jour. Entom. and Zool., 40:72-92.

Keen, A. M., and D. L. Frizzell, 1939. *Illustrated Key to West North American Pelecypod Genera.* Stanford University Press, 28 pp.

MacGinitie, G. E. 1941. On the Method of Feeding of Four Pelecypods. Biol. Bull., 80:18-25.

Miller, R. C. 1924. The Boring Mechanism of *Teredo.* Univ. Calif. Publ. Zoöl., 26:41-80.

Newell, I. M. 1948. Marine Molluscan Provinces of western North America; a Critique and a New Analysis. Proc. Amer. Phil. Soc., 92:155-166.

Oldroyd, Ida S. 1924. The Marine Shells of the West Coast of North America. I. Pelecypoda. Stanford Univ. Publ. Geol. Sci., 1:1-247, 57 pl.

Schenck, H. G., and A. M. Keen. 1936. Marine Molluscan Provinces of Western North America. Proc. Amer. Phil. Soc., 76: 921-938.

———. 1937. An Index-Method for Comparing Molluscan Faunules. *Ibid.*, 77:161-182.

Sommer, H., and K. F. Meyer. 1935. Mussel Poisoning. Calif. and Western Med., 42:1-11.

Yonge, C. M. 1923. The Mechanism of Feeding, Digestion, and Assimilation in the Lamellibranch *Mya.* Jour. Exp. Biol., 1:15-63.

———. 1926. Structure and Physiology of the Organs of Feeding and Digestion in *Ostrea edulis*. Jour. Mar. Biol. Assoc., 14:295-386.

———. 1928. Structure and Function of the Organs of Feeding and Digestion in the Septibranchs, *Cuspidaria* and *Poromya*, Phil. Trans. Roy. Soc. London, B, 216:221-263.

———. 1939. The Protobranchiate Mollusca; a Functional Interpretation of Their Structure and Evolution. *Ibid.*, 230:79-147.

———. 1948. Formation of Siphons in Lamellibranchia. Nature, 161: 198-199.

———. 1949. On the Structure and Adaptations of the *Tellinacea*, Deposit-feeding Eulamellibranchia. Phil. Trans. Roy. Soc. London, B, 234:29-76.

Yonge, C. M. 1951. Marine Boring Organisms. Research, 4:162-167.
——. 1951. Studies on Pacific Coast Mollusks. I. On the Structure and Adaptations of *Cryptomya californica* (Conrad). Univ. Calif. Publ. Zoöl., 55:395-400.
——. 1951.——. II. Structure and Adaptations for Rock Boring in *Platyodon cancellatus* (Conrad). *Ibid.*, 55:401-407.
——. 1951.——. III. Observations on *Hinnites multirugosus* (Gale). *Ibid.*, 55:409-420.
——. 1952.——. IV. Observations on *Siliqua patula* Dixon and on Evolution within the Solenidae. *Ibid.*, 55:421-438.
——. 1952.——. V. Structure and Adaptation in *Entodesma saxicola* (Baird) and *Mytilimeria nuttallii* Conrad, with a Discussion on Evolution within the Family Lyonsiidae (Eulamellibranchia). *Ibid.*, 55:439-450.
——. 1952.——. VI. A Note on *Kellia laperousii* (Deshayes). *Ibid.*, 55:451-454.

Prosobranchia

Bullock, T. H. 1953. Predator Recognition and Escape Responses of Some Intertidal Gastropods in Presence of Starfish. Behaviour, 5:130-141.
Coe, W. R. 1936. Conditions Influencing Change of Sex in Mollusks of the Genus *Crepidula*. Jour. Exp. Zool., 77:401-424.
Crofts, D. R. 1929. *Haliotis*. Liverpool Mar. Biol. Comm. Mem. No. 29: 1-174.
Dall, W. H., and P. Bartsch. 1909. Monograph of West American Pyramidellid Mollusks. U. S. Nat. Mus. Bull. 68:1-258. 30 pl.
Demond, J. 1952. The Nassariidae of the West Coast of North America between Cape San Lucas, Lower California, and Cape Flattery, Washington. Pacific Science, 6(4):300-317.
Dimon, A. C. 1905. The Mud Snail, *Nassa Obsoleta*. Cold Spring Harbor Monographs, No. 5:3-48.
Fisher, W. K. 1904. The Anatomy of *Lottia gigantea* Gray. Zool. Jahrb., Abt. f. Anat. u. Ontog. d. Thiere, 20:1-66.
Fretter, V., and A. Graham. 1949. The Structure and Mode of Life of the Pyramidellidae, Parasitic Opisthobranchs. Jour. Mar. Biol. Assoc., 28:493-532.
Gowanloch, J. N., and F. R. Hayes. 1926. The Physical Factors, Behaviour and Intertidal Life of *Littorina*. Contr. Canad. Biol. Fish., 3:135-165.

Keen, A. Myra, and J. C. Pearson. 1952. *Illustrated Key to West North American Gastropod Genera*. Stanford University Press. 39 pp.

Moore, H. B. 1936. The Biology of *Purpura lapillus*. I. Shell Variation in Relation to Environment. Jour. Mar. Biol. Assoc., 21:61-89.

———. 1938*a*. II. Growth. *Ibid.*, 23:57-66.

———. 1938*b*. III. Life History and Relations to Environmental Factors. *Ibid.*, 23:67-74.

———. 1937. The Biology of *Littorina litorea*. Part I. Growth of the Shell and Tissues, Spawning, Length of Life and Mortality. *Ibid.*, 21:721-742.

———. 1940. Part II. Zonation in Relation to Other Gastropods on Stony and Muddy Shores. *Ibid.*, 24:227-237.

Orton, J. H. 1912. The Mode of Feeding of *Crepidula* . . . and Some Remarks on the Mode of Feeding in Gastropods and Lamellibranchs. Jour. Mar. Biol. Assoc., 9:444-478.

Strong, A. M. 1928. West American Mollusca of the Genus *Phasianella*. Proc. Calif. Acad. Sci., Ser. 4, 17:187-203.

———. 1930. Notes on Some Species of *Epitonium*, Subgenus *Nitidiscala*, from the West Coast of North America. Trans. San Diego Soc. Nat. Hist., 6:185-196.

Yonge, C. M. 1947. The Pallial Organs in the Aspidobranch Gastropoda and Their Evolution throughout the Mollusca. Phil. Trans. Roy. Soc. Lond., B. 232:443-518.

Opisthobranchia

Bergh, R. 1879-1880. On the Nudibranchiate Gastropod Molluscs of the North Pacific Ocean with Special Reference to Those of Alaska. Proc. Acad. Nat. Sci. Philadelphia, Pt. I, 31:71-132; Pt. II, 32: 40-127.

———. 1894. Die Opisthobranchien. *From* Reports on the Dredging Operations off the West Coast of Mexico, and in the Gulf of California in charge of Alexander Agassiz, carried on by the U. S. Fish Commission Steamer "Albatross" during 1891. Bull. Mus. Comp. Zool., Harvard, 25:125-233.

Cockerell, T. D. A. 1915. The Nudibranch Genus *Triopha* in California. Pomona College. Jour. Entom. and Zool. 8:228-229.

Cockerell, T. D. A., and Sir C. Eliot. 1905. Notes on a Collection of Californian Nudibranchs. Jour. of Malacol., 12:31-53.

Costello, D. P. 1938. Notes on the Breeding Habits of the Nudibranchs of Monterey Bay and Vicinity. Jour. Morph., 63:319-343.

MacFarland, F. M. 1906. Opisthobranchiate Mollusca from Monterey Bay, California and Vicinity. Bull. U.S. Bureau of Fisheries, 25:109-157.

MacFarland, F. M. 1912. The Nudibranch Family Dironidae. Zool. Jahrb., Suppl. 15:515-536.

———. 1923. The Morphology of the Nudibranch Genus *Hancockia*. Jour. Morph., 38:65-104.

———. 1923-1925. Opisthobranchiate Mollusca. Proc. Calif. Acad. Sci., 13:389-413.

———. 1926. The Acanthodoridae of the California Coast. Nautilus, 39:49-65, 94-103.

———. 1929. *Drepania*, a Genus of Nudibranchiate Mollusks New to California. Proc. Calif. Acad. Sci., 18:485-496.

MacFarland, F. M., and C. H. O'Donoghue. 1929. A New Species of *Corambe* from the Pacific Coast of North America. Proc. Cal. Acad. Sci., 18:1-21.

MacGinitie, G. E. 1934. The Egg-laying Activities of the Sea Hare *Tethys californicus* (Cooper). Biol. Bull., 67:300-303.

O'Donoghue, C. H., 1921. Nudibranchiate Mollusca from the Vancouver Island Region. Trans. Roy. Canad. Inst., 13:149-209.

———. 1922. Notes on the Nudibranchiate Mollusca from the Vancouver Island Region. Pts. I, II, III, *Ibid.*, 14:123-168.

———. 1924. Pt. IV, *Ibid.*, 15:1-34.

———. 1927. Pt. V, *Ibid.*, 16:1-12.

———. 1926. A List of the Nudibranchiate Mollusca Recorded from the Pacific Coast of North America with Notes on their Distribution. *Ibid.*, 15:199-247. (Bibliography for Pacific North American forms fairly complete to 1926.)

———. 1927. Notes on a collection of Nudibranchs from Laguna Beach, California. Pomona College Jour. Entom. and Zool. 19:77-117.

Oldroyd, I. S. 1927. The Marine Shells of the West Coast of North America. Stanford Univ. Publ., Geol. Sci., 2(Pt. 1):23-52.

Pulmonata

Ingram, W. M., and Carol Lotz. 1949. Land Molluscs of the San Francisco Bay Counties. Jour. Ent. and Zool., 41(2): 3-22; 1949, 41(3): 39-48; 1950, 42(1): 5-27; 1950, 42(2):20-39.

Mead, A. R. 1943. Revision of the Giant West Coast Land Slugs of the Genus *Ariolimax* Mörch (Pulmonata: Arionidae). Amer. Midl. Nat., 30:675-717.

BRYOZOA AND ENTOPROCTA

Bassler, R. S. 1922. The Bryozoa or Moss Animals. Ann. Rept. Smithsonian Inst., 1920: 339-380, pl. 1-4.

Canu, F., and R. S. Bassler. 1923. North American Later Tertiary and Quaternary Bryozoa. Bull. 129, U. S. Nat. Mus., pp. 1-302.

Hincks, T. 1880. *A History of the British Marine Polyzoa.* 2 vols., London: Ray Society.

O'Donoghue, C. H. 1926. A Second list of Bryozoa (Polyzoa) from the Vancouver Island Region. Contr. Canad. Biol. and Fish., new ser., 3:47-132.

O'Donoghue, C. H., and E. O'Donoghue, 1925. List of Bryozoa from the Vicinity of Puget Sound. Publ. Puget Sound Biol. Sta., 5:91-108.

Osburn, Raymond C. 1950. Bryozoa of the Pacific Coast of America. Part 1, Cheilostomata-Anasca. Allan Hancock Pacific Exped., 14:1-269.

———. 1952. Part 2, Cheilostomata-Ascophora. *Ibid.*, 14:271-611.

———. 1953. Part 3, Cyclostomata, Ctenostomata, Entoprocta, and Addenda. *Ibid.*, 14: 613-841.

Robertson, Alice. 1900. Studies in Pacific Coast Entoprocta. Proc. Calif. Acad. Sci., Ser. 3, Zool., 2:323-348.

———. 1905. Non-encrusting Chilostomatous Bryozoa of the West Coast of North America. Univ. Calif. Publ. Zoöl., 2:235-322.

———. 1908. The Encrusting Chilostomatous Bryozoa of the West Coast of North America. *Ibid.*, 4:253-344.

———. 1910. The Cyclostomatous Bryozoa of the West Coast of North America. *Ibid.*, 6:225-284.

Swazey, L. H. 1933. Ordinal Classification of the Bryozoa. Micropaleo. Bull., 4: 23-43.

PHORONIDEA

Cori, C. J. 1939. Phoronidea, *in* Bronn, *Klassen und Ordnung des Tierreichs*, 4:4:1:1.

Rattenbury, J. C. 1951. Studies of Embryonic and Larval Development in California Phoronidea. Unpublished Ph. D. dissertation, University of Calif., Berkeley.

———. 1953. Reproduction in *Phoronopsis viridis*. The Annual Cycle in the Gonads, Maturation and Fertilization of the Ovum. Biol. Bull., 104: 182-196.

BRACHIOPODA

Hertlein, L. G., and U. S. Grant, IV. 1944. The Cenozoic Brachiopoda of Western North America. U.C.L.A. Publ. in Math. and Phys. Sci., 3: 1-236.

Oldroyd, I. S. 1924. The Marine Shells of the West Coast of North America. Stanford Univ. Publ. Geol. Sci., 1. 221-232.

ECHINODERMATA

General

Bather, F. A., J. W. Gregory, and E. S. Goodrich. 1900. The Echinoderma. Part III of Lankester, *Treatise on Zoology.* London: Adam and Charles Black. 344 pp. (In this older work the emphasis is rather largely on skeletal parts.)

Bush, Mildred. 1918. Key to the Echinoderms of Friday Harbor, Washington. Publ. Puget Sound Biol. Sta., 2:17-44. (This includes descriptions.)

———. 1921. Revised Key to the Echinoderms of Friday Harbor. *Ibid.*, 3:65-77. (This brings the previous key up to date, but should be used in conjunction with the 1918 paper, since descriptions are not repeated unless altered.)

Fell, H. B. 1948. Echinoderm Embryology and the Origin of Chordates. Biol. Rev., 23:81-107.

Grassé, P. (ed.). 1948. *Traité de Zoologie. Anatomie, Systematique, Biologie. Tome 11: Échinodermes, Stomocordés, Procordés.* Paris: Masson. 1077 pp. (This is the only recent general work on echinoderms, and is especially good for comparative morphology and development.)

Kindred, J. E. 1924. The Cellular Elements in the Perivisceral Fluid of Echinoderms. Biol. Bull., 46:228-251.

Mortensen, T. 1921. *Studies of the Development and Larval Forms of Echinoderms.* Copenhagen: G. E. C. Gad. 261 pp., 33 pl.

Smith, J. E. 1937. The Structure and Function of the Tube-feet in Certain Echinoderms. Jour. Mar. Biol. Assoc., 22:345-357.

Asteroidea

Agersborg, H. P. K. 1918. Bilateral Tendencies and Habits in the Twenty-rayed Starfish *Pycnopodia helianthoides* (Stimpson). Biol. Bull., 35:232-255.

Bullock, T. H. 1953. Predator Recognition and Escape Responses of Some Intertidal Gastropods in Presence of Starfish. Behaviour, 5: 130-141.

Chadwick, H. C. 1923, *Asterias*. Liverpool Mar. Biol. Comm., Mem. on Typical British Marine Plants and Animals, no. 25.

Fisher, W. K. Asteroidea of the North Pacific and Adjacent Waters. Bull. 76, U. S. Nat. Mus. Part 1, Phanerozonia and Spinulosa, 419 pp., 122 pls., 1911. Part 2. Forcipulata (part), 245 pp., 81 pls., 1928. Part 3. Forcipulata (concluded), 356 pp., 93 pls., 1930.

Jennings, H. S. 1907. Behavior of the Starfish *Asterias forreri* de Loriol. Univ. Calif. Publ. Zoöl., 4:53-185.

Moore, A. R. 1939. Injury, Recovery and Function in an Aganglionic Nervous System. Jour. Comp. Psychol., 28:313-333.

———. 1941. Dysfunction in Righting and Locomotion in a Starfish (*Patiria*) with Supernumerary Rays. *Ibid.*, 32:483-487.

Smith, J. E. 1937. On the Nervous System of the Starfish *Marthasterias glacialis* (L.). Phil. Trans. Roy. Soc. London, B, 227:111-173.

———. 1944. The Role of the Nervous System in Some Activities of Starfishes. Biol. Rev., 20:29-43.

———. 1946. The Mechanics and Innervation of the Starfish Tube Foot—Ampulla System. Phil. Trans. Roy. Soc. London, B, 232: 279-310.

———. 1947. The Activities of the Tube Feet of *Asterias rubens* L. I. The Mechancis of Movement and of Posture. Quart. Jour. Micr. Sci., 88:1-14

Osterud, H. L. 1918. Preliminary Observations on the Development of *Leptasterias hexactis*. Publ. Puget Sound Biol. Sta., 2:1-15.

Verrill, A. E. 1914. Monograph of the Shallow-water Starfishes of the Northern Pacific Coast from the Arctic Ocean to California. Smithsonian Inst., Harriman Alaska Series, 14: Part 1, text, 408 pp.; Part 2, plates 1-110. (Much of nomenclature obsolete.)

Ophiuroidea

Clark, H. L. 1911. North Pacific Ophiurans in the Collection of the U. S. National Museum. Bull. 75, U. S. Nat. Mus. 302 pp.

McClendon, J. F. 1909. The Ophiurans of the San Diego Region. Univ. Calif. Publ. Zoöl., 6: 33-64.

MacGinitie, G. E. 1949. The Feeding of Ophiurans. Jour. Entom. and Zool., 41: 2-4.

May, R. M. 1924. The Ophiurans of Monterey Bay. Proc. Calif. Acad. Sci., ser 4, 13: 261-303.

Echinoidea

Clark, H. L. 1925. *A Catalog of the Recent Sea Urchins (Echinoidea) in the Collection of the British Museum (Natural History)*. 250 pp., 12 pl.

Johnson, M. W. 1930. Notes on the Larval Development of *Strongylocentrotus franciscanus*. Publ. Puget Sound Biol. Sta., 7:401-411.

Moore, H. B. 1934. A Comparison of the Biology of *Echinus esculentus* in different habitats. Part I. Jour. Mar. Biol. Assoc., 19: 869-886. Part II. *Ibid.*, 20: 109-128, 1935. Part III. *Ibid.*, 21: 711-720, 1937.

Mortensen, Th. 1928-1951. *A Monograph of the Echinoidea*. 5 vols. Copenhagen: C. A. Reitzel; London: H. Milford.

Parker, G. H., and M. A. Van Alstyne. 1932. Locomotor Organs of *Echinarachnius parma*. Biol. Bull., 62: 195-200. (On an eastern species comparable to *Dendraster.*)

Weese, A. O. 1926. Food and Digestive Processes of *Strongylocentrotus dröbachiensis*. Publ. Puget Sound Biol. Sta., 5: 165-179.

Holothuroidea

Clark, H. L. 1901. The Holothurians of the Pacific Coast of North America. Zool. Anzeiger, 24: 162-171.

———. 1924. Some holothurians from British Columbia. Canad. Field Nat., 38: 54-57.

Cowles, R. P. 1907. *Cucumaria curata, sp. nov.* Johns Hopkins Univ. Circular no. 195: 1-2, pl. 2 and 4.

Dawbin, W. H. 1949. Auto-evisceration and the Regeneration of Viscera in the Holothurian *Stichopus mollis* (Hutton). Trans. Roy. Soc. New Zealand 77: 497-523.

Deichman, E. 1938. New Holothurians from the Western Coast of North America, and Some Remarks on the Genus *Caudina*. Proc. New England Zool. Club, 16: 103-115.

———. 1941. The Holothurioidea Collected by the *Velero III* during the Years 1932-1938. Part I. Dendrochirota. Allan Hancock Pacific Exped., 8(3):61-195.

Filice, F. P. 1950. A study of some variations in *Cucumaria curata* (Holothuroidea). Wasmann Jour. Biol., 8: 39-48.

Heding, S. G. 1928. Synaptidae. Papers from Dr. Th. Mortensen's Pacific Expedition 1914-16, no. 46. Vidensk. Medd. Dansk Naturhist. Foren., Copenhagen, 85: 105-323.

Kille, F. R. 1935. Regeneration in *Thyone briareus* Lesueur Following Induced Autotomy. Biol. Bull., 69: 82-108.

———. 1939. Regeneration of Gonad Tubules Following Extirpation in the Sea-cucumber, *Thyone briareus* (Lesueur). *Ibid.*, 76: 70-79.

Wells, W. M. 1924. New Species of *Cucumaria* from Monterey Bay, California. Ann. Mag. Nat. Hist., ser. 9, 14: 113-121.

HEMICHORDATA

Bullock, T. H. 1945. The Anatomical Organization of the Nervous System of Enteropneusta. Quart. Jour. Micr. Sci., 86: 55-111, pl. 2-8. (Excellent anatomical figure in pl. 2.)

Davis, B. M. 1908. The Early Life-history of *Dolichoglossus pusillus* Ritter. Univ. Calif. Publ. Zoöl., 4: 187-226.

Dawydoff, C. 1948. Stomocordés. In *Traité de Zoologie*, P. Grassé, ed. Paris: Masson. pp. 367-532.

Horst, C. J. van der. 1934-1939. Hemichordata. *In* Bronn, *Klassen und Ordnungen des Tierreichs*, 4:4:2:2: pp. 1-737.

Ritter, W. E., and B. M. Davis. 1904. Studies on the Ecology, Morphology, and Speciology of the Young of Some Enteropneusta of Western North America. Univ. Calif. Publ. Zoöl., 1: 171-210.

CHORDATA

Urochordata

Abbott, D. P. 1953. Asexual Reproduction in the Colonial Ascidian *Metandrocarpa taylori* Huntsman. Univ. Calif. Publ. Zoöl., 61:1-78.

Berrill, N. J. 1950. *The Tunicata. With an Account of the British Species.* London: Ray Society, 354 pages, 120 figs.

———. 1951. Regeneration and Budding in Tunicates. Biol. Rev., 26: 456-475.

Berrill, N. J., and D. P. Abbott. 1949. The Structure of the Ascidian, *Pycnoclavella stanleyi* n. sp., and the Nature of Its Tadpole Larva. Canad. Jour. Res., D, 27: 43-49.

Brien, P. 1948. Embranchement des tuniciers. Morphologie et reproduction, In *Traité de Zoologie*. Tome 11: *Echinodérmes, Stomocordés, Procordés*, Grassé, P., (ed.). Paris: Masson. 1077 pages.

Hopkinson, J. 1913. *A Bibliography of the Tunicata 1469-1910*. London: Ray Society. 288 pages.

Huntsman, A. G. 1912. Holosomatous Ascidians from the Coast of Western Canada. Contrib. Canadian Biol. 1906-1910: 103-185, pls. 10-21.

MacGinitie, G. E. 1939. The Method of Feeding in Tunicates. Biol. Bull. 77: 443-447.

Ritter, W. E. 1900. Some Ascidians from Puget Sound, Collections of 1896. Ann. New York Acad. Sci., 12: 589-616, pls. 18-20.

———. 1913. The Simple Ascidians from the Northeastern Pacific in the Collection of the United States National Museum. Proc. U. S. Nat. Mus., 45: 427-505, pls. 33-36.

Ritter, W. E., and R. A. Forsyth. 1917. Ascidians from the Littoral Zone of Southern California. Univ. Cali. Publ. Zoöl., 16: 439-512, pls. 38-46.

Thompson, H. 1948. *Pelagic Tunicates of Australia*. Melbourne, Australia: Commonwealth Council for Scientific and Industrial Research. 196 pages, pls. 1-75, figs. 1-19.

Van Name, W. G. 1945. The North and South American Ascidians. Bull. Amer. Mus. Nat. Hist., 84: 1-476, pls. 1-31, figs. 1-327.

See also Skogsberg and Vansell, 1928 (Amphipod references) and the appropriate sections of H. G. Bronn, *Klassen und Ordnungen des Tierreichs*, and Kükenthal and Krumbach, *Handbuch der Zoologie*.

Fish

References below include common fishes of offshore waters as well as those of the intertidal zone.

Barnhart, P. S. 1936. *Marine Fishes of Southern California*. University of California Press, 209 pp., illus.

Bolin, R. L. 1944. A Review of the Marine Cottid Fishes of California. Stanford Ichthyol. Bull., 3(1): 1-135, illus.

Hubbs, C. L. 1927. Notes on the Blennioid Fishes of Western North America. Papers Michigan Acad. Sciences, Arts, and Letters, 7: 351-394.

Hubbs, Clark. 1952. A Contribution to the Classification of the Blennioid Fishes of the Family Clinidae with a Partial Revision of the Eastern Pacific Forms. Stanford Ichthyol. Bull., 4(2): 41-165, illus.

Jordan, D. S., and B. W. Evermann, 1896-1900. *Fishes of North and Middle America.* Bull. 47, U. S. Nat. Mus., 4 vols, illus.

Rodel, P. M. 1948. Common Marine Fishes of California. Calif. Dept. of Fish and Game, Fish Bull. 68: 150 pp. illus.

Schultz, L. P. 1936. Key to the Fishes of Washington, Oregon, and Closely Adjacent Regions. Univ. Wash. Publ. Biol. 2(4):103-228, illus.

Walford, L. A. 1931. Handbook of Common Commercial and Game Fishes of California. Calif. Dept. of Fish and Game, Fish Bull. no. 28. 181 pp. illus.

Birds

Grinnell, J., and A. H. Miller. 1944. The Distribution of the Birds of California. Pacific Coast Avifauna, no. 27. Cooper Ornithological Club.

Hoffman, Ralph. 1927. *Birds of the Pacific States.* New York and Boston: Houghton Mifflin. 353 pp., illus.

Peterson, Roger Tory. 1941. *A Field Guide to Western Birds.* New York and Boston: Houghton Mifflin. 240 pp., illus.

Marine Mammals

Bonnot, Paul. 1928. Report on the Seals and Sea Lions of California. Calif. Dept. of Fish and Game, Fish Bull. No. 14·61 pp.

———. 1951. The Sea Lions, Seals and Sea Otter of the California Coast. Calif. Fish and Game, 37: 371-389.

Scheffer, V. B. 1948. The Whales and Dolphins of Washington State with a Key to the Cetaceans of the West Coast of North America. Amer. Midl. Nat., 39:257-337.

PLANTS

Smith, G. M. 1944. *Marine Algae of the Monterey Peninsula.* Stanford University Press. 622 pp. illus.

Index

(Italicized numbers indicate figure references.)